Advances in Particulate Materials

Advances in Particulate Materials

Animesh Bose
Parmatech Corporation
Petaluma, California

Butterworth–Heinemann

Boston   Oxford   Melbourne   Singapore   Toronto   Munich   New Delhi   Tokyo

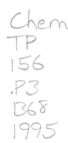
Library of Congress Cataloging-in-Publication Data

Bose, Animesh.
    Advances in particulate materials / Animesh Bose.
        p.   cm.
    Includes bibliographical references and index.
    ISBN 0-7506-9156-5 (acid-free)
    1. Particles.   I. Title.
TP156.P3B68   1995
620'.43—dc20   94-44281
                                        CIP

British Library Cataloguing-in-Publication Data

A catalogue record for this book is available from the British Library.

Butterworth–Heinemann
313 Washington Street
Newton, MA 02158-1626

10 9 8 7 6 5 4 3 2 1

Printed in the United States of America

**To my parents Mrs. Bani Bose and Mr. Amal
Kumar Bose, my wife Prarthana, son Pinaki
and daughter Shree**

Contents

Contents

Preface

The challenges in the area of advanced materials have resulted in the widespread use of processing techniques that are based on particulate materials (P/M). With the increasing demands of modern technology, P/M has emerged as the leading material-shaping process that is flexible enough to fabricate near net-shape products of extremely difficult-to-machine materials and yet attain very high rates of production. The advantages of this unique process are currently being exploited by material scientists and enginers to consolidate new and advanced materials for the demanding aerospace, automotive, defense, chemical and medical industries.

It is important to realize that individual particulates serve as the rudimentary building blocks for the fabrication of useful components from these advanced materials. This book introduces some of the basic powder production approaches and some of the underlying scientific principles associated with them. Though some of the commonly used powder processing and consolidation methods are briefly mentioned, detailed discussions dwell only around the more recent advances in this critical area. The main emphasis of the book will be to provide readers with an in-depth coverage of a broad segment of the new and exciting developments in the area of particulate materials.

I have divided the book into six chapters and a short introductory section. The introductory section describes the contents of the various chapters in detail and also acknowledges the important segments that could not be discussed. The first chapter introduces the attributes, unique characteristics, problems, and numerous applications of this technology and also briefly touches on some of the safety regulations when using materials in the particulate form. The next two chapters deal primarily with two popular powder production approaches, namely melt atomization and chemical powder production approaches. The fourth chapter on mechanical alloying deals with the power production, consolidation, and some properties of these nonconventional materials. The fifth chapter entitled "Intermetallic Compounds" deals with this specific class of material that has exploded into the area of advanced materials since the 1980s. The last chapter discusses in detail the most significant development in the area of material-shaping technology, namely particulate injection molding. The chapter provides the basic science and technology that is pertinent to near net-shape processing of advanced materials using a process similar to plastic injection molding. This chapter also covers the basics of the conventional sintering processes used for material consolidation (both solid- and liquid-phase sintering).

It is my belief that the wide range of topics that have been covered in this book will provide the interested reader with a broad overview of several advanced topics in the area of particulate materials in sufficient depth. Though this book tries to cover a section of the new and exciting developments in the area of particulate materials, it by no means claims to have covered all the "new and exciting" topics in the P/M area. Due to the finite volume of the book, it was necessary to select the topics that are covered in great detail. The aim of this book is to cover a fairly broad cross section of the new and rapidly developing areas of particulate materials in sufficient depth that readers with a little background in material science can get a very good grasp of the exciting possibilities that particulate materials have to offer.

I would like to take this opportunity to thank several individuals for their help in getting the complete manuscript together. Special thanks are due to Professor Randall M. German, Pennsylvania State University, for reviewing the text and providing excellent guidance. The author would also like to acknowledge the work of Mr. Victor Hernandez, SwRI, in preparing some of the excellent artwork. Special thanks are due to Mr. Karl Zueger, Parmatech Corporation, for providing me with the opportunity to complete this enormous task by creating conditions that are ideal for such intellectual exploration. I would like to acknowledge the help provided by the dedicated staff of Butterworth–Heinemann, especially Ms. Aliza Lamdan, Mr. Alexander Greene, and Mr. Frank Satlow. The stimulating discussions with Dr. Malay Ghosh, Alcon, provided the motivation for undertaking a project of this nature. I would also like to acknowledge the support provided by Mrs. L. Galapate, Parmatech Corporation, Dr. J. Strauss, HJE, Dr Richard Page, SwRI, and Mrs Mary Bannon Stoner, MPIF. Lastly, the book would be incomplete without mention of the assistance provided by Mrs. Prarthana Bose both in terms of encouragement to go ahead with the book and also for her tremendous help in the manuscript preparation.

Aminesh Bose
Petaluma, CA

Introduction

Particulate materials (P/M) are increasingly playing a vital role in the processing of advanced materials. The unique advantages provided by the P/M processing route are currently being exploited by material scientists and engineers to consolidate new and advanced materials for the ever-demanding aerospace, automotive, defense, and structural components. With increasing demands of modern technology, P/M technology has emerged as the leading metalworking process that is flexible enough to fabricate near net-shape products of materials that are extremely difficult to machine and yet attain very high rates of production.

For the readers to get a reasonably good understanding of the P/M processes and how the particulate materials are consolidated into useful bulk shapes, it is important to realize that individual particulates serve as the rudimentary building blocks for the fabrication of these advanced parts. This book will, therefore, introduce the basic powder production approaches and some of the underlying principles associated with them. Though some of the commonly used powder processing methods will be briefly mentioned, detailed discussions will only dwell around the more recent advances in this critical area. Similarly, traditional consolidation techniques such as cold compaction and sintering will not be treated as separate chapters. Some of the cold consolidation techniques will be briefly touched in the first introductory chapter and some of the conventional sintering methods will be included in Chapter 6 on Particulate Injection Molding. Numerous pressure-assisted hot consolidation techniques will, however, not constitute a part of this book, primarily due to their sheer volume. The main emphasis of the book will be to provide readers with an in-depth coverage of a broad segment of the new and exciting developments in the area of particulate materials. The book will be divided into six chapters, which are outlined below:

1. Introduction to Particulate Materials

2. Chemical Powder Production Approaches
3. Melt Atomization
4. Mechanical Alloying
5. Intermetallic Compounds
6. Particulate Injection Molding

The introductory chapter first initiates the readers to the term "particulate materials," and provides some of the reasons why it is necessary to use the term "particulate materials" instead of the more commonly used name of "powder metallurgy." The next section of that chapter introduces the readers to different types of particulate materials that are used for processing advanced materials and some of their important powder characteristics. A brief discussion on several aspects of safety and regulations when using materials in the particulate form, especially when it is in the submicrometer range, should be of interest. The toxic effect of some powders on the human system is also briefly touched in this chapter. The first chapter also briefly covers the cold consolidation techniques of die compaction, cold isostatic pressing, and the CONFORM process. Lastly, the chapter deals with some of the applications of this unique P/M process.

The next few chapters provide detailed coverage of a number of powder production methods. The basic premise is that almost all materials can be reduced to powders provided there is input of sufficient energy. Very large energy consumption will result in an increased price tag for the powder, which in turn will reduce its use. However, if there is sufficient value added to the part, the higher price of the powders can be justified. In the particulate injection molding (PIM) process, where complex shapes can be produced easily (which provides a value-added component) the added price due to the fine powder requirement is acceptable as long as the total process economics is comparable or better than the competing processes. In some cases, however, P/M processing is the only possible route for fabricating the required component. In such cases, high powder production costs are often justified. Similarly, high-temperature refractory metals, the majority of the structural ceramics, dispersion-strengthened composites, etc., are good examples where higher powder production costs can be justified.

The basic powder fabrication approaches are discussed under several broad chapters, such as chemical powder production approaches, melt atomization, and mechanical alloying. Some of the processes discussed are already producing powders that are being used extensively in fabricating advanced materials. Some of the processes discussed in these chapters are presently laboratory curiosities, but possess the attributes of becoming extremely important powder fabrication routes in the near future. However, it should be mentioned that well-known processes for powder production, such as electrolytic processing, mechanical milling, beneficiation of ores followed by metal extraction, and common chemical approaches such as that used in the production of tungsten and molybdenum powders (e.g., reduction of tungsten oxide obtained from ammonium paratungstate) will not be covered in this book. A detailed description of the majority of the common powder processing approaches has been very well covered in most of the powder metallurgy textbooks, both old and new. This book will attempt to cover some of the recent developments in powder fabrication. However, there will still be a number of novel processes that will not be adequately covered. As a case in point, the process of chatter machining to produce powders with high aspect ratios will not be discussed.

It should be mentioned that though the chapter on mechanical alloying will deal extensively with powder production, this chapter will also discuss the consolidation and some of the properties of these nonconventional materials. Thus, the chapter cannot be classified as dealing only with powder production, as in reality its coverage includes the making of the powders, their consolidation, and also some of the properties of these advanced materials. This chapter also includes discussions on amorphous or glassy metals and alloys, nanocrystalline materials, and new materials with extended solid solubility. The chapter on melt atomization, for the most part, deals exclusively with powder production, except for a small section on spray forming, which in reality is a combination of powder production and powder consolidation techniques rolled into one. In contrast, the chapter on chemically produced powders deals almost exclusively with powder production techniques.

The fifth chapter, entitled "Intermetallic Compounds," offers coverage of this rapidly developing class of materials with very interesting property combinations. This chapter, in a true sense, deals with a specific class of material. Intermetallic compounds have exploded into the area of advanced materials since the 1980s. It was decided that this class of materials was so unique and has become so important that it warrants a separate chapter by itself. The chapter starts with a description of what intermetallic compounds are and their importance in the materials arena, later goes on to discuss specific compounds such as nickel aluminides, titanium aluminides, iron aluminides, silicide-based intermetallic compounds, and several other classes of intermetallic compounds.

The last chapter, on "Particulate Injection Molding," discusses the key steps associated with this net-shape processing technique that in recent years has shot into the limelight. Some of the exciting developments in the area of particulate injection molding such as the formation of aligned fiber-reinforced composites, injection molding of thixotropic materials that uses particulates as the starting material, development of several new binder systems and the processes adopted to remove the binders, freeze compression molding of materials mixed with water based binders, etc., are discussed in detail. It was felt that this material shaping process was so important that its individual processing steps such as feedstock formation, injection molding, debinding, and sintering deserved detailed coverage. This provided the opportunity of covering some of the conventional sintering processes used for material consolidation (both solid and liquid phase sintering). It should be remembered that the process of sintering is also applied to densify green shapes that can be attained by other P/M methods such as conventional die pressing and cold isostatic pressing that are briefly described in the first introductory chapter.

Though this book tries to cover a section of the new and exciting developments in the area of particulate materials, it by no means claims to have covered all the "new and exciting" topics in the P/M area. This is an appropriate place to mention some of the topics that could not be covered within the scope of this book, and thereby acknowledge their importance in the area of particulate materials. Some of these topics are: explosive compaction, laser glazing, solid free-form processing, smart processing of particulate materials, dynamic consolidation, ceramic materials in general, roll compaction processes, high-temperature sintering (as applicable to ferrous materials), secondary operations, application of computers in particulate material technology, developments in compaction presses and sintering furnaces, microwave sintering, rate-controlled sintering, and numerous pressure-assisted hot consolidation techniques such as rapid omnidirectional compaction (ROC), hot isostatic pressing, hot pressing, sinter+HIP, CERACON^{TM}, Quick HIP, bioactive ceramics and other particulate-based biomaterials, and others. Due to the finite volume of a book, it was necessary to select the topics that would be covered in great detail, and in the process to be forced to leave out a number of other important topics.

When writing a book on the new and exciting developments in any processing area that covers a very broad range of topics it becomes extremely difficult to pick and choose the areas that will be covered in any great detail. If an attempt is made to cover almost every exciting development in the area of particulate materials in any meaningful detail, the information can in no way be incorporated into the structure of just one or two books, simply owing to its sheer volume. To provide a case in point, the topic of particulate injection molding (whose subsections are known as metal injection molding, or ceramic injection molding, or powder injection molding) has been the subject of several books and conference proceedings; thus, if any extensive coverage in that area is attempted, the topic by itself would probably require at least two or three large books. There will always be a need for books dedicated entirely to such new, exciting, and rapidly developing areas as rapid solidification, particulate injection molding, melt atomization, mechanical alloying, etc. However, that was not the purpose of this book — as is clearly evident from the title "Advances in Particulate Materials." The aim of this book is to try to cover a fairly broad cross section of the new and rapidly developing areas of particulate materials in sufficient depth, so that readers with a little background in material science can get a very good grasp of the exciting possibilities that particulate materials have to offer. Attempts have been made to briefly provide the fundamentals of topics covered in any detail, and then to discuss the development of the technology based on the fundamental principles. For readers who wish to delve at greater depth into one particular topical area, this book provides a significant number of references that can suitably guide them toward that goal. Unfortunately, to provide a broad coverage with sufficient depth it has been necessary to make some difficult choices with regard to the topic selections. The readers would appreciate that within the scope of one book, it would have been impossible to cover all topics that can be deemed as "new, exciting, and rapidly proliferating." Though the choice of the topics is a reflection of my personal preferences, it is expected that the topics discussed cover a significantly broad area that will be of interest to a large cross section of readers.

An alternate way of attempting to write a book of this nature, and yet keep it within some manageable magnitude, was to try to cover nearly all the exciting topics (most of which could be several books by themselves) of certain interest within a few pages. However, that coverage would have been so superficial in nature that the readers would not have gained any understanding of the process or the material whatsoever. Thus, a difficult choice was made to limit the coverage of the topics to the chapters that have been outlined earlier. It is expected that the broad range of topics that have been covered in reasonable detail will aid a relatively broad spectrum of readers in their endeavor to gain a more in-depth understanding of what P/M technology can offer and also to give some an idea about the recent developments in their particular area of interest. Also, the references, around 80% of which are after the 1980s, should provide the reader with a recent reference source covering the developments in the areas of particular interest.

With this brief preview of the chapters and the basic philosophy of this book, I would like to conclude this brief introductory section. It should be remembered that the initial success of P/M technology was rooted in the ability of this process to produce unique materials in very high volumes and at significantly lower cost. Self-lubricating bronze bearings and WC-Co-based cermets are a couple of examples that epitomize the early P/M successes. Obtaining the controlled porous structure that could retain the lubricating oil was a unique attribute that could be provided only by P/M processing. The press-and-sinter method was suitable for very high production volumes and the copper and tin powders provided an inexpensive pool of raw material. This initiated the proliferation of P/M into metal shaping technology. That same urge to process unique and difficult-to-fabricate advanced materials into near-net shapes is expected to be the prime driving force in the expansion of this unique technology in the years to come.

An extremely important and recent development that has not been covered in this book is the new single compaction process known as ANCORDENSE™. This process, recently unveiled by Hoeganaes Corporation, has taken the conventional press and sinter industry by storm. The process uses a heated powder and a heated die, together with a new lubricant system. The powder system is warm compacted and can provide conventional ferrous-based parts with densities in excess of 7.45 g/cm^3 by a single press and single sinter operation. It is my belief that the potential of this new warm compaction process will be utilized all over the world to produce parts that will have superior performance compared to their conventional press and sintered counterparts.

Chapter 1

Introduction to Particulate Materials

At the heart of all powder-processed materials are the tiny individual building blocks known as "particulates" or "powders." These powders serve as the rudimentary building materials from which a number of advanced high-performance parts are fabricated. This introductory chapter will initiate readers into the topics of powder characterization and some of the important consolidation techniques that are not covered as separate chapters in this book. However, the very first task of this chapter will be to introduce the readers to the term "particulate materials" and provide a brief explanation for the choice of the term "particulate material" over the conventional terminology of "powder metallurgy."

Particulate Materials

The term "powder metallurgy" has become extremely popular, and has been used extensively by the industry to generally categorize any process that starts with materials remotely in the size range of "powders." The start of the book is perhaps the best place to point out that, in spite of its popularity, the term "powder metallurgy," does not convey the broad sense that this process presently encompasses. The term "metallurgy" is usually associated only with metals and alloys. Thus, a discussion in the area of "powder metallurgy" would convey that the discussion will be confined to processes dealing with powders and parts produced from powders only of metals and metallic alloys. Present-day advanced materials, however, are seldom made of only metals and metallic alloys, as they often incorporate ceramics, ceramic fibers, and intermetallic compounds. In fact, some of the advanced materials often do not contain any metals or alloys, such as whisker-reinforced ceramic matrix composites or oxide dispersion-strengthened intermetallic compounds. Thus, the term "metallurgy" should be replaced by "materials" in order to involve all the aspects that will be covered during the course of this book.

It is also felt that the term "powder" has a narrow connotation and is generally associated with fine spherical or to some extent irregularly shaped "powdered" materials. Present-day advanced materials, however, utilize various forms of fibers and whiskers that often have aspect ratios as high as 1000. It is difficult to categorize these whiskers and fibers as "powders." A more all-encompassing term accounting for the size and shape of the materials presently used would be "particulates" instead of "powders." Thus, a more apt term to replace "powder metallurgy" would be "particulate materials." During subsequent sections of this book, the term "powder metallurgy" can be considered to imply "particulate materials."

In a similar vein, some mention should be made of the chapter entitled "Particulate Injection Molding." This title is also a deviation from the conventional titles such as "metal injection molding" (MIM) or "ceramic injection molding" (CIM). Recently, the trend has been at least to use the name "powder injection molding" to cover both ceramic and metal injection molding. However, it was felt that even that change did not go far enough, as at present advanced materials that contain whiskers or chopped fibers are being processed by injection molding. Thus, "Particulate Injection Molding" was selected as the title of the chapter to represent the broader coverage that the process presently encompasses.

Lastly, to provide readers with a broader coverage of advanced materials, this book has been entitled "Advances in Particulate Materials." It is felt that this reflects the general broadening of the horizons of "powder metallurgy," and is quite apt as the coverage of the book includes intermetallic compounds, ceramic matrix composites that are reinforced with fibers and whiskers, ceramic particle-reinforced intermetallic compounds, and a host of other materials that are far removed from the "metallurgy" that one is familiar with.

Characteristics of Particulate Materials

Although this book deals exclusively with powder-processed materials, no separate chapter has been devoted to the issue of powder characteristics, which is perhaps one of the most important factors that play a vital role in the selection of the subsequent processing methods. Thus, this introductory chapter provides the opportunity to briefly discuss some of the important powder characteristics and also to touch on how these powder characteristics affect the choice of the subsequent processing steps. In fact, the following material on powder characterization will to some extent lay the foundation for the subsequent

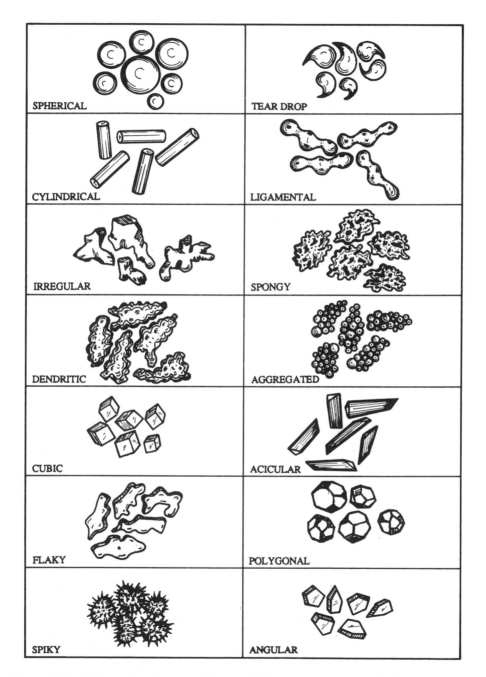

Figure 1.1: *Schematic of some of the possible powder shapes and their qualitative descriptions.*

chapters, which are mainly concerned with advanced processing of particulate materials and several methods of producing particulate materials.

1. Particle Shape

The shape of the particles is greatly influenced by the fabrication methods used to produce the particulates. The powder particles can exhibit a wide variety of shapes depending on the fabrication process. Some of the shapes and their rough qualitative names are outlined in Figure 1.1 and a few scanning electron microscope pictures of some actual powders are illustrated in Figure 1.2.

It can be observed from the two figures that a wide variety of powder particle shapes and sizes can be produced by different processes. Also, by manipulating the processing conditions, it is possible to obtain different powder particle shapes even though the essential processing route is the same. For example, variations in the oxide reduction conditions can produce both irregular (porous spongy) or polygonal (dense) particles. The particle shape and other characteristics such as size, size distribution, chemistry, and surface condition all play a crucial role in determining the consolidation process and to some extent the final properties.

To elaborate this point, let us consider the classic press-and-sinter approach, which is still the most popular route in powder metallurgy fabrication. In this method, loose powder is poured into a die orifice, and uniaxial pressure is applied on the mass of powder by two moving punches. This applied pressure results in a green body whose shape is determined by the shape of the die. Generally, the shape of the die is quite simple, and the compacted sample is in the form of a cylindrical disk or a sample with a square or rectangular cross-section. Shapes that have axial symmetry (such as gears) can also be die compacted. The process is unsuitable for producing true complex shapes from powdered materials.

Irregular, sponge, and dendritic types of powders, which are often "soft" (usually elemental), are generally the best choice for the press-and-sinter approach. The mechanical interlocking provided due to the particle shape results in excellent green strength (which is the strength of the as-pressed part), which is necessary for handling the material before sintering. Compacting spherical powders in a conventional die, on the other hand, is extremely difficult, and the compacted material offers very little handling strength after it is ejected from the die. This is generally true, unless the powder used is of a very soft material, such as aluminum or lead, which due to their low yield strengths can be compacted to very high green densities. Compaction of hard spherical powders in a die where only uniaxial pressure is applied results in literally no shape retention once the compact is ejected from the die. However, spherical powders have the best packing characteristics, and will often pack to 60% of the theoretical density, while the irregular shaped powders tend to pack very poorly.

Let us contrast the powder requirements for a classic press-and-sinter process and the particulate injection molding (PIM) process. In PIM, very little pressure is applied on the powder particles. However, for PIM it is extremely important that the powders pack to high densities, as this aids the final densification and reduces excessive shrinkage. This requirement makes primarily fine spherical particles the unanimous choice for particulate injection molding applications. However, a small amount of irregular powder is often added to promote some strength in the parts when they have been debound. In contrast, spherical powders (especially prealloyed) are generally not the preferred shape for the classic press-and-sinter route.

From the above, readers can get a feel for the importance of particle shape in the processing of advanced particulate materials. However, the particle shape, though very important, is not the only criterion that affects the subsequent processing. In fact, powder particle size and size distribution, and the powder surface area, play an equally important role in influencing the subsequent processing steps.

2. Particle Size, Size Distribution, and Surface Area

The particles often used in powder metallurgy processes are relatively small in size, usually in the range of 10 μm to 1 mm. However, both smaller and larger particle sizes are also used for powder processing. The particulate materials usually possess a large surface area to volume ratio, which increases as the particle size decreases. To produce these particulate materials it is necessary to supply energy that goes to create the extra surface area. Generally, the finer the particulate size the greater is the required energy that is necessary to produce the powders.

The specific surface area of the powder, usually expressed in m^2/g, provides an average measure of the external surface area of the powder particles within the same batch of powder of the same shape. It can easily be visualized that coarse spherical powders will have a lower specific surface area compared to a finer powder. Irregular powders will have a larger surface area compared to spherical powders. A powder particle having totally interconnected porosity will exhibit increased specific surface area compared to its totally solid counterpart of the same size, while powder particles with closed porosity will not be affected. Finer powders of similar shape will exhibit a larger specific surface area.

The discussion in the previous section on particle shape illustrated that spherical shaped powders are the natural choice for PIM applications. However,

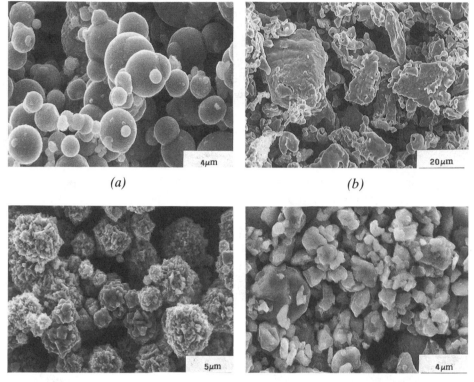

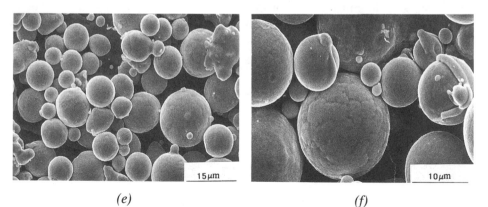

Figure 1.2: *Scanning electron micrographs showing examples of different powder shapes and sizes: (a) spherical iron powder (carbonyl process); (b) spongy iron powder (oxide reduction); (c) spiky nickel powder (carbonyl process); (d) angular chromium powder; (e) spherical prealloyed $Ni_3Al + B$ powder (inert gas atomization); (f) spherical prealloyed $Ni_3Si + B$ powder (inert gas atomization);*

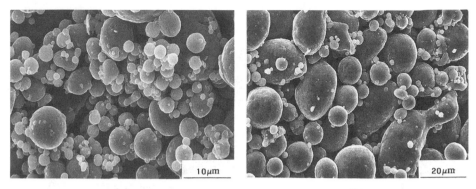

(g)

(h)

(i)

(j)

(k)

(l)

Figure 1.2 contd. (g) spherical Al-Si alloy (inert gas atomization); (h) spherical aluminum powder (inert gas atomized); (i) angular tantalum powder (chemical process); (j) polygonal tungsten powder; (k) cylindrical fibers of alumina; (l) partially alloyed Fe₃Al powder (milling of reacted elemental powder).

large spherical particles (greater than 50 μm) by themselves are not suitable even for PIM applications. This becomes clear if the reader tries to imagine glass beads that are held together into a certain shape by a glue. Once the glue is removed, the glass beads can no longer hold the shape as they will simply slide over one another. This happens because the spherical contours provide very little internal friction. Thus, shapes made by PIM of coarse spherical powders will invariably slump once the binder holding it together is removed. This problem of slumping in coarse spherical powders has been discussed by Bose and German [1]. The same paper also illustrates the fact that using small volume fraction of a much finer powder (around 3 to 5 μm) results in better shape retention in the PIM material after debinding. This illustrates the importance of particle size and the size distribution.

Most PIM processes routinely use spherical powders. These spherical powders are quite small and are usually in the 5 to 20 μm size range. The question then arises why these powder do not create the slumping problem that is associated with coarse spherical powders. The reason is, in fact, quite simple. As the powder particle size decreases, the interparticle friction tends to increase. Alternately, a small amount of irregular powder can also be blended with coarser spherical powders to achieve similar ends. This interparticle friction can prevent the slumping of the parts. The aspect of interparticle friction is discussed in more detail in a later section of this chapter.

It is also important to point out that the sintering kinetics is significantly altered depending on the powder particle size. The finer the powder particle size, the greater is its surface area. This increased surface area provides a greater driving energy for a compacted powder mass as it tries to reduce its surface energy when heated to elevated temperatures. Thus, to obtain higher densities in a particular material when the sintering conditions are the same, it is necessary to use a finer powder. A word of caution that should be introduced at this point is that it is often impractical to use very fine powders (in the submicrometer size range) in spite of their faster sintering kinetics. Very fine powders with very high surface areas are often pyrophoric in nature, and therefore create tremendous problems during the powder handling stage. The problem becomes especially acute in case of reactive materials such as aluminum, titanium, hafnium, etc., which have a great affinity for oxygen. Even relatively nonoxidizing materials such as nickel can become pyrophoric in nature if the particle size is extremely fine. Thus, the finer the powder (especially finer than submicrometer), the greater is the explosive hazard of the material. Very fine and reactive powders can, however, be handled in specially designed glove boxes under a protective atmosphere. This definitely increases the cost of the final product.

The role of powder particle size distribution becomes important in providing

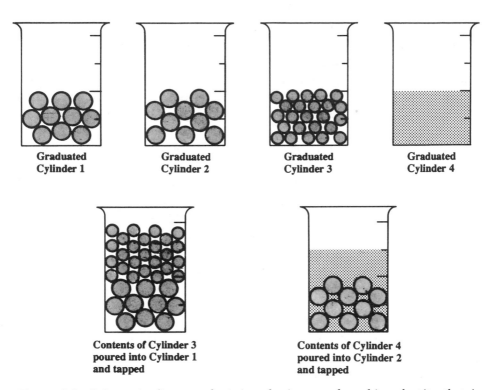

Figure 1.3: *Schematic diagram depicting the increased packing density that is possible by suitable adjustment of the powder particle sizes.*

higher packing densities and for better alloying and faster homogenization in the case of mixed powders where one of the powders is a minor constituent. Consider a certain volume filled with large spherical particles. If we could have another particle size range where the smaller particles fill in the interstices of the large particles, we could theoretically have more solid filling of the same volume, thus increasing the packing density. An effective and easy demonstration of this principle can be obtained by a simple experiment. Consider two large, exactly similar, graduated cylinders about halfway filled with monosized marbles (around 12 mm in diameter). Let us obtain two other graduated cylinders of exactly the same size as the ones before and fill one of them halfway with monosized glass beads whose diameter is only slightly smaller than the marbles used in the first cylinder (say around 10 mm), and half fill the fourth graduated cylinder with fine-grained sand (average size approximately 10 μm). Now pour the glass beads into the first cylinder containing the marbles and tap the cylinder a number of times. Even though the glass beads are smaller than the size of the marbles, they are not small enough to fit into the interstitial voids between the marbles. Thus, in spite of the tapping, the glass beads will remain on top of the marbles and the combined volume of the marbles and the glass beads will almost completely fill up the graduated cylinder. Next the fine-grained sand is poured into the second graduated half filled with marbles and tapped a number of times. The sand particles, which are small enough to fit into the voids between the marbles, gradually fill up the gaps and crevices with continued tapping. Finally, a stage will be reached when no more sand can move down as most of the interstitial spaces between the marbles have been filled up with the sand particles. In this case, the total volume of the graduated cylinder that is occupied by the marble–sand mixture will be much less than the total volume of the cylinder, though each of the two materials taken separately fill up half the graduated cylinder. Thus, in the second case, with the marbles and the sand particles, a much higher solid volume loading has been accomplished by choosing the particle sizes suitably. The concept is depicted pictorially in Figure 1.3. This concept has also been described in detail by German [2]. Theoretically, by tailoring the particle size distribution, it is possible to attain packing densities that are in excess of 80%.

The role of powder particle size ratio between two powders in a mixture is also critical when one of the constituents is present in a small proportion and forms a liquid or is reacting or alloying with the major constituent. The particle size of the minor constituent should be smaller, with the optimum particle size being dependent on the volume fraction. Large particle size of the minor constituent that forms a liquid can leave behind pores and cause other processing problems.

A good example that demonstrates the above statement is illustrated in the experiments conducted on the reactive sintering of nickel aluminide [3] from an elemental powder mixture of nickel and aluminum. Elemental nickel and aluminum powders were mixed in the desired stoichiometric ratio to form the compound Ni_3Al. The aluminum in this case was the minor constituent (34 vol.%) which melts and rapidly reacts to form the desired compound when a compacted mass of the mixture is heated. Experiments were conducted with a nickel powder that had an average particle size of $3.3\,\mu m$ and aluminum powders of varying powder particle sizes. When a very coarse aluminum powder (mean size around 95 μm) was used with the fine nickel powder, the as-reacted material was porous with large pores that had been left behind by the aluminum particles. However, when finer aluminum powders were used, a dense (around 95% to 97% theoretical density) compound of Ni_3Al was formed. Thus, in materials that are reacting to form a compound, or where quick homogenization is necessary during alloying with an additive that is a minor phase, or where it is necessary to have a minor phase forming a fully interconnected network, or in the case where the minor phase is mainly required along the grain boundary region, it is necessary to have the minor phase particle size sufficiently smaller than the particle size of the material forming the main constituent. In fact, in most of these cases, it is almost ideal if the minor phase additive forms a nearly continuous network around the major phase particles. The microstructure of a reactively sintered Ni$_3$Al where the starting elemental powder of aluminum, which is the minor phase in the compound, was much larger than the particles of the major phase, nickel, is shown in Figure 1.4a. When they were reacted together, they formed the compound Ni$_3$Al, but the resultant microstructure exhibited very large porosity. In comparison, using much finer aluminum powder resulted in a microstructure that was almost fully dense, as shown in Figure 1.4b. The black dots observed in Figure 1.4b are mainly etch pits and not porosity.

The powder particle size distribution can also alter the sintering kinetics in a subtle way. Let us consider a powder mass that has a bimodal size distribution such that the finer powder particles can easily fit into the interstices formed by the larger ones. This type of powder size distribution should have faster sintering kinetics compared to a powder mass that is made up only from the coarser powder. Let us now consider another bimodal powder size distribution where the finer powder particles are just large enough that they do not fit the interstitial gap formed by the larger powder particles. In this case, however, the finer particles could conceivably hinder the particle–particle contacts between the larger powders, thereby resulting in poorer sintering characteristics even when compared to the powder that was made up of only the coarser-sized powders. However, these concepts have to be experimentally tested, and further investigations are needed to verify the role of powder size distribution on the sintering kinetics of different materials.

3. Particle Chemistry

The role of powder chemistry is also of great importance in the P/M process. To illustrate this, consider the example of a low-alloyed steel product that is produced by powder metallurgy techniques. Let us assume that the alloy contains small quantities of additive elements like nickel, carbon, molybdenum, and chromium. There is more than one way by which the material of the desired composition can be produced from powders.

One obvious method that suggests itself is to have the additives added during the powder production step, thus producing a hard prealloyed powder in which each individual powder particle will be represent the final alloy chemistry. This approach could, however, lead to processing problems during subsequent press-and-sinter operation. To provide sufficient green strength in the as-pressed

(a)

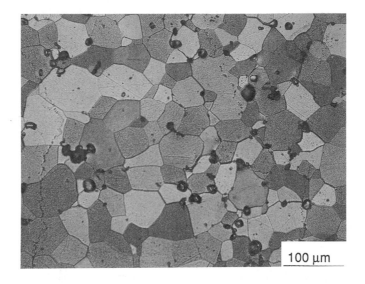

(b)

Figure 1.4: *Microstructure of reactively sintered Ni₃Al, showing the effect of particle size on the reactive sintering process: (a) large aluminum powder particle size; (b) small aluminum powder particle size.*

parts, it would be necessary to press at higher pressures, which, coupled with the hard powder particles, would result in excessive die and tool wear. To prevent this, softer elemental powders forming the desired composition can be directly pressed into the green shape and sintered. Due to the high sintering temperatures, the alloying additives can rapidly diffuse in to yield the desired alloy composition. Again, care has to be taken to select a smaller powder particle size for the minor alloying additive. For press-and-sinter type operations it is often desirable to have the softer elemental powders rather than prealloyed powders having the final alloy chemistry.

The above example can be contrasted with processes that rely on the rapid application of pressure. Some of the recent consolidation processes such as rapid omnidirectional compaction (ROC), CERACON™, Q-HIP™, etc., depend on rapid application of high pressures at the consolidating temperature. In such cases the material is exposed to high temperature only for a short period of time. The process of homogenization, which is largely dependent on the diffusion kinetics, and hence on the exposure time at elevated temperatures, is therefore much lower in these rapid pressing operations. Thus, in these processes, it is important to have prealloyed powders where individual powder particles have almost the same chemistry as the final material. The use of mixed elemental powders in such cases will result in an inhomogeneous chemical segregation in the final product, since the time of exposure at high temperatures will not be sufficient to produce a homogeneously alloyed material.

Powder chemistry, in some instances, also dictates the final properties of the materials processed by the P/M route. Very small amounts of impurities like sulfur or phosphorus in tungsten heavy alloys (W–Ni–Fe or W–Ni–Cu based alloy with usually 90 wt.% or more tungsten) can cause embrittlement of the material due to the impurities preferentially segregating to tungsten–matrix interfaces. Thus, it is extremely important to have powders with the proper chemistry and low content of harmful impurities. At the other extreme is the rare case of beneficial segregation of boron in polycrystalline Ni_3Al. Without the addition of the boron, the polycrystalline Ni_3Al has practically no ductility. With very small amounts of boron additions (around 0.02 to 0.05 wt.%), the ductility of polycrystalline Ni_3Al can be increased to almost 40%. It is found that the boron is segregated to the grain boundaries and this segregated boron is believed to increase the ductility of the material. Thus, in Ni_3Al-based intermetallic compounds, the use of prealloyed powder chemistry is considerably advantageous since the alloying of boron at a later stage could result in the inhomogeneous distribution of boron in particular areas.

The discussion on powder chemistry will be incomplete without a brief mention of coated powders. In the preceding discussion it was brought to light that it is ideal if the minor additive could almost coat the major phase of the system. If this argument is extended a bit further it becomes obvious that to obtain a very homogeneous distribution of the additives it is often best to provide a uniform coating of the minor additive on the major phase. For example, an Fe–2Ni alloy material is extensively used in the P/M industry. Particulate injection molding of the material is generally used to form complex shaped products from the material, and the starting material is usually mixed elemental powders. It is relatively simple to coat nickel on iron by various coating processes. A coated Fe–2Ni alloy powder will provide a much better homogeneous starting material compared to the mixed powder approach as the diffusion distances for proper homogenization will be greatly decreased in case of the coated powders.

Another example where coated powders could be used to great advantage is probably in the W–Hf composite system. It is presently being postulated that in order to impart the desired properties in tungsten alloys so that its penetration properties become on a par with depleted uranium, it is necessary to make the tungsten alloys undergo shear localization and failure. Pure hafnium has been known to promote shear localization at high strain rates and its physical properties are such that it could even be prone to adiabatic shear at high strain rates. Thus, a W–Hf based composite is expected to promote the desired shear localization phenomena. However, in case of W–Hf-based composites, although the volume percentage of tungsten is approximately 75%, it is desirable to have the hafnium forming a continuous network, preferably surrounding the tungsten. To achieve that end, it is necessary to obtain very fine hafnium powder. In reality, it is extremely difficult to obtain hafnium powders that are sufficiently fine to form the desired microstructure. A W–Hf composite formed by the mixed elemental powder approach results in a final microstructure in which the hafnium particles are unfortunately surrounded by tungsten, instead of the other way around [4]. This situation can be altered if the tungsten particles can be coated with hafnium, resulting in a consolidated material where the tungsten is embedded in hafnium. However, care should be exercised to prevent contamination in the case of coated powders, as these contaminations could result in a large deterioration of the final properties.

Various processes can be used to produce these powders with the special chemistry where the additive material, instead of being alloyed and homogeneously distributed in the powder particles, is predominantly present as a coating on the surface of one material. Chemical vapor deposition and the carbonyl process using a fluidized bed are some of the ways by which these coated powders can be produced. Extreme caution should be used during the coating process to minimize the level of impurities, which can totally nullify the advantages that one can obtain from the use of these coated powders. In the chemical vapor deposition process the contamination could result from the chemical precursor and in the carbonyl process the contaminant could be the carbonyl group. Oxygen contamination is always a problem, especially during the coating of reactive materials such as hafnium or titanium. However, this unique powder chemistry can serve as a precursor for the processing of various advanced materials.

4. Particle Pore Structure

Another important aspect of particulates, which is also dependent on the powder processing technique, is the internal pore structure of the individual powder particles. For most applications, the preferred powder type will consist of fully dense individual particles. However, under a variety of conditions, individual powder particles may have both closed and open porosities.

As a case in point, some variations of inert gas atomization processes could result in the entrapment of a gas that has practically no diffusivity through the material. This could result in a powder particle with a pore that is enclosed on all sides with the solid material. This type of porosity is difficult to remove during sintering. Thus, the use of these powders as the starting material results in a final part that has retained porosity due to the inability of the sintering process to remove these initial pores in the powder particles. If the part produced from these powders is to be used in high-performance applications, the retained porosity could serve as a defect, causing premature failure of the part itself.

The other type of pore structure is interconnected in nature. Often, oxide reduced sponge powders, due to the large volume change associated with low temperature reduction, exhibit powder particles with interconnected porosity. In this case, the pore channels usually lead to the surface of the individual powder particles. This form of porosity could be useful when it is being infiltrated with a molten liquid that has good wetting characteristics. For example, a compact produced by pressing a mixture of reduced sponge iron powders with copper can be densified by infiltrating the iron with molten copper, where the fine pore channels in sponge iron will exert a capillary force on the molten liquid. Powder particles can also have a mixture of both open and closed porosity. A schematic view of powder particles that are fully dense, with only open pores, with only closed pores, and with both open and closed pores is shown in Figure 1.5. The aggregated powder shown in Figure 1.1 can also be considered as a powder particle with interconnected porosity. However, if the aggregated particles are de-agglomerated, the individual particle characteristics will be dependent on the type of powders originally forming the agglomerate.

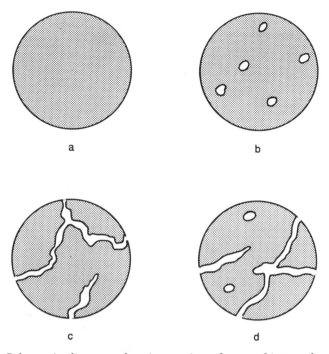

Figure 1.5: *Schematic diagram showing various forms of internal pore structure in a powder particle: (a) dense powder; (b) powder particle with closed pores; (c) powder particles with interconnected pores; (d) powder particles with both closed and interconnected pores.*

5. Apparent Density, Tap Density, and Interparticle Friction

Several other important characteristics of particulate materials that are to some extent dependent on the other characteristics discussed earlier are apparent density, tap density, and interparticle friction. Tests to determine most of these powder characteristics have been described in detail in established P/M standards and have also been described in numerous P/M textbooks.

The apparent and tap densities are important powder characteristics that provide information about the ability of a particular batch of powder to pack without any pressure. The apparent density of a batch of powder is obtained by allowing the powder to fall freely from a fixed height through a special funnel (known as a Hall flowmeter), into a cup of specific volume. The cup is allowed to overfill and the excess powder is carefully removed. The weight of the powder mass is determined. Since the volume of the cup is known, the powder density is easily determined by dividing the weight of the powder held in the cup by the volume of the cup. This density is termed the apparent density. To find the apparent density of powders that do not flow very well, as well as to correctly simulate the flow of loose powder from the feed shoe to a die cavity, a different form of apparatus known as an Arnold density meter is employed. In this apparatus, a bushing sits on a steel block that has a powder collection orifice with a volume of $2 \times 10^{-5} \, m^3$. One hundred grams of powder is weighed and poured into the bushing. The powder-filled bushing is then slid on the steel block until the bushing is directly over the orifice in the block. The powder in the bushing drops into the powder collection hole, and when the bushing is again slid back to its original position, the excess powder is removed. The amount of powder retained in the collection hole is carefully weighed and that weight when divided by the known volume of the orifice provides a measure of the apparent density.

The powder density that is determined by vibrating the powders to afford better packing and removal of powder bridges is termed the tap density. The more spherical the powders the closer will be the tap and apparent density values, since the spherical powders are already well packed without any vibration. However, irregular or dendritic powders, especially the finer powders, will have a low apparent density, which will rapidly increase with tapping or vibration. Thus, even though the ultimate tap density of irregular or dendritic powders will generally be lower than the tap density of spherical powders of the same material, the ratio between the tap and apparent density is much larger for the irregular powders due to larger interparticle friction.

Interparticle friction is another important powder characteristic that often has a profound effect on the processing of the powders. Often the ratio between the tap and apparent density, known as the Hausner ratio, is used to get a rough idea about the interparticle friction of the powder. It has been determined that this interparticle friction plays a very important role in particulate injection molding. Materials with very low interparticle friction will have less problems during the injection molding process, but will usually slump and distort during the debinding stage. To visualize the process of debinding powders with very low interparticle friction, one can imagine the case of spherical ball bearings that have been glued together by a synthetic resin to form the simple shape of a disk. Now if the glue holding the metallic balls together is removed, it is almost certain that the spherical balls will not retain the simple disk shape. The reason is the extremely low interparticle friction between the ball bearings. Thus, for PIM applications, it is necessary to have some degree of interparticle friction in the powders so that slumping and shape distortion can be prevented. It should be realized that this type of coarse spherical powder does not pose problems in case of processes such as hot isostatic pressing, hot pressing, CERACON™, etc. The above example really goes to demonstrate the vital importance of powder characteristics on the final processing that the material has to undergo.

A measure for the interparticle friction can be obtained from a simple test in which a fixed mass of powder is allowed to flow from a fixed height on to a flat surface. The powder mass is expected to take up a conical shape. Depending on the interparticle friction, the angle that is formed between the flat base surface and the edge of the cone formed by the powder mass can vary. This angle, often termed the angle of repose, provides a measure of the interparticle friction, which in turn influences the shape-retention capability of PIM parts [5]. The lower the interparticle friction of the powder, the larger the base diameter of the cone the powder will form, and the powder will have a lower angle of repose.

6. Powder Compressibility and Flowability

The powder compressibility and flowability are two important factors that directly affect the processing of P/M components. The powder flowability and compressibility have a direct bearing on the conventional P/M technique of die-pressing. The flowability of the powder affects the productivity, while the powder compressibility provides an indication of the pressure that would have to be applied to produce a sound green compact of the desired density. Powders with poor compressibility can be responsible for increased die wear and therefore in a sense also affect the productivity to some extent.

The powder compressibility provides a measure of the ability of the material to be compressed at various pressures. Let us consider two different powders of the same material that are pressed into small cylindrical disks using a pressure of 250 MPa. The green density (as-pressed density) of the disks can then be measured. If the green density of the first batch of powders is greater than that of the second batch, the first batch of powder can be said to have a higher compressibility. This is an important parameter, especially for press-and-sinter type operations. The compressibility of a particular kind of powder is expected to differ under the following conditions:

1. A batch of powder that has been annealed will have a greater compressibility compared to a work-hardened powder of the same material.
2. Particles having a spherical shape will have a lower compressibility compared to irregularly shaped particles.
3. A powder batch produced from elemental powder mixes will generally have a higher compressibility compared to a prealloyed powder of the same composition.

The powder flowability is a measure that provides a rough idea whether the powder mass has the ability to flow easily from the hopper into the die during high-production conditions. The flowability is measured by passing a fixed amount of powder through a standardized flow apparatus and measuring the amount of time required for the total mass of powder to flow through the apparatus. It has been determined that the higher the interparticle friction, the lower is the powder flowability. Empirically, powders with a Hausner ratio greater than 1.5 generally do not flow freely. It has also been observed that slightly oxidized samples always flow better than their non-oxidized counterparts (probably due to their increased hardness). Agglomerated powders can also create a great deal of problems with the powder flow. All the above properties can be affected by the powder particle size, which in turn can be influenced by the powder production technique.

7. Particle Crystallinity

Another aspect of particulate materials that can significantly influence the properties of the final consolidated material is the nature of the powder particles in terms of their crystallinity. Recent advances in powder production have resulted in powders that can be crystalline and, strangely, even amorphous in nature. The idea of "metallic glasses" was a totally alien phenomenon before the finding that rapidly quenched metallic materials could lose their crystallinity and exhibit amorphous behavior. These rapidly quenched metallic alloys even exhibited a glass transition or devitrification temperature above which the crystalline nature of the "metallic glasses" was restored.

The use of rapid solidification could also produce some hitherto unknown powders due to the extension of the solubility limit, which represents a major deviation from the equilibrium conditions. Recently, mechanical alloying has also been used to produce amorphous or glassy metallic materials. Chemically produced powders that use various salts or other organometallic precursors also have the ability to produce amorphous metallic powders of different materials. The chemically produced powders have the added advantage of being able to produce atomic-scale mixtures of materials that do not normally have any solubility for each other. The above processes also have the ability to produce nanocrystalline powders. Some of these advances will be discussed in detail in later chapters.

The crystalline state of the particles that are used as the starting materials for processing P/M parts often dictates the choice of the subsequent consolidation steps. For example, a normal crystalline powder may be sintered at temperatures that are often as high as 75% of its melting point. However, when trying to consolidate "metallic glasses" that have an almost amorphous structure, it would essentially defeat the purpose if the sintering were carried out at temperatures as high as 75% of the melting point of the material. The amorphous structure, which is a nonequilibrium one, will tend to revert to its more stable crystalline form if heated above what could be termed the devitrification (or glass transition) temperature. Thus, for this class of material the consolidation process should use the simultaneous application of temperature and pressure, with the temperature being on the low side as well as the time of exposure at the elevated temperature being very short. Hot consolidation techniques such as CERACON™ (a trade mark of Ceracon Inc.), rapid omnidirectional compaction (ROC), or explosive compaction could be used to consolidate these materials (metallic glasses) to try to preserve as much as possible their amorphous characteristics.

8. Hazards Posed by Particulate Materials

One aspect of P/M technology that is increasingly coming under closer scrutiny, especially when fine powders are used, is the potential hazard that is posed by the powder itself. Risk to human beings can be divided into several categories as outlined below.

1. A very fine material that can be pyrophoric in nature, and therefore be a fire or explosive risk. If a powder is very fine, it will generally burn spontaneously when exposed to air.
2. Materials that are reactive and can oxidize very easily are generally pyrophoric in nature and produce an increased risk of fire and explosion.
3. The material may be carcinogenic.
4. The material in its very fine form (0.5 to 5 μm) being able to penetrate into the alveoli and cause fibrosis of the lungs.
5. Large quantities of very fine materials present in the environment can be inhaled during normal breathing. The resulting damage to the lungs would depend primarily on the material itself.

Fine whiskers of certain materials, due to their high aspect ratio and size, are often extremely harmful if present in the air above a certain level. These materials may be biologically inert in nature, but their shape and size makes them extremely dangerous. Thus, the use of very fine whiskers of several inert materials is presently being restricted due to the ability of this material to cause severe damage in the pulmonary area. Other particulate materials, especially those identified as potential carcinogens, are extremely harmful due to their adverse interaction with the human body.

The fire and explosive hazards should be taken very seriously in the work environment. Unfortunately, the other hazard of working with particulate materials that is often not considered by many is the toxicity of the material itself. This area is now being heavily investigated by various organizations all over the world, and there is a strong coordinated effort to introduce so called "tough" regulations to protect the health of individuals exposed to various kinds of particulate materials. However, this area goes into uncharted waters and it is extremely difficult to ascertain with certainty what is and what is not harmful or toxic to the human system. There is also the question of what level of particulate material in the environment can be tolerated and also of what is the critical particle size above which a so-called nontoxic material is no longer considered a health hazard.

A very brief discussion of what fine particles of certain materials can do to the human system, and why they do what they do, should provide more insight into this extremely serious issue of the health hazard that is posed by particulate materials. The major human organ that is affected by the inhalation of particulate materials is the lungs. The human lungs are responsible for supply of oxygen and removal of carbon dioxide from the blood. This exchange between the inhaled air that we breathe in and the circulating blood occurs across the thin membranes in the lungs' alveoli. Generally, the dust that is present in the air, when inhaled, is filtered out in the bronchial channels. However, airborne particulate materials having the right shape and size (usually in the range of 0.5 to 5 μm) can go straight into the alveoli, where they are taken up by the microorganisms known as microphages. This uptake of certain foreign particulates by the alveolar microphages can cause the release of chemicals that cause scar tissue in the lungs, which in turn makes the lung stiff and severely impairs its usual exchange function. Well-known materials that elicit such response are silica, hard metals, and asbestos. The reader can therefore appreciate the magnitude of the problem posed by particulate materials in general.

It is imperative that all the trade organizations that deal with particulate materials combine their resources and share their knowledge to suggest guidelines and regulations for individuals who are constantly exposed to particulate materials. Presently, in the United States, some basic guidelines and regulations are provided by the Occupational Safety and Health Act (OSHA) and American Conference of Governmental Industrial Hygienists (ACGIH). Similarly, the effort by the European Community, which had designated 1992 as the European Health and Safety Year, is extremely commendable. In Europe the legislation is formed in the Directorate General (DG) bodies, and after their development the regulations are published in the "Official Journal of the European Community." These regulations come into force once published in the official journal. It would be interesting to briefly cover the part that is relevant to the P/M industry is discussed in the review article by Tracey [6]. The directive that is of direct importance to the European P/M community is the directive 67/548EEC that has been modified 12 times since its conception in 1967. The introduction of ECOTOX labelling which requires manufacturers to label substances and preparations that can be classified as dangerous to the environment, is currently in a proposal state.

Several directives that do concern the P/M industry are listed below.

67/548 (Annexe 1):   List of dangerous substance in order of atomic number.

67/548 (Annexe 2):   List of appropriate symbols for labeling to indicate danger. This section also provides the phrases to be used to denote the risk.

90/394EEC: Establishes materials that are specified as categories 1 and 2 carcinogens.

91/155EEC: Defines specific information for data sheets for dangerous preparations.

91/322EEC: New directive that establishes Threshold Limit Values (TLVs) at the workplace and uses the data established by ACGIH on 40 substances.

It is imperative that the companies involved in the P/M industry try to adhere to the guidelines and also try to keep themselves updated on the changes and developments that occur in the safety regulations. It is also important that several preventive and good practices are strictly enforced in the workplace. Some of the workplace practices that are quite common and hazardous are outlined below.

1. Not using the appropriate respirators or masks when weighing, mixing, or milling particulate materials.
2. Eating and/or drinking in the area where particulate materials are handled.
3. Areas next to ball mills being left open and powders dusting into the atmosphere.
4. Nonfunctional extraction vents and inefficient powder removal ducts.
5. Handling of reactive powders without the proper precautions.

One of the major problems in establishing any definite safety guidelines for the amount of particulate material dust that can be tolerated in the work environment is the fact that very little is known about the hazards posed by the materials. A case study of the two commonly used particulate materials, nickel and cobalt, will serve as excellent examples that will alert the reader to the magnitude of the problem that is faced by the industry in making suitable regulations.

The concern over nickel in the particulate form as a cancer-causing agent first arose in the 1920s with the prevalent occurrences of nasal and pulmonary cancer in individuals exposed to nickel dust. At first, the prime suspect was the carbonyl process that was being used to produce the nickel powder. The blame soon shifted to the roasting of sulfide ores as similar incidence of cancer was reported from Canada and Russia (which had similar practices of roasting the sulfide ores). The actual cause remained uncertain. Finally in 1974, the National Institute of Occupational Heath proposed a safe nickel dust limit of 15 μg Ni/m^3 air. This proposal was considered highly controversial as numerous personnel using nickel over a number of decades were not afflicted as were the people in the refining industry. Over and above that, it was considered impossible to achieve a nickel dust level of less than 15 μg Ni/m^3 air. Thus, it was considered that the proposal was not applicable to metallic nickel. Any firm decision in this regard was postponed until a committee entitled "International Committee on Nickel Carcinogenic in Man" completed and published its study in 1990. They concluded that more than one form of nickel, such as oxides and sulfides of nickel, gives rise to lung and nasal cancer. Also, soluble nickel increases the risk, especially when the individual is associated with exposure to the less soluble form. The quantification guidelines stated a limit of 1 mg Ni/m^3 air when the nickel is in the soluble form and 10 mg Ni/m^3 air for nickel in the insoluble form. The committee also concluded that the risks are extremely small for individuals exposed to the very low levels of nickel concentration in the ambient atmosphere. However, at the same time, a study conducted by the International Agency for Research on Cancer, a World Health Organization Agency, concluded that there was sufficient evidence that sulfate, sulfide, and oxide of nickel encountered in the nickel refining industry were carcinogenic in the human system; but there was inadequate evidence that metallic nickel or its alloys were carcinogenic in the human system though their carcinogenicity was sufficiently established in animals. Based on the above studies, in the United States, metallic nickel is labeled as "May Cause Cancer," and the European Community designates it as a Category 3 Carcinogen with the label "Possible Carcinogen."

The dermatological problems associated with nickel were first reported from Canada. Since then, knowledge in this regard has increased significantly. The exposure of skin to nickel salts can result in nickel dermatitis, which is, however, low in case of nickel powders. The result is a rash that appears first at the point of contact and then may spread to other parts of the body. However, good engineering practice can significantly reduce this problem. Thus, in the United States nickel powder does carry a warning label that the material may cause skin irritation.

It is often very difficult to know the exact interaction of a particular material with the human body. As a case in point, it has been recently reported that cobalt powder by itself is not necessarily as toxic as it was believed to be. Cobalt was the prime suspect in the affliction commonly known as "hardmetal disease." Cobalt in hardmetal was believed to be responsible for both asthma and fibrosis of the lungs. Cobalt was the prime suspect although it is the minor constituent in the hardmetals, since similar lung problems have also been reported from other industries that use cobalt as the binder phase. However, there was a growing suspicion that pure cobalt was being wrongly blamed as there were no reports of lung problems from the cobalt industry in general. This suspicion was confirmed at a recent HMP 92 International Conference by Swennen [7]. Comparative studies carried out on groups in the cobalt and the hardmetal industry showed that, irrespective of the exposure rate, there were no lung problems for those who were exposed only to pure cobalt or pure tungsten carbide. However, mixed WC/Co powders were found to be as toxic as silica. It was concluded that cobalt was the casual agent and some kind of abrasive attack on the lung surface was necessary to promote the toxic effects.

From the above it can be concluded that the toxicity of a material and its true interaction with the human body are extremely difficult to predict, but are factors that should never be neglected. In fact, a far greater emphasis should be placed on rapidly determining the effects of various particulate materials on living cells; and with the recent finding, one should also consider the effect of mixed powders that are actually used by the industry. The total industry and the national and international health organizations as a whole have to support these efforts, as the health of the individuals exposed to these materials should be of prime concern. In the meantime, companies involved in the P/M business should ensure good engineering and housekeeping habits. Threshold limit values, as outlined by OSHA or ACGIH for foreign particles that are tolerable without detrimental effects to the individuals working in the environment, should be known by the management and should be carefully monitored. Good work habits should be inculcated in employees at the very beginning. Practices such as eating and drinking in the work area should be strictly prohibited. Use of respirators in the appropriate conditions should be strongly enforced. It should be remembered that the effectiveness of respirators is influenced by handling, cleaning, and storage practice. Respirators should be inspected after each use and at least bi-weekly if not in use. Spent filters should be discarded and replaced. All these good work habits can add up and create a much cleaner and safer work environment.

It would be best to conclude the brief discussion on powder characteristics at this point. To gain a complete picture of the powder or powder mixture, it is extremely important to know the powder particle shape, size, size distribution and size ratio, the surface area, the apparent and tap density (which give an idea of packing), powder flow characteristics, chemical composition and chemical gradients in the powder structure, internal structure of the powder particle (which could determine whether the powders have entrapped pores), and the presence of surface films (such as oxides that could hinder densification). The powder fabrication process often influences most of these properties, which in turn reflect on the final properties after processing. Various methods of characterizing the powders have been described by German in his book *Powder Metallurgy Science* [8]. A brief discussion on the health hazards, especially the toxic effects of particulate materials on the human system, has been also included under the heading of powder characteristics. It is felt that this topic will be of importance to the P/M industry.

Cold Consolidation Techniques

The previous section introduced the different types of particulates and their important characteristics. The next obvious question is: How are these particulates produced and how are they processed into useful bulk consolidated materials?

There are, in fact, a large number of ways in which these particulate materials can be produced and then consolidated into useful materials. Detailed descriptions of the common powder processing and consolidation approaches have been given by various authors including German, Lenel, Hausner and Mal, Goetzel, and Jones [8–12]. Several important powder processing methods and some of the advanced consolidation techniques are covered in the subsequent chapters of this book. Since one of the main purposes of this book is to cover some of the advanced topics in particulate materials, no separate chapter has been dedicated to the conventional cold compaction techniques. However, to date, the press-and-sinter operation is responsible for the fabrication of over half of the P/M products that are being used for commercial applications. It was felt that one of the best places to introduce the important topic of cold compaction was in this introductory chapter. Thus, the next part of the discussion will serve as a brief introduction to the process of green part formation by cold compaction. Since cold compaction techniques in general will not be covered in the subsequent chapters, it was felt that a brief discussion about a relatively new cold consolidation technique, known as the CON-FORM^{TM} process, should be given in this chapter.

1. Die Compaction

The process of compaction is used to provide some green shape to the particulate materials. These green, as-compacted shapes usually serve as precursors to further densification processes such as sintering. The most conventional pressing approach most extensively used in the powder metallurgy industry is "die pressing" of powdered materials. In this process, the desired powder is introduced into a die cavity and is generally pressed from both the top and the bottom by punches that fit closely in the die cavity. Pressing from both sides (top and bottom) is usually preferred because pressing from only one direction results in excessive density gradients within the green compact.

The process of compaction first involves the free filling of the die cavity with the powder, usually from the top surface through a feed shoe or a powder feed hopper, with the bottom punch inserted in the die cavity. The die cavity at this juncture is completely filled with a predetermined amount of powder. The lower punch is then moved downward to allow some space at the top of the powder-filled die cavity, where the top punch is then inserted. Pressure is then applied through the motion of both the top and the bottom punches. This results first in the rearrangement of the powder particles and then a gradual deformation of the individual particles to form interparticle bonds. These interparticle bonds are responsible for the green strength of the as-pressed part. Once the green shape has been attained by pressing the part at the desired pressure, the top punch is moved upward and out of the die cavity. At this juncture, either the bottom punch may be moved in an upward direction with the die remaining stationary, or the die may be moved downward with the lower punch remaining stationary. Both the processes results in the ejection of the green part from the die cavity. The feed shoe then moves into place for the next round of powder insertion into the die cavity. As the feed shoe moves into place, it knocks the as-pressed green part into the collection bin, and the above processing cycle is repeated.

Several types of presses have been used for imparting the desired green shape to the powder particles. They include hydraulic, mechanical, rotary, and isostatic types of presses. It is not very difficult to imagine that the powder characteristics will play a very significant role in determining the pressure–density–strength relationship during the formation of green shapes. The powder characteristics can be classified as intrinsic or materials-related (which is usually independent of the powder production technique) and extrinsic (which is dependent on the powder fabrication method). The materials-related properties include the yield strength and modulus of the material, the strain hardening rate of the material, and the tendency for chemical bond formation between the individual particles; the extrinsic factors could be the powder particle shape, size, size distribution, and surface contamination.

A "softer" material is expected to compact very well and attain very high densities at low pressures, with the green compacts exhibiting high strengths. Examples of such materials can include aluminum and lead. A hard and brittle material is, however, not expected to show any significant green strength or densification with increasing pressure, though some degree of densification may occur due to the fragmentation of the brittle particles. A powder with a spherical shape will result in poor densification with applied pressure and is expected to exhibit the worst green strength. A powder that has a layer of surface oxide will hinder the development of chemical bonds, and thereby reduce the green strength of the part. A powder with a desirable size distribution can result in relatively high initial packing density and hence a higher final density after pressurization (true for irregular powders). A fine powder is also expected to result in lower densification compared to its coarser counterpart due to the high interparticle friction of these powders. Ideally, pressurization in a die cavity should result in a decrease in the volume fraction of porosity, an increase in the number of particle–particle contacts over a fixed surface area, an increase in the total contact area over a fixed volume, and a steady increase in the strength of the green part.

One of the principal concerns during the process of conventional die compaction of powders is the die wall friction. The pressure that is applied to the top of the powder body is not totally transmitted to the bottom layer. A significant portion of the energy that is applied is lost in overcoming the friction between the die wall and the powder, and the pressure loss increases as the distance from the surface where the pressure is applied increases. This results in die wear and an increase in the density gradient as the height of the part is increased. To combat this problem, powder metallurgists have resorted to the use of admixed lubricant with the powder. These lubricants reduce the die wall friction and greatly increase the life-span of the pressing dies. However, the use of lubricants can lower the green strength of the part as well as create problems during the sintering stage. It can be argued that, instead of using admixed lubricants with the powder (which is a source of contaminant), if the die wall itself can be lubricated the problem associated with the removal of the lubricant from the green part and its detrimental effect on the green strength, can be significantly reduced. However, this is easier said than done. It should be realized that the process of conventional die pressing is a high-volume production process. Thus, interrupting the pressing cycle each time to apply some form of die wall lubricant to the die surface will result in a tremendous decrease in the productivity. Thus, admixture of lubricants with the powders will continue until some means of very rapid application of lubricants to the die wall can be devised so that the process does not compromise the high productivity that is the hallmark of the conventional pressing operation. Usually the lubricants are stearic acid or other forms of organic fatty acids that can usually be burned out at relatively low temperatures without leaving behind harmful residues.

2. *Cold Isostatic Pressing (CIP)*

One of the important parameters that have been outlined in the earlier discussions is the length to diameter ratio of the green part. The higher the ratio, the greater is the density gradient within the part, which translates into nonuniform shrinkage of the part. This problem becomes so acute above a certain ratio of length to diameter (greater than 5), that the process of conventional die compaction in a hard tool is not feasible. Another limitation of the conventional die compaction method is the relative simplicity of the shape of the desired part. In order to be die compacted, the part should exhibit some form of axial symmetry. Thus, other forms of die compaction are usually necessary. It has been found that the use of cold isostatic pressing in a soft and flexible mold can overcome some of the problems of the conventional die compaction process discussed above. For a more detailed discussion on the conventional process of compaction, readers are referred to the books by German [8], Lenel [9], Goetzel [11], and Jones [12].

The process of cold isostatic pressing (CIP) involves the packing of the desired powder or powder mixtures in a soft flexible mold often known as the "bag." The mold is usually made out of polyurethane or rubber. In this case the pressure is applied to the powder-filled soft "bag" through a fluid medium, which is usually a water-soluble oil. The pressure applied in CIP is isostatic in nature, in contrast to the uniaxial pressure applied in case of conventional die pressing. Thus, long rods or even tubes can easily be fabricated by cold isostatic pressing. The use of soft and flexible rubber or polyurethane molds also allows the fabrication of complex shapes that cannot be achieved by conventional die pressing methods. The flexible mold is usually oversized to allow for the shrinkage of the part after compaction.

Figure 1.6 shows the photograph of two complex rubber molds (black in color) representing the faces of a rabbit and a monkey. These rubber bags were filled with aluminum powder and the ends were sealed with a rubber cap, which was then wrapped with electrical tape to ensure a good seal that would prevent the leakage of the pressurizing fluid into the mold. These powder-filled rubber bags were introduced into the CIP cavity, which was then sealed with a metallic plug. An external pump is generally used to pressurize the fluid to the desired

Figure 1.6: Soft rubber molds used in the process of cold isostatic pressing and actual complex shapes produced from aluminum powder by cold isostatic pressing.

pressure at a slow rate. Since pressure is applied on a fluid medium, the pressure is transmitted through the fluid to the surface of the flexible bag. The applied pressure is isostatic in nature and is applied uniformly over the entire surface of the bag. Once the desired pressure has been attained, the pressure is maintained for a short period of time (around 60 s) and then released. The powder-filled bag is removed from the CIP cavity and the as-pressed green part is carefully removed from the rubber bag. The as-pressed green part is a smaller version of the rubber mold. It can be seen in Figure 1.6 that the as-pressed shape faithfully reproduces all the intricate details of the rubber mold (the monkey's face has been hand-polished to give it the shiny appearance). It would be almost impossible to produce such complex shapes in the conventional die pressing operation. The same figure also shows a 125-mm-long polyurethane mold with a top plug of the same material. This mold is used to produce cylindrical rods from powders.

It is also possible to process hollow tubes by CIP. Two possible processing approaches can be adapted to fabricate the hollow shapes. In one case, the powder can surround a hard machined mandrel (solid core) and the pressure is applied to the flexible mold that surrounds the powder. The reverse process of using a flexible bag to press the powder outward (commonly known as the dilating bag technique) against a surrounding hard tool material can also be used to produce a hollow shape.

If the length-to-diameter ratio of the part to be pressed is very high, it is often necessary to provide an external metallic support for the isostatic bag in order to keep it fairly straight. The support is a metallic tube with a number of holes drilled in it (around 6 mm in diameter) to allow the fluid to freely enter the inside portion of the tube. The powder-filled flexible mold, after being sealed, is placed inside the metallic tube and the complete setup (the metallic tube with the powder-filled bag) is lowered into the die cavity and then pressurized by the fluid. The outer diameter of the soft mold is such that it just fits into the inner diameter of the support metallic tube.

There are primarily two types of cold isostatic pressing techniques, namely the dry bag and the wet bag technique. The dry bag technique is the faster of the two. In this technique the bag is essentially built into the pressure cavity and the powder filling and the ejection of the as compacted specimen is achieved without removing the bag from the cavity, while the sealing is achieved by a punch. In case of the wet bag technique the pressurizing fluid is in direct contact with the bag and the bag is removed from the pressure cavity to extract the compacted specimen from the bag. This process, which has been reviewed by various authors [13,14], provides the opportunity of processing somewhat complex shapes and rods and bars with high length-to-diameter ratios and even tubes with high length-to-diameter ratio.

3. CONFORM^{TM} Process

The development of the CONFORM^{TM} process, which can use not only the conventional fine to medium-sized (10 to 250 μm) powder particles but also coarse beads, machined chips, etc., as the starting material, has captured the interest of the powder processing world. The CONFORM^{TM} continuous extrusion process can be used to produce wires and small solid or hollow sections from nonferrous powders, granules, droplets, chopped wire pieces, and machine swarf. This unique process, invented by the UK Atomic Energy Authority Springfields Laboratories in 1971 [15], has the ability to use a wide variety of feedstock as the starting raw material. The materials that can be used in this process, however, are confined to the soft (low yield strength) nonferrous materials such as aluminum, copper, silver, and their alloys. The process has the capability to add tremendous impetus to the efforts in recycling or recovery of metals.

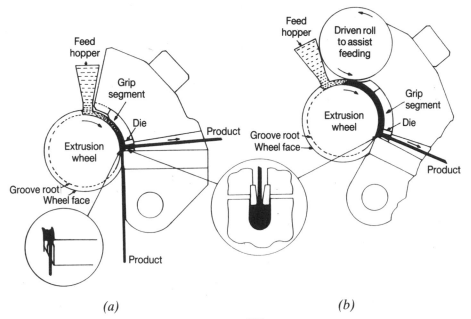

(a) *(b)*

Figure 1.7: *Schematic of the CONFORM^{TM} machine arrangement for use with particulate feeds: (a) radial input without any roll precompaction; (b) radial input with a coining roll for precompaction. (Reprinted with permission from Ref. 16.)*

The principle of the CONFORM^{TM} process is shown schematically in Figure 1.7a. In this process the feedstock (powders, machined swarf, chopped wires, granules, droplets, etc.) is gravity fed directly into the wheel groove. The rotating wheel has a circumferential groove and a pivoted section overlapping a portion of the wheel surface. The pivoted shoe also incorporates the tooling. The tooling consists of a grip segment and an abutment. The tooling serves as the "mini-extruder" through which the loose material is gradually gripped, deformed, and heated due to frictional heat. The feed material gradually fills the chamber as it continues to move forward until its motion is stopped by the abutment. A separate die segment is often positioned between the grip and the abutment and material is finally squeezed through an orifice to form the dense product. The large compressive stresses that build up before the die can be as large as 1000 MPa. The resulting temperatures can reach as high as 723 K (450°C).

It is generally a good idea to precompact the particulate material within the wheel groove [16]. The precompaction of the feed material can be achieved by introducing a coining roll after the feed hopper and before the tooling. This becomes mandatory when, due to certain feed characteristics (spherical powders, high-strength material, etc.) proper grip conditions for extrusion cannot be attained. Even in cases where the feed characteristics do not require precompaction, it is beneficial to have the coining roll in the system as it minimizes air entrapment, which results in blisters on the product surface. A schematic diagram showing this arrangement is depicted in Figure 1.7b [16].

Some of the materials that have been successfully fabricated by the CONFORM™ process include, aluminum and its alloys, copper and its alloys including dispersion-hardened copper with alumina, silver and silver–cadmium oxide electrical contact materials, zinc, and magnesium. The flexibility of this process enables it to produce small-diameter (2 mm) wires to larger solid or hollow sections. Small amounts of feed material can be extruded by this process. The process provides the opportunity to retain the beneficial effects of rapidly solidified materials, since the temperatures associated with the processing are not very high. Thus, this can be classified as an extremely potent process from the point of energy saving and recycling of scrap materials as well as for producing quality materials with properties comparable to ones processed by other routes. The process will, however, be limited to soft nonferrous materials since the materials need to be deformed and welded together without the aid of large external heating.

Conceptually, it may be possible to apply the CONFORM™ process to consolidate some of the higher-strength materials if the particulates of the materials can be subjected to external heating. This may be possible as the strength of most materials (exceptions being several intermetallic compounds) tend to decrease with elevated temperatures. However, it will be necessary to maintain a protective atmosphere to prevent oxidation of the powders.

This essentially concludes the discussion on several of the cold compaction techniques that can be used to produce green shapes. These green shaped compacts are subsequently heated to the desired sintering temperatures in designated furnaces having the proper protective atmosphere (where necessary) to obtain the final densified products. Conventional cold pressing, usually followed by sintering, is still the most popular and widely used process for producing P/M components in the industry. However, cold-pressed billets or rods may also be used as the precursor for several hot consolidation techniques.

With this basic introduction to particulate materials, it would be natural to make a transition into the area of applications of these particulate materials. Thus, the remaining section of the chapter will dwell briefly on some of the applications of particulate material technology. This section will cover topics such as applications of P/M in the transportation, biomaterials, magnetic, electrical and electronics fields, etc. This section will also deal with specific materials such as intermetallic compounds, hard materials, ceramics, refractory metals, light metals, etc., and point out some of their applications.

Applications

The previous sections of this introductory chapter provide readers with a brief coverage on the types of particulate materials and their characteristics, and several means of cold-consolidating the powders into useful green shapes. Most of the subsequent chapters will deal with the means and methods of producing the particulates, some of their consolidation processes (not including pressure-assisted hot consolidation), and also means of producing near net-shaped parts from them. However, all of this exercise will be futile unless the reader can at least get a feel for the potential applications of particulate materials and realize their potential in servicing the demands for advanced materials. During the course of some of the chapters, some of the potential uses of particulate materials are briefly alluded to. However, the use of particulate materials is so widespread that the coverage of all its uses could only be accomplished in a number of books. Thus, this chapter will also briefly point to some of the broad classes of applications of particulate materials without attempting to get into the details of each individual application.

It is almost impossible to trace the exact century when the applications of particulate materials first evolved. Some of the earliest known applications of particulate materials were found in the ancient religious scripts of India. Use of particulate materials dates farther back in time than what is commonly known in the P/M literature. An interesting paper on this by Dube [17] entitled "Aspects of Powder Technology in Ancient and Medieval India" provides a detailed account of what can be considered as the earliest known uses of particulate materials. In *Atharva Veda*, one of the earliest known Hindu texts, which probably dates back to 4500 BC, there is reference to lead that was probably used as granules or shots. According to Dube, it was reasonable to assume that the lead "granules" were prepared by some form of powder atomization technique. Another ancient Hindu script, *Manasollasa*, which dates back to AD 1131, indicates the process of making gold powders for use in paints. According to this, which is given in detail in the paper of Dube, the process of making the gold powder is as follows: Gold pieces together with a fragrant grass (Virana Grass or *Andropogon muricatus*) are slowly ground on a stone slab, using an instrument with a sharp tip. The powder that is produced is placed in a bronze vessel, and water is poured into it. The mixture is continually stirred and the water is panned to remove lower-density impurities. This results in the separation of pure gold powder having the color of the rising sun. The gold pulp is used with a binder to produce the paint.

Perhaps the most cited application of particulate material in the open literate is the Delhi Iron Pillar of India, which probably dates back to AD 319. The total length of the pillar is approximately 7.2 meters, of which 6.7 meters are above the ground. The diameter at the bottom is 0.42 meter and it tapers off at the top to an approximate diameter of 0.29 meter [17]. The pillar is covered with a black enamel-like coating, which has not corroded even though it has been exposed to the elements for over 1600 years. The average composition of the pillar [18] and its coating [19] (referenced from Dube's paper) are

Pillar: Fe–0.15C–0.05Si–0.05Mn–0.25P–0.005S–0.05Ni–0.03Cu–0.02N

Coating: $67.0Fe_3O_4$-$13.1FeO$–$14.8H_2O$-$1.7FePO_4$–$3.1SiO_2$–$0.2MgO$–$0.1CaO$

It is postulated that the pillar was manufactured by successive hot forging of porous sponge iron pieces obtained by the reduction of selected iron ore. According to Dube, ancient Indians, rather than making the preforms from the iron/steel powders, used the ores directly as the starting material, thereby combining the steps of powder production, powder consolidation, and sintering.

For some reason, this ancient science remained in the shadow of the more illustrious metalworking technique of melting and casting. However, the ability of this process to fabricate materials with unique properties, such as materials with interconnected porosity, and also to process materials with hitherto unknown material combinations soon resulted in the explosion of this technology into the modern materials arena. Presently, increasing systems requirements necessitated by the rapid advancement of modern technology have resulted in a great demand for new and improved materials performance. The new materials are increasingly moving away from equilibrium microstructures, and the new processing evolution adopted to fabricate such materials is being spearheaded by particulate material processing routes. Applications of particulate materials have proliferated and will be briefly pointed out in the following parts of this chapter.

1. Amorphous Metals, Metglass, Nanocrystalline and Microcrystalline Materials

The market is rapidly heating up for the rapidly cooled materials. The amorphous metals, metallic glasses, nano- and microcrystalline materials can all be produced by one or more of the following processes:

1. Rapid solidification processes

2. Chemically produced powders
3. Mechanical alloying

With cooling rates around 10^5 to 10^6 K/s, amazing changes in the microstructures and properties of the materials are observed. The crystalline nature exhibited by all metals and alloys can be suppressed and they can behave like noncrystalline or glassy materials. This has opened up a whole new dimension in the world of metallurgy. Recent reports have demonstrated that amorphous materials can also be produced by mechanical alloying, which does not require the melting and subsequent rapid cooling.

The metallic glasses or amorphous alloys do not exhibit any long-range crystallographic order, but they can and do display local short-range ordering like glassy materials. The list of such materials is long, and new ones are constantly being added to the already impressive list: a few examples are Ni_{63}-$Cr_{12}Fe_4B_{13}Si_8$, $Fe_{50}Ta_{50}$, and $Al_{50}Hf_{50}$. The materials, due to their glassy nature, exhibit very interesting and useful properties. However, retaining the glassy properties in bulk materials is extremely difficult. As the powders are heated to form the bulk material, the amorphous nature of the material is lost. Thus, nonconventional techniques like dynamic compaction or explosive shock compaction seem to hold promise for producing bulk amorphous alloys.

Nanocrystalline materials (also termed nanophase materials, nanocrystals, nanostructures or microcrystalline alloys) are a relatively new generation of materials in which the crystal sizes are often in the order of few nanometers. The extremely fine dimensional scale of the microstructure leads to novel and enhanced properties. Mechanical alloying can produce an equiaxed nanocrystalline structure in a number of systems. A few examples of nanocrystalline structures formed by MA are Fe–W–Ta alloy, Ti–Ni–C cermet, Al–Fe, Fe–Ta, and Ag–Fe. Mechanically alloying AlN with Al in liquid nitrogen resulted in the formation of nanocrystalline structures. An alternate way of producing nanocrystalline materials is by using the rapid solidification technique to produce non equilibrium or amorphous powders, which on subsequent processing can be devitrified to yield the microcrystalline structures. Excellent examples of this process have been demonstrated in numerous systems, such as Co–Cr and Fe–B. In almost all cases very high hardness compared to normal polycrystalline materials is observed. A new and evolving technique of chemically produced powders also has the capability of producing nanocrystalline powders.

The potential uses of these new class of materials are still being developed. These materials have already found applications in extrusion dies, small glass-blowing dies, novel magnetic materials, and so on. However, applications of these materials are still in their infancy and are expected to grow at an extremely rapid pace.

2. Refractory and Hard Metals

The refractory metals comprise elements that have high melting points and include metals such as osmium, platinum, iridium, tantalum, tungsten, molybdenum, niobium, etc. These metals and their alloys have found numerous applications in industry and are characterized by very high melting points and high densities. Though most of the refractory metals fall in the same section of the periodic table, each them possesses unique properties. These unusual properties of refractory metals provide engineers with a pool of materials that can potentially be used in numerous applications, some of which have been outlined in a booklet published by the Metal Powder Industries Federation [20].

One of the most extensively used and widely available refractory materials is tungsten. One of the most common applications of tungsten that every reader will be familiar with is in filaments for light bulbs. Tungsten oxide, unfortunately, is volatile above 811 K (538°C), so to use it at high temperatures the material has to be coated, or used in a protective atmosphere or in vacuum. Tungsten is also extensively used in welding electrodes, heating elements, and evaporation boats. The recent finding that doping tungsten with very small amounts of iridium can produce a desired softening effect should lead to the development of new alloy systems for use as very high-temperature materials.

Composites of tungsten such as W–Ag or W–Cu can be used as heavy-duty electrical contact materials, and they are used in light switches, automotive ignition systems, and voltage control devices. Another unique application of tungsten is in the two-phase metal matrix composite known as tungsten heavy alloys. This class of material provides a unique combination of strength, hardness, ductility, and high density. These properties, coupled with good corrosion resistance and easy machinability, make this alloy an extremely desirable material for numerous applications such as aircraft and helicopter counterbalances, radiation shields to attenuate gamma-rays and X-rays, containers for radioactive cobalt-60, X-ray targets in radiation therapy for cancer patients, vibration damping devices, and kinetic energy penetrators. To improve the strength and hardness of these alloys, they are generally subjected to deformation followed by aging. Recent results on alloying of conventional tungsten heavy alloys with molybdenum, tantalum, and rhenium have resulted in heavy alloys with increased strength and hardness without the added step of deformation. The resulting alloys exhibit a refined grain size. Thus, these materials can now be processed into complex and near net shapes.

Applications of molybdenum developed slowly at first, and then very rapidly after the middle of the twentieth century. Its first use was as an armor plate. Primary uses of molybdenum include alloying additives for various ferrous and nonferrous alloys, heat sinks, die casting dies and cores, extrusion and special forging dies, and high-strength high-temperature alloys. An alloy having a composition of Mo–0.5Ti–0.08Zr has a strength that is almost twice that of unalloyed molybdenum in the temperature region around 1273 K (1000°C) and is being considered for structural applications. Molybdenum disulfide powder is also known as one of the best high-temperature solid lubricants and is used to enhance the lubrication provided by oils and grease and is also effective in vacuum environments. Another compound of molybdenum that is being heavily investigated for its potential use as high-temperature material is molybdenum disilicide. This material is presently an excellent heating element, having good oxidation properties at temperatures above 1773 K (1500°C). Attempts are currently underway to try to use this compound in very high-temperature structural components. Other uses of molybdenum include high-precision grinding wheel spindles, supports and backing for transistors and rectifiers, electrodes and stirring equipment in glass manufacturing, heat shields, and nuclear reactor control rod production.

The refractory element rhenium has a very high density and melting point, and is unique among refractory metals because it does not form carbides. The effect of rhenium additions on the softening behavior of both molybdenum and tungsten is presently being used to process these high-temperature refractory alloys into bulk shapes with some ductility. Rhenium alloyed with molybdenum and tungsten is often used in grid heaters, cathode cups, and special filaments. Rhenium alloys are also gradually being used in applications such as semiconductors, electronic tube components, tungsten–rhenium thermocouples, gyroscopes, and thermionic converters. The primary use of rhenium, however, is not in the bulk form but as a chemical with important applications in catalysts.

Niobium and niobium-based alloys are also gaining popularity. The reasonably light weight and high melting point of this material make it very attractive. Some interesting alloys based on niobium are Nb–1Zr, Nb–10Ta–10W, and Nb–10W–10Hf–0.1Y. These alloys find applications in various high-temperature rocket components. The ductility of niobium allows the addition of other refractory alloying elements while retaining high strength at both room and elevated temperature as well as fabricability. For example, sheets made from an alloy having the composition Nb–10Hf–1Ti–0.7Zr are being used as fabricated liners in jet engines. When this material is coated with silicides, it can withstand temperatures from 1143 K to 1700 K (870 to 1427°C) for 288×10^4 to 360×10^4 s (800 to 1000 hours). This alloy is an improvement over the best cobalt-based alloy and it satisfies most of the rocket engine criteria up to a temperature of 1720 K (1447°C) [21]. Work is also underway to process high-temperature intermetallic compounds based on niobium. Extensive investigations on niobium aluminides and niobium iridites are in progress. Due to their bioinertness, niobium-based alloys are also being investigated for applications in the biomaterials arena.

The use of tantalum has also gradually increased over the decades. Tantalum aluminide-based intermetallic compounds are being investigated for possible high-temperature applications. TaC is being used in the hardmetal industry. Tantalum is one of the most corrosion-resistant materials that is available. Tantalum, which forms a very stable anodic oxide film, is extremely useful in electronic capacitors. Thus, it is not surprising that the major use of tantalum still remains in the capacitor industry where it is used in computer hardware, color television sets, and automobile radios. Another interesting use of tantalum might be in the area of shape charges. Generally, shape charges are made of soft ductile copper, but use of tantalum, which also is quite ductile and soft, could provide a tremendous boost in terms of density.

Other applications of tantalum include heat exchangers, vacuum tube filaments, bayonet heaters, and chemical process equipment. Another interesting area where tantalum is increasingly being used is that of biomaterials. The bioinert nature of tantalum, especially in body fluids and tissues, has made it useful for applications in surgical sutures and wire gauzes for abdominal muscle support after hernia repair.

The hard metals, on the other hand, are a separate class of materials that have extremely high hardness and wear resistance. They usually consist of various kinds of oxides, carbides, borides, and nitrides of various materials including those of most of the refractory metals. Other forms of hard metals include diamond–metal composites, various cermets, new oxide-based composites, oxy-carbo-nitrides, etc.

The major volume of the hard metal business is still controlled by the cemented carbide industry. Cemented carbide-based cermets, still one of the most important classes of cutting tool materials, continue to be improved. The improvements of conventional WC/Co-based hard metals are brought about by additives such as VC, TiC, TaC, and NbC replacing a portion of the WC. All these hard phases are held together by a cobalt binder. The alternate route that has been used to improve the performance of these materials is to use thin layers of very high-hardness materials. At present various types of coated carbides are being used by the multibillion-dollar cutting tool industry. The coatings include

TiC, TiN, titanium oxy-carbo-nitrides, and even alumina. Coatings have significantly increased the performance level of these tools. Starting with relatively simple single-layer coating of TiN, modern cutting tools can use a combination of coatings like TiC, TiN, and Al_2O_3. Development of TiCN coatings by physical vapor deposition has also resulted in better tool performance. The WC–Co tools have also undergone considerable development, including the addition of other carbides. A recent development in which the outermost layer of the WC is textured to make full use of the higher hardness of the flat prism face of the WC could prove to be extremely interesting.

Cermet technology, which developed as a means of circumventing the patent problems of the WC–Co-type cutting tools, has come a long way from its modest beginning. An area of rapid development in the carbide industry is the use of new binders replacing cobalt. New binder materials are being actively studied and carbides bonded with high-speed steels are rapidly developing as a new generation of cutting tool. Recent-generation cermets based on Ti(CN) often contain MoC, TiC, Ti(CN), VC, WC, TaC, NbC additives and use a Ni/Mo binder. These cermets are excellent materials for tools used in turning, milling, and grooving operations. Powder injection-molded shapes made from these materials are presently being investigated for applications in corrosive and abrasive environments.

Metal-bonded diamond tools are extensively used in cutting cements, drilling rocks, and wire drawing dies. Diamond powder is mixed with the bonding material (which can vary from soft bronze to the hard WC powder) and is hot-pressed in graphite dies. In a new method, cobalt binder infiltrates the diamond powders during high-pressure sintering and at the same time the diamond tool is bonded to a WC–Co substrate. It is expected that the use of metal-bonded diamond tools will continue to grow. Some of the not so well known applications of these bonded diamonds are in the construction arena, where one of the major uses of these bonded diamond pads brazed on to large-diameter metallic wheels is to cut grooves in the asphalt.

3. Intermetallic Compounds

Intermetallic compounds form the basis for the next generation of high-performance, high-temperature materials. A recent surge of research in the intermetallic compounds has taken place as ceramics have failed to live up to their promise and superalloys have apparently been exhausted. Among the intermetallic compounds, the aluminides have attractive characteristics of high strength at elevated temperatures, good corrosion and oxidation resistance and low density. Some of the aluminides have very high melting points, which makes them attractive as high-temperature alloys. Most of these compounds, however, are extremely brittle at room temperature. A number of intermetallic compounds exhibit a unique characteristic of increased strength with increasing temperature. Some of the intermetallic compounds exhibit excellent shape memory characteristics and are therefore extensively used as shape memory alloys, like TiNi.

Presently the aluminide-based intermetallic compounds, especially the Ni_3Al-based compounds, are being used commercially. It was expected that the intermetallic compounds would first find applications in the aerospace industry, but their initial applications have been in other areas; some of the applications include diesel engine components, powder metallurgy products, resistance heating wire, and aircraft fasteners. Since the advanced Ni_3Al alloys do not provide any exceptional advantages over the nickel-based superalloys, it would seem unlikely that Ni_3Al-based alloys will see direct applications in the aerospace industry in the near future except for noncritical components. However, it is expected that this material will be used in radiant heaters, dies and molds, pump impellers, hot furniture for furnaces, land-based steam turbines, oil and gas well components, and hot components in automotive applications.

Other interesting aluminides that are being extensively studied include NiAl, Fe_3Al, FeAl, Ti_3Al, TiAl, and other compounds with alloying additions to the above. The iron aluminides have exceptional sulfidation resistance and are being considered for applications in fossil fuel conversion systems. The titanium aluminides, and composites based on titanium aluminide matrices, however, are being seriously considered for aerospace applications, especially in advanced aircraft engines.

The other interesting class of intermetallic compound is the silicides. This class of compound has received much less attention compared to its aluminide counterpart. However, due to the ability of the silicides to form a stable silica film that is extremely resistant to oxidation and sulfidation, it is expected that this class of materials will find numerous commercial applications.

Other types of intermetallic compounds that could have potential applications are the beryllides and the various ternary and quarternary alloys of the other intermetallic compounds. One application of intermetallic compounds that has declined to some extent is in superconductors. Prior to the development of ceramic-based superconductors, the intermetallic compound of niobium–tin was extensively used. This intermetallic compound was also processed by P/M route. For the more exotic intermetallic compounds, the primary applications are expected to be in the new generation of hypersonic aircraft. A detailed coverage of intermetallic compounds has been carried out in Chapter 5.

4. Magnetic, Electrical and Electronic Applications

A host of P/M products are used in magnetic, electrical, and electronic applications. The applications include new magnetic materials, electrical contacts, electronic packaging materials, and superconductors. The developments in the various areas are so extensive that it would be an impossible task to cover all the applications, so only some of them will be discussed here.

Magnetic materials play a crucial role in modern electrical and electronic devices. The hard or permanent-magnet market is estimated to be around 1 billion US dollars. The energy product expressed in MGOe determines the power of the permanent magnets. The first magnets were hardened steels with a energy product of 1 MGOe; these were followed by the Alnico family of magnets and ferrites, which had an energy product in the range of 6 to 7 MGOe; these were quickly followed by the rare-earth–cobalt-based permanent magnets such as SmCo, which provided energy products of up to 25 MGOe; and recently, the new class of hard magnetic material $Nd_2Fe_{14}B$ in the fully dense condition can yield energy products in the range of 20 to 40 MGOe [22]. The rare-earth–cobalt magnets (RE_2Co_{17}) have found numerous applications in stepping motors, loudspeakers, headphones, voice coil motors, optical pickup heads, and coreless motors. The recent development of the high-performance permanent magnetic material based on Fe–Nd–B alloys and plastic-bonded composites of this alloy is also an area that should be carefully followed by the P/M industry. Recently there has been a sharp rise in the use of such magnets in actuators, motor sensors, electrically-assisted powder steerings, electrically controlled suspensions, air bags, breaking systems, medical devices, flexible gasket magnets for door seals in automobiles to reduce noise, drives of CD players, hub spindle drives for computer hard disks, oil burners, pumps, centrifuges, and electronically commutated DC motors for driving handling devices, positioning systems, etc.

Soft magnetic components have expanded from pure iron to Fe–P, Fe–Si, Fe–Ni, Fe–Co, and ferritic-grade stainless steels. The development of Fe–P, and Fe–Ni alloys as competitors for the Fe–Si alloys for magnetic applications constitute an interesting development in the P/M area. More exciting magnetic applications stem from metallic glasses like the iron–silicon–boron and iron–silicon–boron–carbon containing materials that display low magnetic hysteresis and eddy-current losses. Soft magnetic materials are being used in the following applications: magnets for relay, stators, yokes, pole pieces, print heads in computer printers, printer actuator mechanisms for both line and dot-matrix printers (Fe–Si, Fe–P, or Co–Fe); disk drive bearing supports; ABS sensor rings (ferritic stainless steel); triggers for electronic ignition (Fe–P); sensors for regulator system for gasoline fuel injection (Fe–Ni); and cores for gas burner safety valves (Fe–Ni). Some of the magnetic materials are required in complex shapes, which has resulted in an increase in PIM-processed soft magnetic parts.

An interesting process of producing fully dense iron strips from high-grade ores found in India has been discussed in a paper by Govila and Knopp [23]. A very fine-grained pure iron ore known as "blue dust" is first reduced at 1393 K (1120°C) and the iron powder obtained is then roll-compacted and densified by repeated sintering and cold compacting. These strips are being considered for DC electrical equipment requiring a high degree of magnetic stability.

With the advent of "high-temperature" superconducting materials, the issue of fabrication has become a prime concern. Again, powder processing seems to be a viable route for producing bulk forms from this material. Superconductors in the RE–Ba–Cu–O system (where RE represents most rare earths) having a transition temperature of 90–100 K, have been studied for a couple of years. The recently discovered Bi–Sr–Ca–Cu–O and Tl–Ca–Ba–Cu–O based superconducting materials have transition temperatures approaching 125 K. Potential uses of these superconductors are in powder generation and storage, magnets, motors, and electronics. Thus bulk shapes, wires, and tapes are required, and are being processed by a variety of P/M processes.

The electrical and electronics industry has also used P/M products quite extensively. The majority of the electrical applications have been discussed in the applications of refractory materials. However, a whole series of Cu–X alloys (where X may be any BCC refractory metal) with an extremely good combination of strength and electrical conductivity have been developed by using a P/M process of heavily deformed in-situ composite. These new materials have an exceptional combination of electrical conductivity and strength at elevated temperatures. P/M copper is also extensively used in heat sinks, shading coils for contacts, components for fuse blowouts, components for switch boxes, and armature bearing blocks.

P/M-processed, controlled thermal expansion materials such as invar (Fe–Ni) and kovar (Fe–Ni–Co) have been used extensively as packaging materials. Several newly developed composites such as W–Cu and Mo–Cu are also expected to find applications in heat sinks and various kinds of electronic packages. W–Ag and W–Cu contacts have been used for electrospark machining and electrodes for resistance welding. Various tungsten-based heavy alloys are used for heavy-duty electrical contact materials. A variety of ceramic materials have been used in a variety of electronic applications such as IC substrate materials. $PbTiO_3$ and other ceramics with a perovskite structure that exhibit ferroelectric properties are also being used as piezoelectric vibrators, thermistors, varistors, and capacitors.

5. *Light Metals and Alloys*

These are an extremely important class of materials due to their widespread application in the aerospace industry. Powder processing of this class of materials is being actively pursued. Materials in this class can be divided into categories that include aluminum-, titanium-, and magnesium-based alloys.

In all the three class of materials, rapid solidification technology plays a vital role. In aluminum-based systems, three classes of alloys have played a vital role. The precipitation-hardened high-strength alloys produced by gas atomization of the 2000, 5000 and 7000 alloy series are important aerospace materials. The intermetallic and silicide phases are very fine and the combination of precipitation hardening and grain refining (Hall–Petch) results in very high mechanical properties. The second generation of these alloys is produced by extruding an alloy of Al, Zn, Cu, Mg, Zr, and Ni. They exhibit 30% improvement in strength and 40% improvement in toughness over the ingot metallurgically produced counterpart. The dispersoid-strengthened aluminum alloys are for use in high-temperature structural parts. These alloys are impossible to produce by the ingot metallurgy route and are produced either by rapid solidification or by mechanical alloying. Some of the compositions are Al–8Fe–2Mo, Al–4Cr–3Zr, Al–5Fe–3Ni–6Co, and the lightweight Al–Li or Al–Li–Be alloys. Also of interest in the aluminum-based materials are the lightweight foams, aluminide-based intermetallic compounds, and whisker-reinforced composites.

Mechanical alloying has also been used to process a variety of aluminum alloys. A dispersion-strengthened aluminum alloy known as DISPAL is already in use as piston preforms, high-temperature electrical conductors, and interferometers. The development of a mechanically alloyed Al–6Ti material has also shown excellent combination of elevated temperature strength, creep resistance, thermal stability, stiffness, and corrosion resistance when compared to the conventional series of 2xxx and 7xxx aluminum aerospace alloys produced by ingot metallurgy. This new mechanically alloyed material is expected to find applications in the skin, wheels, and structure of advanced aircraft. Production scaleup for commercial use of this material is in progress.

The drive for the use of magnesium alloys was kept alive by the lure of significant weight saving that can be gained by using these materials. However, the magnesium-based alloys have poor resistance to corrosion, especially in salt water or salt air environments. The ability of rapid solidification P/M processing approaches has been viewed as one of the best ways to process these alloys owing to the possibility of extended solid solubility, improved chemical homogeneity, and refined grain structures. A P/M alloy having a composition of Mg–6Zn–0.77Zr (wt.%) has been used in support beams for the floors of transport aircraft.

To fully utilize the potential of rapid solidification processes, it is necessary to produce fine thermally stable dispersoids that have limited solubility in magnesium. Zinc, aluminum, and silicon can provide precipitation hardening. Recent investigations on Mg–Zn–Al–Si systems have also demonstrated the capability of producing high-strength magnesium alloys by a combination of rapid solidification and powder processing approaches. The material has excellent formability at the low temperature of 473 K (200°C), and can be formed into complex shapes in a closed die by forging at low rates. These alloys provide high strength and hardness. Similarly, corrosion-resistant magnesium alloys have also been developed by the rapid solidification technique. In magnesium-based systems, alloys containing rare-earth elements like ytterbium and neodymium result in materials with a good combination of strength, ductility, and corrosion resistance.

Interestingly, magnesium has also been used as a supercorroding alloy. Mechanically alloyed Mg–X alloys (X = Fe, Cu, Ti, etc.) react rapidly when placed in an electrolyte like seawater. The rapid corrosion (reaction) produces heat and evolves hydrogen gas, which is being used as a heat source in deep-sea diving and a gas generator for buoyancy control, fuel cells, and fuel in hydrogen-based engines. Since various combinations of magnesium with other noble metals corrode at different rates, they can be used in different applications, for example Fe and Cu will be more desirable when generating hydrogen gas or underwater heat; but, for applications like release of deep-sea equipment from the ocean surface, it would be beneficial to use a slowly corroding link made up of Mg–Ti.

Applications of P/M titanium alloys are very well developed. A number of books and conference proceedings have been devoted to the applications of this light metal. The classic alloy that has been used extensively for a variety of applications is Ti–6Al–4V. This alloy has been used extensively in the aerospace industry as well as in the biomedical area. Plasma sprayed titanium coatings or vacuum sintered porous coatings on artificial bone implants have provided the means of anchoring the artificial bone by the ingrowth of the natural bone tissue into the open pore structure. Presently, titanium aluminide-based composites are being considered for applications in high-temperature aerospace components.

6. Advanced Ceramics

Today's advanced ceramics have developed at a pace that is beyond one's imagination. Powder processing has played a large role in this development. The major problem faced by ceramic materials was extreme brittleness and various toughening approaches have been taken to improve the fracture toughness of advanced ceramics. Some of these approaches include increasing the sintered density by proper sintering procedures, decreasing the grain size, and reducing the size and concentration of crack initiators (such as pores or microcracks that exist after sintering).

Colloidal processing of the ceramic powders has led to better homogeneity in the green state, which translates into better sintered products with fewer inherent flaws. This is achieved by using proper deflocculants and adjustment of the pH so as to produce repulsive forces between particles. Al_2O_3/ZrO_2 composites have been successfully prepared by such colloidal processes. Colloidal-processed and hot-pressed TiB_2–TiC–SiC ceramic composites are considered to have potential as tool and wear resistant materials.

Introduction of various synthetic crack-retarding phases is another popular method that is being applied to combat the problem of low fracture toughness of the ceramics. The incorporation of crack-closing particles (transformation toughened ceramics) has received a great deal of attention. The best example of this is partially stabilized zirconia (PSZ). Two-phase ZrO_2 is produced by partially stabilizing the tetragonal ZrO_2 phase by addition of up to 10% Y_2O_3, MgO, or other oxides. When a crack travels through this material, the tetragonal phase transforms locally into a monoclinic structure, reducing the effective driving force. These materials are being used extensively in a variety of applications that includes dies for wire bonding, cutting tools, and implant materials.

An alternate method of toughening ceramics is by the incorporation of fibers or whiskers in the ceramic matrix. Some of the hot-pressed silicon carbides and silicon nitride reinforced with whiskers are currently of great interest. Alumina reinforced with silicon carbide whiskers is being used as a cutting-tool material. Molybdenum disilicide reinforced with alumina fibers is being investigated for high-temperature structural components. Research on whisker-reinforced composites has mainly been used to improve the fracture toughness of advanced ceramics. The process of colloidal mixing has helped in the attainment of uniform mix of the fiber and whisker. An exciting ceramic application that could be of great help to the P/M industry is ceramic dies for powder compaction. Sialon-based ceramic dies provide extremely beneficial properties. The ceramic material prevents the cold welding of powders to the die walls and provides low ejection forces. As this material has extremely high wear resistance, it can also be used for extended times without significant die wear.

Ceramic-based cutting tools are also becoming increasingly popular. Various ceramic-based cutting tools are presently available in the market. Ceramic-based materials are also being chosen for gas turbines and internal combustion engines as they are expected to provide higher efficiency. Potential candidates for engine applications include silicon nitride, silicon carbide, and partially stabilized zirconia.

A new alumina–zirconia material termed "zirmonite," produced by Diamonite Products, has extremely good flexural strength and wear resistance. Its potential applications could be in chemical plants and raw-material preparation plants. Ceramics that are used for electrical applications include varistor ceramics that have ZnO doped with a variety of different oxides, or PTC ceramic, which is a $BaTiO_3$ doped with a variety of oxides. Some of the ceramics can be used as breeder materials such as lithium aluminates, lithium orthosilicate, and lithium zirconate.

A brief listing of some of the applications of advanced ceramics includes thermistors, IC substrates, capacitors, piezoelectric materials, spark plugs, electromagnetic insulators, cutting tools, wear-resistant materials, high-temperature structural components, sensors in the chemical industry, bioactive ceramics for hard-tissue implants, bioactive ceramic coatings on metallic bone implants for high-stress applications such as hip and knee joints, optical communication fibers, ceramic superconductors, superhard materials, hot-extrusion casting dies, high-temperature furnace insulations, and press nozzles. Though the primary push for ceramics has been for very high-temperature engine applications, their use in other areas has proliferated and the demand for advanced ceramics that are almost always processed by the P/M technique is expected to grow over the next few years. It is felt that the full potential of ceramic materials has not been realized and the recent discovery of ceramic-based superconductors supports this contention.

7. Biomaterials

The use of P/M products in the biomaterials arena has proliferated over the last two decades. P/M techniques have been used extensively to process artificial bone implants that have a thin layer of porous coating over a solid implant. This thin layer of porosity allows the ingrowth of natural bone tissue, and thereby provide anchoring of the implant to the natural bone. This form of anchoring reduces or eliminates the use of bone cements. Inert metallic bio-materials such as titanium- and niobium-based alloys have been processed by P/M techniques to produce useful materials suitable for use within the human body. Powder injection molding has played a key role in the development of numerous orthodontic components. Stainless steel has been used extensively for producing a variety of orthodontic parts that often weigh less than a gram.

However, the greatest use of particulate materials in biomaterials developed as the special properties of bioactive glasses and ceramics became known. A variety of bioactive ceramic materials such as hydroxyapatite, tricalcium phosphates, and ALCAP have been processed by the P/M route. Porous ALCAP ceramic materials produced by the P/M technique have been used as controlled drug-delivery systems. Plasma sprayed hydroxyapatite on titanium implants has promoted bone bonding characteristics of the implants. The development of a new metal–ceramic composite that will provide better bone bonding is presently under investigation. The development of Bio-Glass™ (University of Florida trademark), which has the ability to form strong bonds with the natural bone, led to tremendous development in the area of glass–ceramics that are bioactive. The recent development of a higher-toughness glass–-ceramic material known as A-W.GC has also prompted the development of new artificial implant materials.

Various conventional ceramic materials processed by the P/M technique are also being used as biomaterials. These ceramics, such as alumina and yttrium oxide-stabilized zirconia, are bioinert in nature. Due to their excellent bioinertness, these materials can be used in the human system without triggering detrimental toxic reactions. Alumina has been used extensively in the ball-and-socket joint of artificial hip implants. Alumina has also been used as bone and tooth implants. However, due to the brittleness of alumina, yttria-stabilized zirconia is being investigated. Since the inert bioinert ceramics do not have the ability to bond with natural bone, a porous ceramic coating has been applied. The coating, known as Poral, is deposited on a dense alumina prosthesis by applying several layers of alumina beads (500 to 700 μm diameter) and sintered at 1673 K (1400°C) with the aid of a high-temperature bioglass. This provides a structure with pore sizes in the range of 290 to 500 μm, and animal studies have confirmed that this material does induce bone ingrowth into the pores.

8. Applications in Transportation

The biggest use of P/M products is in the transportation industry. This is a large area that covers materials used in automotive, aerospace, and railroad transportation. P/M parts have heavily infiltrated all these markets and the number of P/M parts for these applications continues to grow at a tremendous rate.

One of the largest users of P/M parts in the transportation sector is the automotive industry. Over the last few years there has been a global increase of P/M components in automobiles. The start of the P/M forged connecting rods in car engines provided a tremendous boost for the P/M industry. A new second-generation P/M steel connecting rod is in production and is going into the engines of several models of Ford cars. The near net-shape ability of P/M forging process has decreased the amount of metal that has to be removed from the connecting rod, and the P/M product provides superior fatigue properties over the conventional wrought product. Other parts that are going into car engines include crank shaft and cam shaft sprockets, data sensing wheels, oil pump generators, and valve seats and guides. Another new product developed is a composite metal camshaft with P/M lobes. A P/M steel inner race for the transmission of General Motors F31 four-speed cars is under production, and a P/M forging will go into the torque converter of a GM vehicle. According to auto industry observers, P/M parts in the modular 4.6 liter V-8 engines will soon be in the 7 to 8 kg range; and for the total car, the net weight of the P/M components will be over the 13.6 kg (30 lb) mark [24].

In his state of the P/M report of 1991, D.G. White reported that though all the three big automakers in North America are very active in developing new P/M parts for engines and transmissions, there has been a significant drop in automobile production. He also added a word of caution that new fuel regulations could present some problems to the P/M industry in the future. The consideration of 85% methanol-containing fuel presents a challenge to the P/M industry as the present alloys are not compatible with the new fuel. In the meantime, P/M parts continue to find applications in many more car models, including those in Europe and Japan. For example, BMW will use a P/M connecting rod in an eight-cylinder engine while Jaguar and Volvo will also start using them. The popular *Saturn* car model already has P/M blocking rings and a hot-forged transmission sleeve. Again a word of caution that needs to be mentioned is the competition that might come about from a new hybrid material connecting rod design that will use carbon-fiber-reinforced plastic with a titanium alloy. This new hybrid material will result in a 30% weight advantage over the P/M connecting rods [25]. However, presently the technology is not suitable for mass production; but P/M technology should be ready for such challenges in the future from other materials and shaping technologies.

Metal injection-molded parts are also gradually infiltrating into the automotive market. Parmatech Corporation, a PIM company in California, is presently injection-molding small parts for the automatic air bag systems. General Motors is also looking at several PIM applications.

Apart from the applications discussed earlier, other automotive parts that use P/M products include a variety of bearings and bushings, tilt levers, steering column locks, etc. Some of the areas where P/M bearings could be used are in the gear bearings for 4-wheel-drive automobiles, electric motor bearings for low temperatures that could be used on electric motors on the cold parts of the car, and cable-linked bearings that endure inclined loads. Other uses of P/M in the automotive industry include various magnetic materials and frictional materials. Some of the new P/M applications being developed for the automotive industry include medium-carbon steel hubs for cooling fans of a Lancia 16-valve engine, a copper–nickel–molybdenum steel connecting rod for the sliding roofs of cars (Peugeot-Citroen group), copper–carbon steel gear box shifting levers (Fiat group), copper–nickel–molybdenum steel double gear for electric windows (Bosch), copper–phosphorus steel timing pulley, copper–nickel–molybdenum steel pinions for an electric window device (Rover), copper–nickel–molybdenum–MnS bushings for the seat back position regulators, pinion and gear for door locks, and speed selector assemblies [26].

Particulate materials have also been used extensively in the aerospace area. Aerospace applications, especially in the critical areas, have one of the most demanding property requirements in terms of high stiffness-to-weight ratio, high strength-to-weight ratio, good elevated temperature creep and strength properties, and good oxidation resistance at elevated temperatures. The R&D efforts for the National Aerospace Plane, or NASP, a futuristic hypersonic aircraft, have fueled interest in new aerospace materials. An excellent review on P/M applications in the aerospace industry has been provided in the presentation by Meetham and Mayo [27].

It is, however, not surprising that the two lightweight materials, aluminum and titanium, are the ones that have found the most extensive applications in the P/M industry. The two nonequilibrium processes of rapid solidification (RS) and mechanical alloying (MA) have been used extensively to fabricate materials for aerospace components. Possible applications of several mechanically alloyed Al–Mg–Li and Al–Mg alloys could be in missile tubing, aircraft control rods, and structural components in aircraft.

Several Al-based alloys that are already commercialized or near the commercialization stage have been listed by Suryanarayana et al. [28]. Some of the high-temperature alloys include Al–8.3Fe–4Ce atomized powders, Al with 6–11% Ti mechanically alloyed materials; low-density mechanically alloyed material having composition Al–1.3Li–4Mg–1.1C–0.4O; and high-strength alloys having the composition Al–9Zn–2.2Mg–1.5Cu–1.4Zr–1Ni, which was atomized, and Al–4Mg–1.1C–0.4O, which was mechanically alloyed. Aerospace parts that are made from aluminum-based alloys include aircraft wheels and fittings, extruded aircraft seat tracks, precision-forged aircraft bulkheads, and channels and angles used in electrical racks.

Sandwich construction is often used in the aerospace industry to produce efficient airframe components. However, these are very difficult to process. The new development of a P/M titanium alloy that can have controlled porosity after the material has been hot isostatically pressed and then carefully given a special thermomechanical treatment is expected to provide an inexpensive substitute for the honeycomb type of structures used by the aerospace industry. The porosity level can be varied depending on the initial partial gas pressure that is applied before sealing the HIP can. The other advantage afforded by this process is the ability to use blended elemental titanium, aluminum, and vanadium powders, instead of the expensive prealloyed Ti–6Al–4V powder. The HIP operation also results in the diffusional homogenization of the elemental powders. This material could serve as a replacement for existing solid Ti–6Al–4V airframe components [29].

Another interesting P/M application in the aerospace industry is the beryllium mast-mounted sights that have been produced into near net shape. The part goes into various scout helicopters, and the low density and very high stiffness requirement makes beryllium a natural choice. An aluminum sensor support that would have weighed 8 kg (18 lb) will weigh only 2.7 kg (6 lb) when replaced by beryllium. The near net-shaped part is obtained by direct HIPing of beryllium powders.

Various intermetallic compounds of titanium and aluminum are also being actively considered for possible use in a variety of aerospace applications. The titanium aluminide-based composites are being considered as hot components for the NASP program.

Newly developed ferrous-based P/M alloys with Cr, Mo, V, and C as the major alloying constituents are being considered as bearing materials for use in cryogenic aerospace engine turbopumps. A new superalloy P/M turbine disk is being used in the engine for French combat aircraft and a silicon carbide ceramic composite is being used for the hot flap of the M88 engine. Various superalloys processed by the P/M technique are gradually finding applications in aerospace engines. It is expected that the drive of the aerospace industry to produce materials that will work at higher temperatures for longer periods of time will lead to the development of new P/M-based materials in the future.

The use of P/M in the rail transportation industry is also well developed. An excellent review on this subject has been recently compiled by Tracey [30]. Friction materials were one of the major applications of P/M products in the railroad industry. Motors, generators, and alternator brushes made from carbon-based materials have been used extensively in electric engine parts. Pantographs and collector systems based on carbon materials impregnated with copper are used in the European railway systems, while the Japanese have developed a Cu–Fe material impregnated with lead for their high-speed bullet train pantographs. Nickel–cadmium batteries produced by P/M technology are also used for diesel engine starting purposes. Electrical contacts are used in large numbers in the rail industry for various trackside signaling purposes. The commonly used electrical contact materials are silver–cadmium oxide or silver–graphite. Magnetic materials are also used to a limited extent in the railroad industry. Their applications, primarily in the areas of braking and warning systems and polarizing relays, include the use of the newly developed Fe–Nd–B based magnets and the extensively used Alnico magnets. One of the new and exciting applications of P/M Fe–Nd–B–based magnetic materials is in magnetically levitated trains. In Germany these magnets are fixed to the train and under the rail to provide the desired levitation. Other applications of P/M materials are in the area of coatings for wear resistance of various components such as diesel engine valve stems, rotary seals, and drive shafts in passenger trains. Thus, the railroad industry also consumes a significant amount of P/M parts, though it is not as well serviced by the P/M industry as are the automotive and aerospace industries.

At this juncture it would be best to end the introductory chapter, which simply tries to initiate readers into the world of particulate materials. This chapter has covered several varied topics such as powder characteristics, some of the commonly used cold-consolidation techniques for producing green shapes, and a novel cold-consolidation technique known as CONFORM^{TM}. This chapter also provides the reader with a brief glimpse of the wide range of applications that is serviced by the P/M industry. The brief list of applications covered in this chapter should definitely not be considered as "almost all the applications of the P/M industry," but only as a list that goes to show the versatility of the P/M industry in producing parts for a wide range of applications. It should be mentioned that though this chapter has briefly discussed several aspects of particulate materials that could not be covered as separate chapters, there are numerous other developments that could not be included within the scope of this book, primarily due to limitations of space.

References

1. A. Bose and R.M. German, *Modern Developments in Powder Metallurgy*, compiled by P.U. Gummeson and D.A. Gustafson, Metal Powder Industries Federation, Princeton, NJ, vol. 18, p.299, 1988.
2. R.M. German, *Particle Packing Characteristics*, Metal Powder Industries Federation, Princeton, NJ, 1989.
3. A. Bose, B.H. Rabin, and R.M. German, *Powder Metallurgy International*, vol. 20, no. 3, p.25, 1988.
4. A. Bose, H.R. Couque, and J. Lankford, Jr., *International Journal of Powder Metallurgy*, vol. 24, p.383, 1992.
5. C.M. Khipput and R.M. German, *International Journal of Powder Metallurgy*, vol. 27, p.117, 1991.
6. V.A. Tracey, *Powder Metallurgy*, vol. 35, p.93, 1992.
7. Reported in *International Journal of Refractory Metals and Hard Materials*, vol. 11, p.iii, 1992.
8. R.M. German, *Powder Metallurgy Science*, Metal Powder Industries Federation, Princeton, NJ, 1984.
9. F.V. Lenel, *Powder Metallurgy Principles and Applications*, Metal Powder Industries Federation, Princeton, NJ, 1980.

10. H.H. Hausner and M.K. Mal, *Handbook of Powder Metallurgy*, Chemical Publishing Company, New York, 1982.
11. C. G. Goetzel, *Treatise on Powder Metallurgy*, Interscience Publishers, New York, 1949.
12. W.D. Jones, *Fundamental Principles of Powder Metallurgy*, Edward Arnold Publishers, London, UK, 1960.
13. W.R. Morgan and R.L. Sands, *Metallurgical Reviews*, vol. 14, p.85, 1969.
14. P. Popper, *Isostatic Pressing*, Heyden and Sons Ltd., London, UK, 1976.
15. D. Green, UK Patent 1370894, filed March 1971.
16. J.A. Pardoe, *New Perspectives in Powder Metallurgy*, Ed. K. M. Kulkarni, Metal Powder Industries Federation, Princeton, NJ, vol. 8, p.77, 1987.
17. R.K. Dube, *Powder Metallurgy*, vol. 33, p.119, 1990.
18. G. Wranglen, *Corrosion Science*, vol. 10, p.761, 1970.
19. M.K. Ghosh, *NML Technical Journal*, vol. 5, no. 1, p.31, 1963.
20. *What are Refractory Metals and How do They Affect Our Lives: Five Metals with Unique Characteristics*, Metal Powder Industries Federation, Princeton, NJ, 1981.
21. E.A. Loria, *Journal of Metals*, vol. 39, p.22, 1987.
22. J. Dickson, *Modern Developments in Powder Metallurgy*, compiled by P.U. Gummeson and D.A. Gustafson, Metal Powder Industries Federation, Princeton, NJ, vol. 18, p.775, 1988.
23. R.K. Govila and W.V. Knopp, *Advances in Powder Metallurgy*, compiled by E.R. Andreotti and P.J. McGeehan, Metal Powder Industries Federation, Princeton, NJ, vol. 2, p.401, 1990.
24. D.G. White, *Advances in Powder Metallurgy*, compiled by E.R. Andreotti and P.J. McGeehan, Metal Powder Industries Federation, Princeton, NJ, vol. 1, p.1, 1990.
25. D.G. White, *Advances in Powder Metallurgy*, compiled by L.F. Pease III and R.J. Sansoucy, Metal Powder Industries Federation, Princeton, NJ, vol. 1, p.1, 1991.
26. E. Mosca, *Powder Metallurgy*, vol. 35, no. 1, p.38, 1992.
27. G. Meetham and D.R. Mayo, *Proceeding of the Conference on Powder Materials in Transportation*, York, October 1991, The Institute of Metals.
28. C. Suryanarayana, F.H. Froes, and W.E. Quist, *Advances in Powder Metallurgy*, compiled by L.F. Pease III and R.J. Sansoucy, Metal Powder Industries Federation, Princeton, NJ, vol. 6, p.15, 1991.
29. R.L. Martin and R.J. Lederich, *Advances in Powder Metallurgy*, compiled by L.F. Pease III and R.J. Sansoucy, Metal Powder Industries Federation, Princeton, NJ, vol. 6, p.361, 1991.
30. V.A. Tracey, *Powder Metallurgy*, vol. 35, p.31, 1992.

Chapter 2

Chemical Powder Production Approaches

Introduction

A large variety of powders can be produced by chemical processes, which by definition will involve some form of chemical reaction during the processing and essentially constitute the fabrication of powders from molecular precursors. Chemical powder production approaches can be adapted to produce elemental, alloyed, or mixed powders. The powder sizes are generally toward the finer side, and the process is especially suited for the production of fine submicrometer powders. The process can also produce powders of both crystalline and amorphous materials. This approach is especially suited for producing powders of advanced ceramics and high-temperature materials. Chemistry and chemical reactions play a dominant role in the production of powders by this method. A large and impressive variety of chemical reactions can be used to produce the same powder with totally different powder characteristics such as powder particle size, powder particle shape, flowability, apparent density, compressibility, etc.

In the chemical approach the physical state of the starting components can include a large variety of combinations of all the three states of matter, namely, solid, liquid, and gas. A few examples are given below to illustrate the point. The Bayer process, an old and well-known process used for producing alumina powder from bauxite, uses liquid as the starting material. Production of $MgCO_3$ from MgO and silicon carbide from CH_3SiCl_3 are examples where the starting components are solid and gas, respectively. The production of silicon nitride by the reaction of $SiCl_4$ and NH_3 (gas–liquid or liquid–liquid), aluminum nitride from alumina, carbon and nitrogen (solid–solid–gas), TiO_2 on alumina from alumina and $Ti(OR)_4$ (solid–liquid), and Al_2TiO_5 from alumina and titania (solid–solid) are examples that serve as illustrations of the myriad possibilities that exist when using the chemical route for powder production [1].

The requirements that the powders must satisfy, the type of chemical reactions, and the powder production routes cannot be generalized. Chemical means of powder production can offer tremendous advantages in the area of high-temperature materials because the melting point of the material is not a factor. Thus, this would be an excellent process for producing powders to be used as very high-temperature materials and composites based on high-temperature matrices.

Chemical powder preparation techniques can be used to great advantage in the production of "amorphous" and "microcrystalline" and/or "nanocrystalline" powders. Conventional preparation of these powders in large bulk quantities is difficult and expensive. The wet chemical synthesis methods offer an inexpensive alternative that is only recently being exploited. This chapter will briefly dwell on some of the new and exciting developments that could become great commercial successes. It will not be possible to cover all the new developments in any depth. Instead, the chapter will attempt to illuminate the vast possibilities that could potentially be realized by the use of chemical processing techniques for producing powders for use in advanced materials. The broad headings used to categorize the different processes also have a considerable degree of overlap. For example, chemical vapor deposition process can be used to produce microcrystalline or even nanocrystalline powders; metallic salts can be used to produce powders that are amorphous; the reaction spray process uses salts of metals in solution. However, for the sake of clarity, the material is divided into smaller subsections.

Powders Using Organo–Metal Chemistry

The use of organometallic precursors to produce metal powders is probably one of the oldest commercial techniques. The decomposition of metal carbonyls to provide the metallic powder is a technique that is still used extensively to produce large quantities of various powders. On the other hand, the use of metal alkyls and metal amides to produce powders, is still in its nascent stage. This part of the chapter will discuss some of the concepts of using organometallic precursors to produce powders.

One of the best, and probably amongst the oldest known, chemical processes for producing powders is a method known as the "carbonyl process," which uses the thermal decomposition of an organometallic precursor. Here the powder is produced by vapor decomposition and condensation of a metal-carbonyl precursor. Although this process was discovered in 1889, it is still an important process used extensively for producing nickel, cobalt, copper, iron and a number of other powders. The recent demand for fine, high-purity powders for powder injection molding has also boosted the use of this century-old process. This process is also being used to produce coated powders, such as nickel-coated graphite or nickel-coated alumina.

The process involves the formation of the desired carbonyl by reacting the metal with carbon monoxide. The metal carbonyl, which normally occurs as a liquid phase close to atmospheric conditions, can be purified by fractional distillation. The liquid is then vaporized under controlled conditions of temperature and pressure and the resultant vapor decomposes to yield fine, high-purity metal powder. A wide variety of powder morphology can be obtained by control of the decomposition parameters. The powders produced by this process can have a spiky or chainlike appearance, rough or smooth spherical appearance, or be of a filamentary type. The microstructure of the cross-sectional area of a carbonyl powder can in some instances have an onion-skin-like appearance. The impurity associated with this process is usually carbon.

Two illustrations briefly describing the production of nickel and iron powders by the carbonyl process will be outlined in this part of the chapter. The production of carbonyl nickel powder as used in the Clydach Refinery [2] is as follows. Nickel oxide, which is reduced to metallic nickel, is reacted with CO at a temperature of 323 to 328 K (50–55°C) to form the nickel carbonyl gas by the reaction:

$$Ni + 4CO = Ni(CO)_4$$

The product gas is filtered to remove the dust particles and is fed to the top of a decomposition unit whose walls are heated to 773 K (500°C). The nickel carbonyl gas instantaneously decomposes into metallic nickel powder and CO gas, which is recycled. The decomposition conditions can be varied to produce powders of various shapes. The powder produced by this process is exceptionally pure and its shape can be varied by varying the reaction conditions. Figure 2.1 shows an SEM photomicrograph of a spiky nickel powder produced by the carbonyl process.

Carbonyl iron powder is another commonly used powder produced by the carbonyl process. This powder is currently in high demand for metal injection-molding applications [3]. Inexpensive molten iron can be used as the starting raw material. The iron is reacted with CO at a temperature of 473 K (200°C) and pressure of 20 MPa to produce iron pentacarbonyl (IPC), $Fe(CO)_5$, a toxic liquid with trace impurities of other carbonyls. The trace impurities are removed by distillation, resulting in extremely pure IPC liquid that boils at 378 K (105°C). The carbonyl vapor is introduced into a decomposition chamber where the temperature is maintained above 523 K (250°C) and more CO gas is added to

Figure 2.1: SEM photomicrograph of a spiky nickel powder produced by the carbonyl process.

the vapor. A complex sequence of reactions result in the formation of iron powder with an onion-skin-like structure formed of alternating layers of iron, Fe_3C, and some amount of free carbon. The typical impurities incorporated are carbon and oxygen. The powder is spherical in nature, extremely hard, and has a mean diameter that varies from 2 to 10 μm. A SEM photomicrograph of a carbonyl iron powder is shown in Figure 2.2. A subsequent hydrogen reduction step is often used to remove the majority of the impurities in the powder and also create a soft iron powder.

The use of the carbonyl process for coating a host of different materials such as glass, oxides, carbides, metals and even fibers with nickel is opening up totally new applications for these powders. INCO has developed a fluidized-bed pilot plant that can coat other powders with 1.7×10^{-3} to 2.8×10^{-3} kg/s of nickel and the coating thickness is extremly uniform [2,4]. The particles to be coated rest on a porous bed designed to allow the passage of a fluid. The flow rate of the fluid is of great importance, as very low flow rates will not result in any fluidization while very high flow rates will blow away the powders from the vessel. Any powder that can be fluidized, and also held at the temperature at which the nickel carbonyl decomposition can occur, can be coated with nickel. The optimum size range of the powders that can be coated lies between 50 and 300 μm. Coating of aluminum and magnesium powders presented the problem

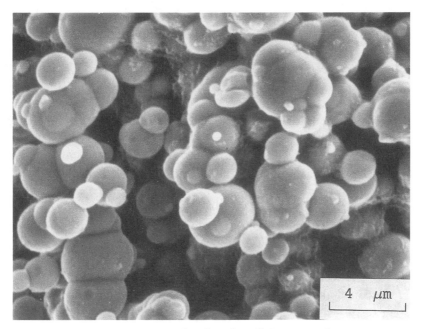

Figure 2.2: SEM photomicrograph of carbonyl iron powder.

of an uncontrolled exothermic reaction. During operation, violent exothermic reaction could occur and fuse the bed into one solid piece. The list of applications for the coated materials include EMI/RFI shielding, conductive fillers, cutting or grinding wheels, abradable seals, fuel-cell electrodes, cutting tools, and a host of other aerospace and automotive applications.

New areas where the carbonyl powder technique can be applied are numerous. According to the present author, this technique could be suitable for producing intimately alloyed fine powders using mixed carbonyl precursors. A few of the possible uses that can be imagined [5] include tungsten-based heavy alloys that are produced by liquid phase sintering of an elemental mixture of tungsten, nickel, and iron powder; nickel–iron–cobalt alloy for magnetic applications; and ceramic powders coated with metal to be used as precursors for processing of novel composites. The metal carbonyls of the desired elements can be mixed together (in the liquid or gaseous state), vaporized, and decomposed to form a very intimate mixture of the desired alloy. The fine homogeneous powder mixture can easily be sintered to very high densities, resulting in improved material properties. The idea is extremely potent and the close involvement of chemists could greatly help the powder metallurgy industry.

The use of metal alkyls and metal amides is also expected to have a potential for producing powders of advanced materials such as intermetallic compounds or dispersion-strengthened materials. This approach is based on the production of electronically or coordinatively unsaturated metal species that combine through metal–metal bond formation to yield large metal-atom clusters or extremely fine powders. The approaches based on the elimination of volatile compounds seem to be the most attractive, since it minimizes the possibility of contamination and maximizes the potential for large-scale economic production of the powders. By varying the reaction conditions, this approach should be capable of forming intimate mixtures of elemental powders or very fine prealloyed powders that could be either crystalline or amorphous in nature.

Several pathways for producing intermetallic powders using metal alkyls and metal amides have been conceived, and they form the basis of an ongoing program sponsored by the Office of Naval Research [6]. The metal powders can be formed by the elimination of volatile amines or hydrocarbons, depending the chosen reaction path. The possible reaction pathways include the reaction of metal alkyls with metal hydrides, reaction of metal amides with metal hydrides, and reduction of mixed metal alkyls with hydrogen.

Several of the above concepts have been investigated and have yielded some interesting results. The usual reaction product is very fine atomic clusters that can combine together to form the powder depending on the reaction conditions. The characterization and handling of such powders is a major problem that currently the P/M industry is not totally geared for. The fine particulates are usually dispersed in a liquid medium (usually within a reaction flask). Attempts to characterize the material included drying of the particulates by driving out the liquid medium (under a protective atmosphere, usually nitrogen), followed by isolation of small amounts of powder from the main reaction flask into small glass crucibles with airtight caps. The transfer of the particulate material into the small glass crucible was carried out inside a glovebox with a protective atmosphere. Thus, the atmosphere over the crucible prevented any spontaneous reaction of the fine particulate material inside the crucible as long as the airtight lid was in place.

The next concern was to get the powder inside the SEM chamber. It was found that the moment the lid over the glass crucible was opened to the atmosphere the particulate material instantly burst into flame due to the spontaneous reaction of the fine reactive material particulates with atmospheric oxygen [7]. Thus a special inflatable glove bag was designed for use over the SEM chamber that would allow the SEM chamber and the glove bag to be flooded with nitrogen before the airtight cap of the glass crucible was opened. The fine particulate is then removed from the crucible on to a small metallic holder, which is then inserted into the open SEM chamber. All this transfer takes place inside the special glovebag, which is flooded with the protective atmosphere. Once transferred into the SEM chamber, the usual steps of

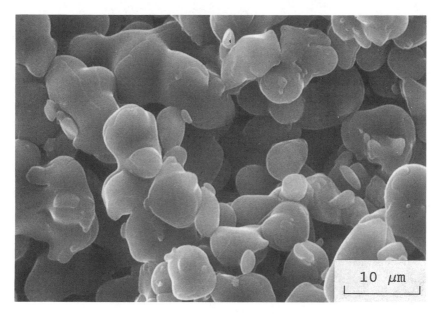

Figure 2.3: *SEM photomicrograph of a chemically produced titanium aluminide powder.*

evacuation and SEM analysis are carried out. An SEM photomicrograph of such a particulate material is shown in Figure 2.3 [7]. For compaction of such materials, pellets have been obtained by pressing the fine particulate material inside the protective atmosphere of the glovebag. The pellets can then be transferred to a glass crucible inside the protective atmosphere of the glovebag, then sealed and moved to the furnace location where the transfer into the furnace chamber can be achieved using a similar process as used to transfer the particulate material into the SEM chamber. The pellets can then be sintered using a suitable sintering atmosphere [7].

Several different reactions can be employed to form different powders of titanium aluminide. Reaction of tetramethyltitanium, $Ti(CH_3)_4$, with aluminum trihydride, AlH_3, at low temperatures such as 195 K ($-78°C$), can yield titanium aluminide powder through the elimination of methane. Alternately, aluminum alkyls could conceivably be reacted with titanium hydride or titanium hydride complexes to yield similar products. Reaction between metal alkyls, aryls, or amides under moderate pressure of hydrogen can also be used to produce the desired titanium aluminide powder through the elimination of alkane or amine. This approach can be extended to mixed aluminum–titanium alkyl complexes. Many of these complexes are currently available due to their role in the Ziegler–Natta process used for polymerization of ethylene. Several compounds containing Ti–H–Al bridges are known [8]. It is expected that titanium aluminide powder can be produced from these bridges by the elimination of alkane. Also, reaction of titanium amide with aluminum trihydride can yield the desired titanium aluminide intermetallic compound.

The above reaction scheme for producing salt-free titanium aluminide has been discussed in detail by Paul and coworkers [9]. In their attempts to produce salt-free titanium aluminide-based intermetallic compounds, titanium amide, $Ti[N(SiMe_3)_2]_3$, taken in a dilute hexane solution, was combined with an excess of aluminum trihydride, AlH_3, at ambient temperature to form a very fine steel-gray particulate material that precipitated after 16 hours of stirring. The particulate material was first washed with ether and then vacuum dried; it is believed to be a loose cluster of atoms of titanium and aluminum with relatively few organic functions and a large number of "dangling bonds." The particulate material was pressed into cylindrical pellets and subjected to an elevated temperature of 1273 K (1000°C) in vacuum for 900 s. The sintered pellet was then crushed into a powder and characterized by X-ray diffraction (XRD) and high-resolution electron microscopy (HREM). Figure 2.4 shows a bright-field micrograph that is composed of agglomerates of particles having a size range of 5 to 25 nm. The XRD pattern of the crushed powder is shown in Figure 2.5 and reveals the presence of the intermetallic compound $TiAl_3$ and small amounts of titanium oxide and Ti_3Al. It is expected that with better control of the reaction conditions it will be possible to process a higher-purity $TiAl_3$ intermetallic compound.

It should be pointed out that the actual titanium aluminide-based intermetallic compounds that are considered to be of practical use do not have simple binary compositions such as Ti_3Al or $TiAl$. For example, Ti–48Al–1V and Ti–24Al–11Nb are two titanium aluminide compounds that are of great commercial interest. Powder production of these ternary compounds containing either niobium or vanadium could be possible by the incorporation of the niobium or vanadium alkyls or amides. The development of different compounds with specific elemental compositions can be achieved with great accuracy by choosing various alternate chemical powder precursor routes. It can also be speculated that the above reaction schemes could be used in conjunction with other organometallic compounds to produce novel dispersion-strengthened materials.

Production of fine powders through the use of polyol liquids is another interesting chemical process that is nearing commercial realization. The patent filed by Figlarz et al. [10] laid the basis for making fine metal powders of easily reducible metals such as Pd, Ag, and Pt and also less reducible metals such as Co, Ni, and Pb. Currently, pilot plants have been started in France for the production of monodispersed, nonagglomerating, spherical nickel and cobalt powders.

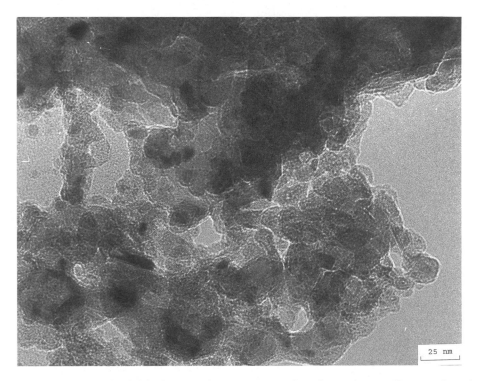

Figure 2.4: Bright-field SEM photomicrograph of a chemically produced titanium aluminide composed of agglomerates of nanosize particles. (With permission from the authors of Ref. 9.)

In this process, nickel hydroxycarbonate or cobalt hydroxide is suspended in a solution containing ethylene glycol or diethylene glycol or a mixture of both glycols. The suspension is heated to the boiling point of the polyol and also stirred simultaneously to form metal powder by the process of nucleation and growth from the solution. At first an intermediate turbostratic phase consisting of precipitates with an incompletely ordered lamellar structure is formed with polyol molecules or the corresponding alkoxy radicals. In the second stage, re-dissolution of the intermediate solid phase occurs and the reduction of the cobalt or nickel species takes place in the liquid phase. Variously sized powders can be obtained with a very narrow size distribution. Very fine submicrometer powder can be produced by seeding the reaction medium with other metal nuclei (such as silver, through silver nitrate solution) to induce heterogeneous nucleation. The process produces pure powders with a guaranteed purity of greater than 99.9 wt.%.

It is possible that one or several chemical routes attempted for producing

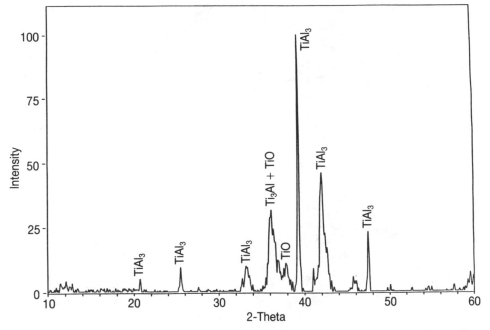

Figure 2.5: *The XRD pattern of the crushed titanium aluminide powder (after heating in the form of a pellet). The pattern reveals the presence of the intermetallic compound TiAl₃ and small amounts of titanium oxide and Ti₃Al. (With permission from the authors of Ref. 9.)*

powders of a particular type of material may not be successful. However, the great advantage of the chemical processing route is the fact that, if one reaction is for some reason unsuccessful, other alternate approaches can still be conceived. Once the feasibility of producing the powders has been established, it is expected that the chemical reactions can be extended to produce powder precursors for advanced materials. Since the melting points of the materials are of no consequence whatsoever, the chemical route will perhaps offer the best powder fabrication route for producing extremely fine powders of ultrahigh-temperature intermetallic compounds.

Powders from Oxide Reduction

One of the best-known and most widely used chemical powder production techniques is the oxide reduction process. This process has been used extensively to produce a variety of metal powders such as iron, nickel, tungsten, and molybdenum. The process is also suitable for reducing mill scales from hot rolling mills and machined swarf. The process relies on reduction of purified metallic oxide using either hydrogen or carbon monoxide. The most critical parameter of the process is the temperature at which the oxide reduction is carried out. The temperature should be high enough to allow the oxide reduction to occur at a reasonable speed that makes the process economically viable. Since the reduction of the metallic oxide to the metal will start from the outside surface and move inward into the oxide powder particle, an outer shell of metal is formed on the oxide particle. The kinetics of further reduction is subject to the slowest step among several diffusion-controlled steps. Since the slowest step is generally diffusion controlled, an increase in the thermal activation will result in a rapid increase in the reduction rate. However, higher temperature will also promote particle-to-particle sintering and result in a coarser powder size. Thus the selected oxide reduction temperature reflects a compromise between the two aforementioned conflicting requirements.

Thermodynamic principles dictate the minimum temperature that should be used to reduce the metallic oxide. The reduction of oxide in hydrogen or carbon monoxide must result in a negative change in free energy for the reactions that are shown below (where M represents the metal).

$$MO \text{ (solid)} + H_2 \text{ (gas)} = M \text{ (solid)} + H_2O \text{ (gas)} \qquad (1)$$

$$MO \text{ (solid)} + CO \text{ (gas)} = M \text{ (solid)} + CO_2 \text{ (gas)} \qquad (2)$$

The equilibrium constants of the above reactions, K_1 and K_2 for equations (1) and (2), respectively, are given below.

$$K_1 = \frac{\text{Partial presure of } H_2O}{\text{Partial presure of } H_2}$$

$$K_2 = \frac{\text{Partial pressure of } CO_2}{\text{Partial pressure of } CO}$$

Thus, the balance between oxidation and reduction will be dependent on the selected temperature and the equilibrium constant established by the ratio of the partial pressures, as shown in the schematic diagram depicted in Figure 2.6. It can be observed that below the equilibrium line the metal oxide is more stable and above it the pure metal is more stable. Thus, for reduction of the oxide, the partial pressure ratio and the temperature have to be adjusted such that the pure metal is more stable than its oxide counterpart (which is below the equilibrium line). Equilibrium charts for individual metals and their oxides can be determined from the thermodynamic literature [11,12].

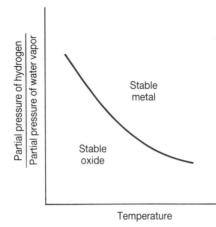

Figure 2.6: Schematic diagram showing the balance between oxidation and reduction, which is denoted by the equilibrium line. Below the equilibrium line the metal oxide will be more stable, while above it the metal is more stable. The temperature and partial pressure ratio have to be adjusted so that conditions favor oxide reduction (above the equilibrium line).

Low-temperature oxide reduction results in a large volume change. Powders obtained by low-temperature reduction are usually porous and spongy in nature, as can be seen in the spongy powder shown in Chapter 1 (Figure 1.2b). On the other hand, the reduction at high temperatures often results in dense polygonal particles, as also seen in Chapter 1 (Figure 1.2j). Tungsten powder, which is usually obtained by the reduction of its oxide at high temperatures, is usually available in dense polygonal form. The reduced powders are mechanically comminuted and often annealed to improve the compressibility of the powders.

The oxide reduction process can also be used to produce quality sponge iron powder from a high grade of naturally occurring iron oxide deposit. An example of this has been demonstrated in India, which has a large deposit of hematite (Fe_2O_3) which is available in a fine dust form. This fine powder known as "blue dust" has low silica and alumina content and is essentially free of impurities such as carbon, sulfur, and phosphorus. Blue dust powder has been reduced to premium-grade sponge iron powder by both hydrogen and carbon. The final properties of the reduced iron powder can be tailored by variations in the post-reduction conditions.

Prealloyed iron powders have also been produced by the coreduction of other oxide powders with blue dust. This naturally occurring resource, approximately

2×10^9 kg (with less than 1% silica level), has tremendous commercial potential for producing high-quality sponge and prealloyed iron powder.

The "blue dust" powder is first subjected to conventional beneficiation treatment by which the silica content is reduced to less than 0.2%. In one process, the blue dust concentrate is reduced by flowing hydrogen in a pusher type of furnace. The reduced sponge forms a cake, which is mechanically comminuted and then followed by a secondary operation. Alternately, the blue dust is reduced by a mixture of coke and limestone. To prevent any contamination, the blue dust and coke powders are arranged in layers separated by a porous refractory felt that allows free passage of gas but prevents ash contamination. Complete reduction can be achieved by heating a multilayered charge for 432×10^2 s (12 hours).

Blue dust powder mixed with the desired proportions of nickel, copper, and molybdenum oxides has been coreduced in dry hydrogen to form an alloy powder cake. This cake is subsequently comminuted to yield the desired powder. Blue dust powder mixed with desired amount of tungsten oxide powder has also been coreduced in dry hydrogen to produce ferro-alloy powders. X-ray results indicated that complete alloying was obtained by reducing in dry hydrogen for 7200 s (2 hours) [13]. Blue dust mixed with the desired amount of tungsten oxide powder, coreduced in dry hydrogen at 1373 K (1100°C) for 72×10^2 s (2 hours), was successful in producing fully alloyed ferro-tungsten. The ferro-tungsten pellets could be directly used for high-speed-steel production.

A large volume of literature exists on this method of producing powders. The process has been used from the very early days, and was probably the basis for the production of a large number of ancient metallic products. It would be outside the scope of this chapter to discuss in detail the variety of materials that are processed by this conventional technique. However, the process was so important and is used to produce powders of so many materials that a brief discussion of the process is necessary. The process is still used today to produce iron powders from mill scales, tungsten and molybdenum powders from their respective oxides, and also numerous other commonly used metal powders.

Amorphous, Microcrystalline, and Nanoscale Powders

The development of specialty materials is extremely dependent on novel microstructures. The horizons of new materials technology have been expanded further by the incorporation of amorphous, microcrystalline, and nanocrystalline materials within its fold. Structures with grain sizes in the scale of 0.01 to 0.1 μm (10 to 100 nm) provide exciting new material properties since now the atoms present in the grain boundaries could be more numerous than those present in the actual grains. Metallic glasses are another class of material that exhibit amorphous characteristics. The production of such microcrystalline, nanocrystalline, and amorphous powders using the chemical approach is gradually becoming the subject of intense research investigations.

Homogeneous nucleation from the vapor phase to produce nanoscale particles is presently being explored. Nucleation from supercooled vapors is achieved by vaporizing carbonyl compounds under 0.0015 MPa residual pressure of argon. Fine copper, cobalt, gold, zinc, and platinum powders have been produced by this process.

Metal vapor synthesis is a recent chemical technique that has been used to produce highly active catalysts and very fine metal powders. The process is based on the generation and controlled reaction of discrete metal atoms within a solvent matrix. Although the production of fine powders has been well characterized for some systems [14] and the potential for producing alloyed powders has been alluded to [15], the utilization of powders produced by metal vapor synthesis in P/M processing of advanced materials is yet to be accomplished.

The process in essence is based on the cocondensation of suitable metal vapor and a solvent. This is accomplished by admitting an excess amount of solvent into a vacuum chamber where the pressure is maintained at 13.33×10^{-10} MPa (10^{-5} Torr) and where the chamber walls are generally cooled to liquid nitrogen temperature while the metal is vaporized by resistance heating or electron beam vaporization. The cocondensation produces an array of individual metal atoms isolated in the frozen solvent matrix. Upon warming in a nonreactive solvent, the metal atoms coalesce to form clusters or very fine particles (in the order of 1.2 nm), depending on the metal, solvent, and the reaction conditions.

The above method has been used to produce approximately 0.006 μm (6 nm) diameter nanocrystalline iron crystals [16]. In the process, pure iron was evaporated from a tungsten heater in 99.9996% helium at a low pressure of 2×10^{-3} MPa, and was accumulated on a vertical cold finger maintained at 77 K. The iron deposited on the cold finger was stripped in high vacuum, and compacted at 70 MPa in a piston-and-anvil device at room temperature to produce a pellet of 5 mm diameter and 0.2 mm thickness. The exciting conclusion of the paper can be summarized as follows.

1. Nanocrystalline solids will not be restricted to only iron, as similar crystals of germanium have also been produced using a similar process.
2. The nanocrystalline systems will not be limited in sample geometry.
3. The system may be extended to multiphase nanocrystalline materials of hitherto incompatible structural and chemical constituents.

The last conclusion opens up large unexplored horizons with practically unlimited potential for producing new and advanced materials.

Physical vapor deposition (PVD) as a means of producing nanocrystalline powders is also being investigated [17]. Here, material is removed from the bulk under vacuum by thermal energy supplied by an electron gun. The technique uses vapor energy in the form of a focused molecular beam that is directed toward a cooled substrate. The vapor flux is contained in a tube with heated walls to prevent condensation on the walls. Carefully controlled amounts of gases that interact with the molecular flux are admitted in the substrate region. The gas absorbs energy from the vapor, which condenses to form very fine particulates. This process is also in its infancy, and whether it will mature to become a viable powder processing route is difficult to predict at this juncture.

Reduction of metallic salts with alkaline borohydrides has been used to produced amorphous metal–metalloid powders whose properties are dependent on the metal and the method of powder preparation [18]. The process consisted of adding 0.2 mol of solid KBH_4 to 100 ml of an aqueous solution of $CoCl_2$, $NiCl_2$, or $FeSO_4$ in an ice bath. The mixing was carried out in a hydrogen atmosphere by vigorous stirring. The resultant powders were washed to remove the excess reagent. The washing step had a strong influence on the chemical composition of the material. When the powders were wet with acetone, they could be handled in air, but the dry powders were pyrophoric due to the extremely fine particle sizes. A safe way to handle the powders is to reduce the surface area by pressing them into pellets before exposing them to air. The X-ray spectra of the Co–B–O and Ni–B–O materials do not show any crystalline phase, while the Fe–B–O show Bragg peaks that are superimposed on an amorphous halo. Annealing at higher temperatures produced the appearance of one or more crystalline phases.

Nanoscale powders have also been produced by hydrothermal synthesis. This novel powder production technique has been used by Tani et al. [19] to produce ultrafine single-phase monoclinic ZrO_2 powders. The powders had a size range that varied from 0.016 to 0.022 μm (16 to 22 nm) depending on the processing conditions.

The initial material used for the process was amorphous hydrated zirconia, which was precipitated from zirconium tetrachloride solution by ammonium hydroxide. The precipitate was washed with distilled water and then dried at 393 K (120°C) for 1728×10^2 s (48 hours). The starting material was taken in a platinum or gold tube (3.3 mm outer diameter, 3 mm inner diameter, and 32 mm length) with 8 wt.% of KF solution. This tube was taken in a test tube-type pressure vessel made from Stellite (6.5 mm inner diameter), heated to a temperature anywhere between 473 and 873 K (200 and 600°C) and pressurized at 100 MPa for 864×10^2 s (24 hours), followed by water quenching. The product was then washed in water and tested. No trace of potassium could be detected by energy-dispersive spectrometry. X-ray powder diffraction confirmed that the powder was monoclinic ZrO_2. Interestingly, the powder particle size, which varied from 0.016 to 0.022 μm for temperatures between 473 and 773 K (200 and 500°C), increased rapidly to 0.07 μm when a temperature of 823 K (550°C) was employed. It has been found that the use of KF solution as the mineralizer yields only monoclinic ZrO_2 while the use of water or lithium chloride solutions produced both tetragonal and monoclinic ZrO_2 under hydrothermal synthesis conditions.

The chemical processing of nanophase WC–Co composite powders has become a commercial reality. The importance of this process, which includes the preparation and mixing of starting solutions, spray drying to produce chemically homogeneous precursor powders, and fluid-bed thermochemical conversion of the precursor powders to nanophase WC–Co powders, warrants detailed discussion during the course of this chapter.

In order to contrast the conventional method of processing WC–Co with the new chemical process, it is necessary to briefly describe the steps involved in the processing of conventional WC–Co. In the conventional process the WC powder is produced by the comminution of WC that is produced by the reaction of tungsten with carbon at around 1773 K (1500°C). The desired powder particle size of the comminuted WC is then mixed with cobalt powder in the desired proportion and ball-milled with paraffin. The mixed agglomerated powder is then pressed to the desired shape, dewaxed and presintered in hydrogen, and liquid-phase sintered at around 1673 K (1400°C) to obtain the fully dense bulk WC–Co part. The grain size of the WC particles is quite large as it cannot be smaller than the comminuted WC powder particle size, which is generally in the 1 to 10 μm range. Often other problems like segregation of inhomogeneous pools of cobalt occur within the specimen after sintering due to the in-homogeneous distribution of the powders.

In contrast, the process for producing WC–Co from the chemical precursor route involves the mixing of H_2WO_4 in ethylenediamine (en) with an aqueous solution of $CoCl_2$, which results in the precipitation of $Co(en)_3WO_4$. This precipitate forms the precursor for the desired compound and has the shape of hexagonal prismatic rods. This precipitate is decomposed in a reducing atmosphere consisting of equal parts of hydrogen and argon to form the reactive intermediate of W–Co, which is nanophase and nanoporous. This high-surface-area reactive intermediate of W–Co is subsequently converted to WC–Co by the carburization reaction with CO_2–CO gas. The final microstructure is determined by the temperature of carburization and the activity of carbon in the gaseous phase. The final powder size varies between 20 to 100 μm and it retains the hexagonal prismatic rod shape of the original precipitate that was obtained from the solution. However, the individual particles are composed of nanophase particles of WC–Co-based composite.

The laboratory-scale development work easily proved the concept of processing these materials [20–22]. However, certain limitations of the laboratory-scale process were immediately apparent. Some of the limitations of the laboratory-scale fixed-bed reactor that have been outlined by McCandlish et al. are outlined below [23].

1. The precursor compound produces a fixed composition of WC–Co where the Co : W is present in equiatomic proportions, which results in a WC–Co with 23 wt.% cobalt. Generally the WC–Co composition varies from 5 to 30 wt.% Co, with majority of the important cutting tool applications being limited to the low cobalt contents. Thus, the fixed WC–Co composition severely limits the versatility of the process in producing WC–Co parts with a wide range of applications.
2. Obtaining large quantities of powder for processing actual bulk parts for testing is difficult. Also the quantity of powder that can be properly carburized is low, as the gas percolation and heat transfer is limited.
3. The shapes of the powders were not very amenable for subsequent handling steps.

A combination of spray drying and fluidized-bed conversion was selected as the processing route for effective production of WC–Co nanocrystalline powders on a large volume basis. One of the major advantages is that both the processes can easily be scaled up to a high-volume industrial-scale processing. The integration of the two processes to produce one particular material was also quite unique. To synthesize the homogeneous precursor powders, the rapid spray drying process is an ideal method, while the fluidized-bed conversion provides a uniform environment with respect to temperature and gas concentration, resulting in uniform conversion of the precursor powder. It has also been demonstrated that various levels of cobalt-containing WC powders can be processed by controlled thermochemical conversion of various solution mixtures of $Co(en)_3WO_4$ and tungstic acid, H_2WO_4, using the rapid spray drying process. Homogeneous precursor powders for the WC–Co-based composites can also be processed from other less expensive starting materials. Presently, precursor powders are being routinely fabricated from other chemical salts that are readily available.

Some of the details of the spray drying process are discussed in detail in the next section, which also deals with a few other specific cases of powders processed by spray drying. This section will only discuss the spray process as

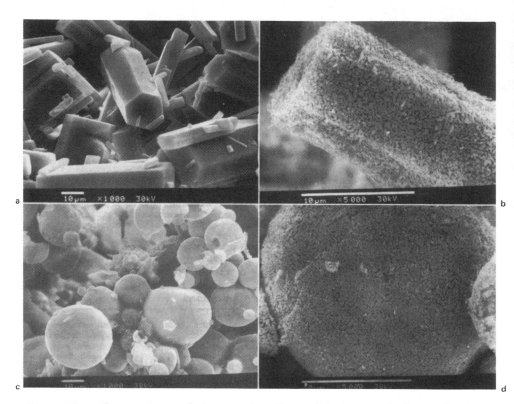

Figure 2.7: Comparison of the powders formed before and after reduction–carburization of Co(en)$_3$WO$_4$: (a) and (b) crystalline powder precipitated from solution; (c) and (d) spray-dried powder. (Reprinted with permission from Ref. 23.)

applied to the processing of the WC–Co composites. In this case, spray drying by a rotary atomization process that provides a narrower range of droplet size was used to process the WC–Co-based composite powders. It was determined that the factors that strongly influence the particle size, bulk density, and powder shape are the atomization energy, feed solution properties, feed rate, air flow rate, and gas temperature. It has been determined that spherical powders can be produced by the reaction spray process. A comparison of the shape and sizes of the powders produced by the simple precipitation of the powders from the solution and from the spray drying process is shown in Figure 2.7. The precipitated powders clearly show the rodlike nature of the powders, while the spray dried powders are spherical in nature.

The spray dried powder of the desired composition is then subjected to a fluidized-bed reactor where the solid particles are levitated in a flowing gas stream. The fluidizing gas velocity at which the particles are just suspended is termed the incipient fluidization velocity. It has been found that the gas velocity can be at least five times the incipient velocity before particle entrainment occurs and the heat transfer is reduced. The gas–solid environment is very uniform. A fluidized bed reactor operating in the dense bed regime is considered to be ideal for rapid scaleup.

These powders can be used for hard-facing applications by low-pressure plasma spraying, laser surfacing, and hypersonic jet spray deposition. Other processes such as press-and-sinter or even particulate injection molding of these powders can possibly be used to produce desired shapes. It has to be realized that a process of rapid liquid-phase sintering has to be used for such powders, since exposing the powders to the liquid phase for a prolonged period of time will result in the coarsening of the WC particles.

Some of the application areas that have been identified include cutting tools, drill bits, wear parts, and bearings. Owing to the unique microstructures that are possible, new applications such as microtome blades, medical scalpels, high-performance saw blades, and so on, are being envisioned.

To end this section on a very positive note, it should be pointed out that this form of processing is not restricted to WC–Co-based cermets. As a matter of fact, this process is not restricted to "cermets," but can conceivably be used to process metal–metal, ceramic–ceramic, and metal–ceramic based composites. A few of the applications where this technique could be successfully used include dispersion-strengthened copper for electrical contacts; super-high-strength alloys of copper reinforced with high volume fraction of tungsten, molybdenum, or other hard ceramic phases for switching gear; high volume fraction of W-based composites for rocket nozzles; and numerous other types of cermets.

Reaction Sprayed Powders

Salts taken in liquid solution, when sprayed into a hot chamber, rapidly lose the liquid and form powders. This process can be used in the formation of composite powders with fine dispersions or mixed homogeneous oxide ceramic powders. A novel reaction spray process (RSP), reported by Haerdtle [24], uses the above principle to produce tungsten-based alloys with oxide dispersions.

The principle of reaction spraying involves the mixing of aqueous or other solutions of salts according to the alloy composition. The liquid solution is sprayed into a hot reaction chamber where the solvent evaporates within a fraction of a second, leaving extremely fine mixed salt particles. The salt particles are finally reduced in hydrogen to obtain the desired powders. Due to the rapid evaporation of the solvent, the homogeneity of the dissolved

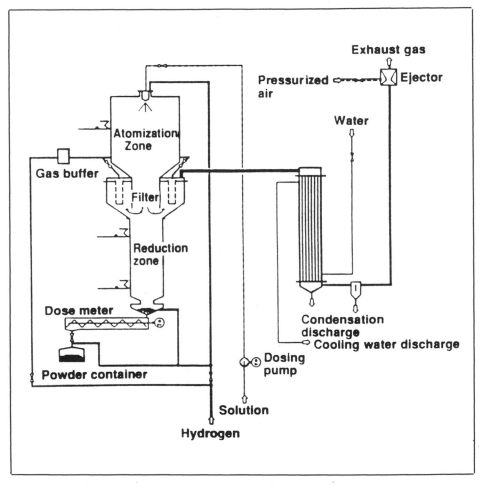

Figure 2.8: Schematic diagram of the reaction spray process. (Reprinted with permission from Ref. 24.)

components translates into powders with uniform composition. The process has been used to demonstrate the production of tungsten-oxide pseudoalloys by producing powders of Y_2O_3, La_2O_3, and ZrO_2 dispersed in tungsten. The oxide content was varied from 0.5 to 5 wt.%. The starting material used ammonium metatungstate with $La(NO_3)_3.6H_2O$, or $Y(NO_3)_3.6H_2O$, or $ZrOCl_2.8H_2O$. Concentrations of 300 kg of metal per cubic meter of solution were used. The powder obtained is spongy in nature with a particle size far below 1 μm. RSP has been applied for preparing hard metal powders, high-speed steels, tungsten heavy alloys, and molybdenum-based alloys. A schematic diagram of the reaction spray process is depicted in Figure 2.8 [24].

The reaction spray process has also been used to produce transformation-toughened alumina [25]. The feed material used in this investigation was again water-soluble salts such as chlorides of the desired components. The salts in the desired proportion are dissolved in water, and additives if necessary are also added in the form of water-soluble salts. The aqueous solution is thoroughly mixed and sprayed into a hot reaction vessel where the water almost instantly evaporates leaving behind individual powder particles that are extremely homogeneous. The powder is filtered, calcined, milled, spray dried, pressed, and sintered. The resultant material can be sintered in air between 1823 to 1873 K (1550 to 1600°C) for 3000 s (50 minutes) to yield an extremely homogeneous structure having 20 wt.% ZrO_2 uniformly distributed in an alumina matrix. The RSP can yield dispersions in the nanometer range. High densities are easily achieved by pressureless sintering due to high sintering activity of the powder.

The reaction spray process provides a large number of advantages over the conventional powder production techniques. The phase homogeneity of one component in another is extremely high and the process is suitable for multiphase materials. It is also a continuous process that can produce reproducible powders in large production quantities. The purity of the final powders is entirely dependent on the purity of the initial salts used in the process. The process is also quite economical with a great potential for producing homogeneously dispersed powders of multiphase materials.

Powders from Salts

Various chemical processes have been utilized to produce powders where the initial starting raw materials are the salts of individual components. Salts can be used to produce elemental powders, alloyed powders, mixed elemental powders, single or mixed ceramic powders, and powders of intermetallic compounds. In fact, some of the processes that have been described earlier use salts of the basic materials as the starting raw material. This part will discuss some of the other processes where salts have been used to produce the desired powders.

Fused salts or metal halides have often been reacted with magnesium to produce sponge powder. This type of process is heavily used in the production of titanium sponge powders. Chloride contamination is often a problem in these powders. In another process [26], tantalum powder is produced by the reduction of the double fluoride salt (K_2TaF_7) by sodium through the reaction

$$K_2TaF_7 + 5Na = 2KF + 5NaF + Ta \quad \text{(exothermic)}$$

The reaction is normally carried out at 1073 K to 1173 K (800–900°C) in inert gas or vacuum atmosphere. Sodium is usually in the liquid or gaseous state when the reduction reaction takes place. The reaction mixture is processed by leaching of the salts and repeated washing of the tantalum powder. This process results in tantalum powders with a narrow grain size above $1 \mu m$.

Reactions and precipitations from liquid provide an alternative approach to powder processing. Metallic salts and halides such as nitrates, sulfates, chlorides, fluorides, phosphates, etc., can be dissolved in a solvent and then precipitated by another compound. Metallic ions can also be reacted with hydrogen to form the powder. Another important variation is the uniform coating of one material on another that can influence the final sintered properties. A case in point is the nickel-activated sintering of tungsten. Very small amounts of nickel (a few atomic layers), when uniformly present on tungsten, can cause very rapid densification. Mixing of nickel powder with tungsten powder does not result in the desired short-circuit diffusion path. The tungsten powder can, however, be mixed with a solution of nickel nitrate in water, dried, and heated to break the nitrate salt into the metallic constituent, which is deposited as a uniform layer on the tungsten powder (the nitrate part is removed as nitrous oxide). The resultant cake can be milled to form the desired powder, which can subsequently be sintered.

An interesting process used to produce very fine powders of $PbTiO_3$, a useful high-temperature ferroelectric material due to its high Curie point (763 K, 490°C), has been described by Awano et al. [27]. Colloidal particles of $(Pb,Ti)(OH)_x$ in organic solvent is dehydrated by azeotropic distillation. $TiCl_4$ is hydrolyzed in distilled water, and $Ti(OH)_4$ is formed by adding aqueous ammonia. $Ti(OH)_4$ is filtered and transferred to aqueous nitric acid to which $Pb(NO_3)_2$ is added. The coprecipitation of $(Ti,Pb)(OH)_x$ is obtained by adding aqueous ammonia and vigorously stirring the mix. The coprecipitated $(Ti,Pb)(OH)_x$ is filtered and transferred into n-butyl alcohol and stirred. The concentration of the sol is controlled in the range of 100 to 300 $PbTiO_3$ mol/m^3. The alcohol sol is distilled and the dehydrated $(PbTi)O_x$ formed is then separated from the alcohol, dried at 383 K (110°C), and calcined at 793 K (520°C) for 3600 s (1 hour). Very pure $PbTiO_3$ powder with particle size ranging from 0.01 to 0.1 μm can be produced by this method.

Salts of aluminum, such as aluminum sulfate, have been used to produce fine alumina powder by urea decomposition in various aqueous media such as pure water, water–glycerol, and water–ethylene glycol [28]. Aluminum hydroxide precipitate was obtained by different methods. The precipitate is filtered, dried and fired at 873 K (600°C). All the powders were later fired at 1473 K (1200°C). It was observed that the precipitation medium had a profound effect on the powder characteristics.

Another use of the metallic salts is in the hydrometallurgical approach that provides an alternate route for producing powder feedstock suitable for plasma atomization [29]. The process involves the use of chemical salts to produce the intimately mixed powder agglomerates. The process has been used to produce tungsten heavy alloy powders (W–Ni–Fe alloy) from ammonium metatungstate, which is dissolved in water to form one solution, and from nickel and iron chlorides which are also taken into a separate solution. The two solutions are mixed and then treated with an ammonium hydroxide solution to increase the pH to 7. The increased pH causes the metallic salts to precipitate out in the form of an intimate mixture of ammonium paratungsteate and nickel–iron hydroxide. The solution slurry is then dried, and hydrogen sintered to provide strong bonds and reduce the salts to the metallic form.

It is noted that the nickel–iron is mostly surrounded by the tungsten, which seems to suggest that the nickel–iron was the first to precipitate and offer nucleation sites for the tungsten to precipitate. The dried material produces natural agglomerates of the powders, where the tungsten, nickel, and iron are intimately mixed in the desired composition. The sinter-reduced agglomerate has been used as the feedstock for processing plasma sprayed powders.

The process of freeze drying is another interesting approach that has been adopted to produce powders from metallic salts. It is an alternate approach to the rapid solidification process. In the freeze drying process, the desired metallic salts taken in the proper solution are rapidly frozen. The metal powders produced by this process are extremely fine and homogeneous. Freeze drying of tungsten and tungsten–rhenium alloys has been discussed and patented [30], and an excellent description of the process has been provided by Schnettler et al. [31]. The process has recently been utilized to produce freeze dried powders of tungsten heavy alloys [32]. Solutions of ammonium metatungstate (AMT) and nitrates or sulfates of nickel and iron were prepared. The AMT has good solubility in water, which allows metal concentrations of 100 to 200 kg of metal alloy per cubic meter. The AMT and the salts of nickel and iron were separately mixed with distilled water, the different solutions were then mixed together, and then flash frozen. The salt solution mixture was sprayed through a 0.5-mm-diameter orifice directly into hexane maintained at a temperature around 203 K ($-70°C$). The frozen microspheres were easily removed from the hexane by sieving. The frozen microspheres were transferred to a freezer where any residual hexane was allowed to evaporate. Drying of the microspheres was achieved by vacuum sublimation [32]. The as-dried powder was decomposed and calcined at 1023 K (750°C) for 7200 s (2 hours) in an air furnace. The calcined oxide powders were reduced in dry hydrogen in the temperature range 1098 to 1223 K (825–950°C). The powder obtained had aggregates of crystallites around 0.1 μm in size. The powder was further granulated to 20-mesh size before compacting and sintering. The compacts exhibited good green strength characteristics. The sintering experiments showed that a sintering temperature of 1513 K (1240°C) with a 7200 s (2-hour) hold was sufficient to obtain full density [32]. Some liquid-phase sintering studies were also carried out on these materials.

The tensile ductility of the alloys processed from the sulfates of nickel and iron was found to be extremely poor. It was found that impurity segregation of sulfur was the main reason for this. Investigations with nitrates were not completed. High sintered hardnesses approaching HRC 48 were reported for materials sintered at 1403 K (1130°C). The relationship between the hardness and the grain size follows the Hall–Petch relationship. Thus, theoretically it can be predicted that using lower sintering temperatures or hot isostatic pressing, hardness levels of HRC 60 can be attained in these materials.

Researchers from Sophia University have also been using a process of freeze drying to produce partially stabilized zirconia with yttria. The powder, having a diameter of around 0.3 μm, is produced by taking aqueous solution of zirconium chloride and yttrium chloride and subjecting the solution to ultrasonic waves (frequency 2.4 MHz). This produces droplets that are freeze dried and pulverized. The pulverized material is fired at 1073 K (800°C) in flowing oxygen for 3600 s (1 hour) to produce the desired powder, which is easy to disperse and has few agglomerates [33].

Salt-based precursor has also been used to produce an interesting bioactive ceramic known as hydroxyapatite, $Ca_5(PO_4)_3OH$ [34]. In this investigation, H_3PO_4 was added by a micropump into a mixed suspension of $Ca(OH)_2$ whose pH was maintained between 8 and 9 through the addition of a 12.5% solution of ammonium hydroxide. The resultant sediment was filtered, dried at 338 K (65°C), ground, and calcined at 1123 K (850°C) for 10800 s (3 hours). X-ray powder diffractometry of the 0.063-mm powder indicated that the powder was predominantly hydroxyapatite with a few percent of tricalcium phosphate, $Ca_3(PO_4)_2$. Porous products processed from this powder if successful in animal experiments will be clinically tested as a material for fillings in bone loss surgery.

Another interesting development in this area was the use of salts to produce high-grade tungsten powder from all kinds of scrap including hardmetals, heavy metals and alloys, grinding sludge, tungsten–copper alloys, and green compacts. In this process, developed by Metek of Israel [35], a mixture of the scrap and an inorganic salt such as sodium nitrite is introduced into a smelter preheated to approximately 823 K (550°C). Sodium tungstate is formed as the primary reaction product. The sodium tungstate is soluble in water. Other reaction products are the oxides of the binders such as cobalt, nickel, and iron, which form insoluble hydroxides. Calcium carbonate is added to the sodium tungstate solution to precipitate calcium tungstate, which is filtered and removed (most other elements remain in solution as calcium salts). Hydrochloric acid is added to the suspension containing the calcium tungstate, resulting in the precipitation of the yellow compound known as tungstic acid. This is followed by separation and washing of the insoluble tungstic acid, followed by drying, calcination, and reduction to provide the tungsten powder. The final tungsten powder particle size is dependent on the particle size of the tungstic acid. This powder can also be used to form WC by reacting with milled carbon black.

Another advance in this area that will not be covered in any detail, but which is certainly worth mentioning, is the process of coprecipitation. Oxide-based ceramic powders can be prepared by coprecipitation in aqueous solutions from the proper mixture of the citrates and oxalates of different materials. The coprecipitated material is filtered and calcined to decompose the salts and form the oxides. In multiphase oxide systems, the key to the success of this process lies in the ability to properly conucleate and precipitate the desired materials. Also, this process requires the use of low solution concentrations, which results in the formation of excessive carbon during the calcination stage.

The Methanol Process

A simple chemical process known as the methanol process (TMP) has been successfully utilized to produce binary or multiple mixed-oxide-based ceramic materials [36]. This process was developed with the goal of producing highly sinterable ceramic oxide powders that exhibit high purity, homogeneity, and fine grain sizes. The methanol process was first used to prepare lithium orthosilicate and lithium aluminate powders for nuclear breeder materials [37]. In this process, an alkaline and an acidic hydroxide are suspended in a primary alcohol. The boiling of the suspension under reflux conditions resulted in the formation of an organometallic complex, which serves as the intermediate phase. The intermediate organic phase, which has not been studied in detail, is separated from the alcohol and decomposed by calcination to form the fine-grained sinterable powders. The metallic hydroxides can be divided into three distinct groups consisting of alkaline, amphoteric, and acidic hydroxides. Materials in the alkaline group include lithium, sodium, potassium, magnesium, calcium, and zinc hydroxides; materials in the amphoteric group include hydroxides of aluminum and yttrium; the acidic hydroxide group consists of chromium, titanium, zirconium, and silicon hydroxides. It has been demonstrated that combinations between different groups are possible. Methanol is usually used as the suspension medium such that the free carbon formation during the calcination process is minimized [36].

The methanol process has been successfully used to produce varistor ceramics, which are essentially zinc oxide doped with a variety of different oxides; mullite;

transformation-toughened ceramics like zirconia–yttria dispersed in alumina; and breeder materials such as lithium aluminate, lithium orthosilicate, or lithium zirconate. The process can be scaled up to produce production quantities of powders. In this process, it is not necessary to work with low suspension concentrations. Usually a concentration of 150 to 200 kg/m^3 of the starting material can be used.

The powders obtained by the methanol process have high flowability. A spray drying–calcination process that can produce extremely fine submicrometer grain sized powders from low concentration solutions often produces very fine spherical powders. The methanol used in the suspension can be reused after removal of the intermediates. It is also possible to convert the alcoholic suspension to an aqueous suspension by simultaneous distillation of methanol and the addition of water. This is definitely a versatile and simple chemical process extremely suitable for the economic production of fine oxide ceramic powders.

Powders by Chemical Vapor Deposition

Powders produced by variations on the chemical vapor deposition (CVD) technique are becoming a popular route for fabrication of fine high-purity powders of reactive and refractory metals. The elimination of the melting step and a crucible-free process ensures the purity of the powder when the feedstock gases are prepurified by vapor distillation. High-purity powders of refractory and reactive metals such as tungsten, hafnium, molybdenum, titanium, zirconium, and niobium can be produced by this route.

Composite powders can also be processed by chemical vapor deposition techniques. Ultramet Corporation of the United States has developed a fluidized-bed powder coating technology over the past few years [38,39]. The company has successfully coated tungsten powder with nickel–iron and nickel–cobalt; alumina with titanium; and silicon and cubic boron nitride with aluminum and alumina. During the CVD process, a gaseous compound of the element to be deposited is passed over a heated substrate, where the thermal decomposition or the chemical reaction of the gas results in the material deposition on the substrate. The initial layer forms at the nucleation sites and then continues to grow on the crystalline face of the deposit. In the fluidized bed CVD reactor, the reactant gas stream is mixed with the powder just before they enter the reactor. The coating species preferentially decompose on the surface of the suspended powder particles. Control of gas velocity at the orifice entrance and in the parallel section over the orifice are of vital importance for the proper coating of the powders. Figure 2.9 shows a schematic diagram of the fluidized-bed chemical vapor deposition equipment.

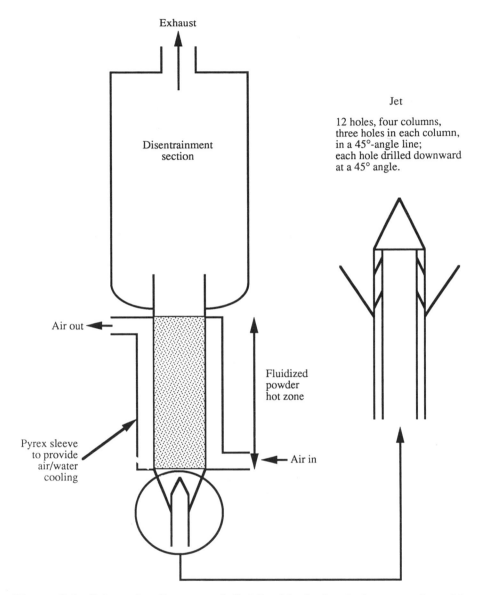

Figure 2.9: *Schematic diagram of fluidized-bed chemical vapor deposition equipment. (Courtesy Ultramet Corporation, Pacoima, CA.)*

Titanium has been used successfully to coat alumina powders by the fluidized-bed CVD technique. The titanium compound tetraisopropyltitanate, $Ti(OC_3H_7)_4$, which boils around 505 K (232°C), thermally decomposes to produce uniform titanium coatings on the surface of irregular alumina powder particles under proper CVD conditions. Recent efforts in producing coated

Figure 2.10: *CVD-coated tungsten that has been coated with a Ni:Co alloy to provide a final composition of around 94.5W, 3Ni, and 2.5Co (wt.%). (Powder provided by Ultramet Corporation, Pacoima, CA.)*

tungsten powders with various materials for penetrator-related applications have given the fluidized bed CVD technique a tremendous boost [40]. Trials of CVD of hafnium-coated tungsten powders are also in progress. This process utilizes the thermal decomposition of hafnium tetraiodide, HfI_4, into hafnium and iodine. One of the key advantages of this process is the very high level of purity that is generally achieved. Under proper deposition conditions, the total carbon and oxygen levels could be lower than 100 ppm. The coatings obtained are dense and adherent. Thus, this unique processing route allows one the flexibility of coating practically any material on another. Examples of powders produced by this coating process are shown in Figures 2.10 and 2.11. Figure 2.10 shows a powder of tungsten that has been coated with a Ni : Co alloy to provide a final composition of around 94.5W–3Ni–2.5Co. Figure 2.11 shows an SEM photomicrograph of hollow mullite microspheres that have been coated with iron. The powder particle in the foreground shows a well-coated mullite powder particle, while just behind it is a clear image of a hollow mullite powder particle that has not been coated.

Ultrafine metal powder produced by laser applications is another important developing area reported recently in *Metal Powder Report* [41]. The report

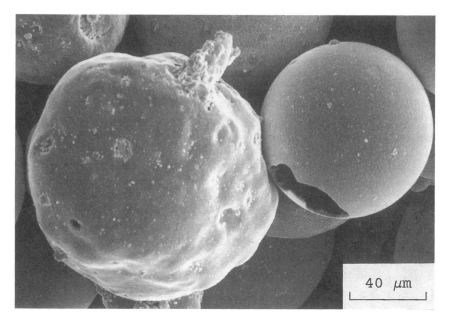

Figure 2.11: CVD-coated hollow mullite microsphere that has been coated with iron. The powder particle in the foreground shows a well-coated mullite powder particle; just behind it is the clear picture of a hollow mullite powder particle that has not been coated. (Powders provided by Ultramet Corporation, Pacoima, CA.)

states that scientists in Japan have been successful in producing ultrafine powders of both metal and ceramic by this technique. The process is based on chemical vapor deposition that incorporates an infrared laser. The energy for the CVD reaction is generated by microsecond pulses of infrared radiation that are produced by a carbon dioxide gas laser. When the energy density of the laser beam exceeds a critical value, ultrafine powders are formed by a phenomenon termed "dielectric gas breakdown." A chain reaction of the feedstock gas results in a plasma from which ultrafine powders are generated.

Titanium diboride powder was obtained from reacting feedstock gases consisting of boron chloride, titanium chloride, and a small quantity of hydrogen. The powder was generally uniform and had an average particle size of 0.1 μm. The powder particle size was said to be dependent on the reaction vessel feedstock gas pressure. The process could become a very important powder-producing route for materials that are difficult to produce since the process is reported to be simple and applicable for mass production for a large number of metal and ceramic powders.

The Future

The adaptation of the chemical powder processing route to suit various materials and the ease with which the process can be scaled up to produce large powder quantities are the greatest advantages of this relatively underexplored powder processing method. Powders produced by this process can be extremely pure since it is generally dependent on the purity of the initial chemicals, which can be obtained in very pure form. The process is suitable for producing powders of very high-temperature materials since the process is not inhibited by the melting point of the material. The chemical process has the ability to produce powders of amorphous or glassy materials as well as microcrystalline and nanocrystalline materials. This process is expected to lead the way in the powder production of these advanced materials.

It is the author's view that the utilization of this extremely potent process has just begun and by the turn of the century the process should become the most important powder processing route. The more recent chemical processes that open up a new spectrum in the area of powder production are sol–gel [42], organosilicon compounds [43], and alkoxides [44]. Readers are referred to the individual references for additional details on these processes. The other important area that could be classified as a chemical powder processing route, but which has not been covered in any detail in this section, is the reactive processing of powders. Reactive processing is discussed in some detail in a couple of chapters of this book. However, fabrication of powders utilizing the reactive sintering process could also be quite attractive. The process would essentially involve the mixing of elemental powders that can be reacted to form the desired compound or composite powder. The as-reacted powder can be comminuted and used as the starting powder for further consolidation.

One of the biggest advantages of the chemical powder processing route is the flexibility that this method offers. To elucidate this point, let us consider the chemical fabrication of silicon nitride and tantalum carbide [45] powders. Theoretically, more than one chemical route can be utilized to produce powders of these important engineering materials, with each process producing powders with special characteristics.

1a Carbothermal nitriding of silicon dioxide:
$$3SiO_2 + 6C + 2N_2 = Si_3N_4 + 6CO_2$$
1b Gas-phase reaction of silicon tetrachloride with ammonia:
$$3SiCl_4 + 4NH_3 = Si_3N_4 + 12HCl$$
1c Nitriding of silicon:
$$3Si + 2N_2 = Si_3N_4$$

1d Thermal decomposition of silicon diimide:
$$3Si + 6NH_3 = 3Si(NH)_2 + 12HCl$$
$$3Si(NH)_2 = Si_3N_4 + 2NH_3$$

2a Reaction with tantalum oxide:
$$Ta_2O_5 + 7C = 2TaC + 5CO$$

2b Reaction with hydride:
$$Ta(H) + C = TaC + \tfrac{1}{2}H_2$$

2c Auxiliary metal bath process:
$$Ta(Fe,Al) + C = TaC$$

2d Tantalum accelerated reaction with tantalum oxide:
$$Ta_2O_5 + Ta + 8C = 3TaC + 5CO$$

2e Gas-phase reaction process:
$$TaX_5 + CH_4 + \tfrac{1}{2}H_2 = TaC + 5HX$$

Thus, a chemist, given the choice of producing powder of a particular material, has a number of alternative chemical routes that he/she can possibly utilize. The factors to be considered are the attainment of proper powder particle sizes, low impurity contents, and an economically feasible manufacturing method. The potential of the chemical processes for generating specific powders is only limited by the imagination of the powder metallurgist or materials scientist and the synthetic skills of the chemist. It can be stated with some degree of confidence that nearly all powders can be produced by utilizing the chemical processing route, and a myriad of powder characteristics can be obtained for the same material by the use of different chemical processing paths. A chemist can control the powder production process in a number of ways, like choosing different starting components where the anions of the salts or the precipitant can be varied; choosing different solvents in which the salts can be dissolved; varying the concentration in the solvents; and changing the reaction kinetics by various means. This flexibility makes it the most versatile process for producing powders. However, lack of proper coordination between chemists and powder metallurgists is the only reason why this process has not become the tremendous commercial success for which the potential is latent. Once the powder metallurgy industry realizes the potential benefits of combining forces with chemists, and when more chemists are made aware of the needs and requirements of the P/M industry, the growth of this area should be phenomenal. The author hopes that this chapter will serve as a catalyst in bringing about the desired marriage between the two disciplines. If proper coordination between the powder metallurgists and chemists does occur, it can confidently be predicted that by the turn of the century the chemical powder processing route will become the most important powder fabrication technique.

References

1. F. Aldinger and Hans-Jurgen Kalz, *Angewandte Chemie Internationale*, English Ed., vol. 26, p.371, 1987.
2. I.J. Mellanby, *Metal Powder Report*, vol.45, p.94, 1990.
3. D.I. Bloemacher, *Metal Powder Report*, vol.45, p.117, 1990.
4. E.L.Ll. Rees, F.W. Heck, and G.A. Dibari, *Modern Developments in Powder Matallurgy*, compiled by P.U. Gummeson and D.A. Gustafson, Metal Powder Industries Federation, Princeton, NJ, vol. 20, p.311, 1988.
5. A. Bose and M. Ghosh, Memorandum of Invention for Patent Application filed, 1993.
6. S.T. Schwab and A. Bose, *Advanced Material Precursors through Organotransition-metal Chemistry*, Funded to South West Research Institute by Office of Naval Research, Contract N00014-91-C-0085, 1992.
7. A. Bose, unpublished results.
8. G.E. Toogood and M.G.H. Wallbridge, *Advances in Inorganic Chemistry*, vol. 25, p.267, 1982.
9. P.P. Paul, Y-M. Pan, S.T. Schwab, S.F. Dec, and G.E. Maciel, paper communicated to *Nano-structured Materials*.
10. M. Figlarz, et al., FR Patent 2 537 898, April 11, 1985; European Patent 0 113 281, December 20, 1983; U.S. Patent 4 539 041, September 3, 1985.
11. F.D. Richardson and J.H.E. Jeffes, *Journal of the Iron and Steel Institute*, vol. 160, 1948.
12. L. Darken and R. Gurry, *Physical Chemistry of Metals*, McGraw-Hill, London, 1953.
13. K.C. Sahoo, D.K. Lahiri, S. Barpanda, G.S. Bhattacharjee, and T.P. Bagchi, *Modern Developments in Powder Metallurgy*, compiled by P.U. Gummeson and D.A. Gustafson, Metal Powder Industries Federation, Princeton, NJ, vol. 20, p.279, 1988.
14. K.J. Klabunde, H.F. Efner, T.O. Murdrock, and R. Ropple, *Journal of the American Chemical Society*, vol. 98, p. 1021, 1976.
15. P.L. Timms, *Proceedings of the Royal Society of London*, vol. 396A, p.1, 1984.
16. R. Birringer, H. Gleiter, H.P. Klein, and P. Marquardt, *Physics Letters*, vol. 102A, p.365, 1984.
17. *Metal Powder Report*, vol. 45, p.26, 1990.
18. A. Corrias, G. Ennas, G. Licheri, G. Marongiu, and G. Paschina, *Chemistry of Materials*, vol. 2, p.363, 1990.
19. E. Tani, M. Yoshimura, and S. Somiya, *Journal of the American Ceramic Society*, vol. 64, p.C181, 1981.
20. L.E. McCandlish and R.S. Polizzotti, *Solid State Ionics*, vol. 32/3, p.795, 1989.
21. L.E. McCandlish, B.H. Kear, B.K. Kim, and L.W. Wu, *Materials Research Society Symposium Proceedings*, vol. 32, p.67, 1989.
22. L.E. McCandlish, B.H. Kear, B.K. Kim, and L.W. Wu, *Protective Coatings: Processing and Characterization*, Ed. R.M. Yazici, TMS, Warrendale, PA, 1990.
23. L.E. McCandlish, B.H. Kear, and B.K. Kim, *Materials Science and Technology*, vol. 6, p.953, 1990.
24. S. Haerdtle, *Metal Powder Report*, vol. 45, p.133, 1990.

25. T. Haug, M. Fandel, and T. Staneff, *Powder Metallurgy International*, vol. 22, p.32, 1990.
26. W. Kock and P. Paschen, *Journal of Metals*, vol. 41, p.33, 1989.
27. M. Awano, K. Nakamura, T. Yamada, and H. Takagi, *Powder Metallurgy International*, vol. 21, p.23, 1989.
28. H.K. Varma, K.G.K. Warrier, and A.D. Damodaran, *Powder Metallurgy International*, vol. 22, p.35, 1990.
29. W.A. Johnson, N.E. Kopatz, and E.B. Yoder, *Progress in Powder Metallurgy*, compiled by C.L. Freeby and H. Hjort, Metal Powder Industries Federation, Princeton, NJ, vol. 43, p.139, 1987.
30. A. Landsberg and T.T. Campbell, *Journal of Metals*, vol. 17, p.846, 1965.
31. F.J. Schnettler, F.R. Monforte, and W.W. Rhodes, *Science of Ceramics*, vol. 4, p.79, 1968.
32. G.D. White and W.E. Gurwell, *Advances in Powder Metallurgy*, compiled by T.G. Gasbarre and W.F. Jandeska, Metal Powder Industries Federation, Princeton, NJ, vol. 2, p.356, 1989.
33. *Metal Powder Report, PM Update*, vol. 47, no. 10, p.8, 1992.
34. A. Slosarczyk, *Powder Metallurgy International*, vol. 4, p.24, 1989.
35. *Metal Powder Report*, vol. 48, no. 7/8, p.28, 1993.
36. H. Wedemeyer and D. Vollath, *Powder Metallurgy International*, vol. 22, p.33, 1990.
37. D. Vollath and H. Wedemeyer, *Advances in Ceramics*, vol. 25, p.93, 1989.
38. J.G. Sheek and J.J. Stiglich, *Coated Tungsten Powder*, ULT/TR-87-4831, Contract DAALO2-86-C-0112, Army Materials Technology Laboratory, Watertown, MA, April 1987.
39. B.E. Williams, J.J. Stiglich, and R.B. Kaplan, *Coated Tungsten Powders for Advanced Ordnance Applications*, Contract DAALO4-88-C-0030, Army Materials Technology Laboratory, Watertown, MA, March 1991.
40. B.E. Williams, J.J. Stiglich, Jr., R.B. Kaplan, and R.H. Tuffias, paper presented at the Annual TMS Meeting, New Orleans, Louisiana, February 18–21, 1991.
41. *Metal Powder Report*, vol. 44, p.574, 1989.
42. D.W. Johnson, *American Ceramic Bulletin*, vol. 65, p.1597, 1985.
43. D. Seyferth and G.H. Wiseman, *Science of Ceramic Processing*, Ed. L.L. Hench and D.R. Ulrich, Wiley, New York, p.354, 1986.
44. H. Okamura and H.K. Bowen, *Ceramic International*, vol. 12, p.161, 1986.
45. E. Schaschel, *Metal Powder Report*, vol. 46, no. 12, p.30, 1991.

Chapter 3

Melt Atomization

Introduction

The last two decades have witnessed the increasing involvement of particulate technology in advanced materials. Atomization is one of the leading forerunners among the most promising and economic techniques for the fabrication of powders. The commonly applied mechanism for atomization involves the disintegration of a liquid into a fine spray of droplets by high-velocity fluids. However, there are other forms of atomization that involve the use of centrifugal forces, vibrational energy, electrical fields, or even the saturation of the molten material with gases to eventually disintegrate the liquid metal into fine droplets.

Essentially atomization involves the disintegration of a film or a stream of molten material by the application of some form of energy. The differences in the various forms of atomization can arise from variations in the manner in which the molten material is formed, the method by which the molten material is contained, and the fashion in which the molten stream or film is disintegrated. The majority of the atomization processes, as mentioned earlier, rely on conventional melting and containment of the liquid in a crucible, followed by the disintegration of a flowing melt stream by the application of energy via high-velocity fluids such as gas or water. An easily conducted experiment to demonstrate the principle of the most popular form of atomization, i.e. gas atomization, involves directing a gas jet on to a thin flowing stream of tap water. The water stream falling freely due to gravity is broken into numerous fine droplets at the point where the gas jet impinges. The atomization of a liquid metal stream follows the same transfer of energy from one fluid (gas) to the other (liquid metal); the equipment needed to direct the liquid metal and to focus the gas jet is only more elaborate and sophisticated in nature. Numerous variations and refinements on this general scheme of forming powders have been adopted to accommodate the necessary melt practices for various alloys and to suit the requirements of the P/M industry.

A large variety of impinging fluids can be used to disintegrate the molten metal stream. They can be mundane materials such as water, synthetic oils, gases like air, nitrogen, argon, helium, and also the more exotic fluids such as liquid nitrogen or argon. Even combinations of two fluids have been recently used to form finer powders. The materials that are produced in large quantities such as steel, copper, etc., still utilize water atomization to a great extent. However, inert gas atomization, commonly referred to as simply gas atomization, is becoming extremely popular and is increasingly being used to produce fine high-purity powders in large quantities.

There are other atomization techniques that are not based on the principle of melt disintegration by impingment with a second fluid. These processes are increasingly being used to produce powders (especially for refractory or reactive metals) for use in advanced materials. Techniques such as melt extraction, rotating consumable electrode, plasma microatomization, melt saturation and explosion, and centrifugal shot casting are some of these processes. Most of these processes can be broadly categorized as centrifugal or rotating atomization techniques, though there are a large number of processes that do not fall into that general category. A few examples of processes that achieve atomization neither by high-velocity fluids nor by the centrifugal atomization techniques are melt saturation and explosion, plasma micro-atomization, and capillary and standing wave atomization.

The primary advantage of the atomization process lies in its flexibility in terms of producing the desired alloy chemistry and particle size. The process is capable of producing small melts for small quantities of experimental powders, but can also be scaled up to produce very large quantities of powder for use in commercial applications. The cooling rates, depending on the processing parameters, can be quite fast and the powders thus produced are quite homogeneous in nature.

The average particle size range varies between 10 and 250 μm, depending on the process and processing parameters. The cooling rate is generally dependent on the powder particle size. The cooling rate can be determined, to some extent, from the secondary dendritic arm spacing in the individual powder particles (unless the cooling rates are fast enough to produce equiaxed microstructures). Still faster cooling, which can be produced by splat techniques, can result in amorphous powders. The cooling rates in atomization processes normally fall in the range of 10^2 to 10^4 K/s, with rates approaching 10^6 K/s in some special cases.

Atomization is extremely suitable for producing powders of pure metals and alloys where the melting point of the material is not too high. Since the process involves melting of the desired metal or alloy and requires some degree of superheating, very high-melting-point materials are not the best choices for atomization. Thus, very high-temperature materials like tungsten, hafnium, and molybdenum are usually not produced by atomization. Also the atomization of niobium aluminide intermetallic compounds where niobium has a very high melting point and aluminum is a low melting additive is difficult, though not impossible. Powders of ceramic materials are usually not produced by melt atomization, except by specialized processes such as plasma atomization.

Contamination could occur due to the crucible in which the atomization melt is contained (usually a ceramic crucible). This problem becomes extremely acute for high-temperature high-performance materials. One of the ways to avoid that problem is to produce within the crucible a thin "skull" of the same material that is to be atomized. The "skull" effectively insulates the molten material from the crucible and thus reduces contamination to a great extent. Since the crucibles in which the "skull" is formed need to be cooled, they are made of high thermally conductive metals, the most popular being copper. Another technique is to have no crucible at all and use the material to be atomized as an electrode that is gradually melted and atomized. Another alternative is to use ceramic filters to filter out large ceramic particles from the melt before atomization. A number of present-day atomization processes use some of the above concepts to produce cleaner powders.

The basic concept in the majority of the common atomization processes is to deliver energy to the liquid melt via the impinging fluid. Variations in the fluid focusing system and the type of fluid used determine the final powder characteristics. Subsequent parts of this chapter will briefly describe various atomization processes that are presently in use and also some of the new processes that are being developed to produce finer powders and powders that have been rapidly solidified. The requirements of the powder injection-molding industry are also providing a large impetus to the atomization processes to provide finer powders that are suitable for the powder injection molding process.

Water Atomization

The process of producing metal powders by the disintegration of the molten metal stream using water is commonly referred to as water atomization. This is one of the most popular techniques for producing nonreactive metal and alloy powders. The rate of heat extraction by this process is quite high.

The apparatus consists of water jets (usually coming through multiple jets or an annular ring) directed on to a molten stream of metal or alloy. The key variables that influence the process are the melt superheat and water pressure,

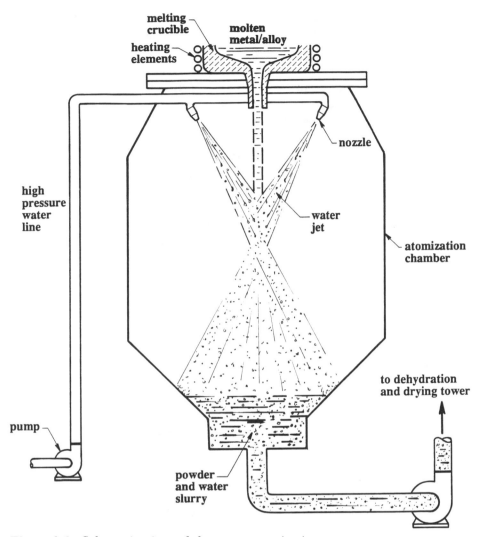

Figure 3.1: *Schematic view of the water atomization process.*

which controls the water velocity. High water pressure results in finer powder particle sizes. Other important parameters are the melt diameter, metal flow rate, and water jet apex angle. The nozzle-to-melt distance is not as critical since water has low compressibility and higher density than gases and thus its energy does not diminish as rapidly. However, this distance should be minimized to maintain the melt superheat.

Due to the rapid extraction of heat, the final particle shape is quite irregular as the particles have less time to spherodize compared to the normal gas atomization process. Thus, it is often necessary to have large superheats for particle shape control. The nature of the environment results in the final powder being slightly contaminated by a surface oxide layer, which under certain circumstances can be reduced by a subsequent hydrogen reduction step. A schematic view of the water atomization process is shown in Figure 3.1.

Water-atomized powders have excellent compressibility, but their apparent density is normally lower than that of gas-atomized powders due to the particle shape. The water pressure has been reported to be the prime controlling factor for the water atomization process. German [1] reported that water-atomized steel powder using a water pressure of 1.7 MPa resulted in powders with a mean particle size of 117 μm compared to a mean particle size of 42 μm when the water pressure was 13.8 MPa.

The need for finer powders (in the 10 μm range) has resulted in a high-pressure water atomization (HPWA) process in which water pressures up to 70 MPa have recently been used to produce powders with an average particle size less than 10 μm [2]. In the same report it has been shown that the water jet nozzle configuration also influences the powder particle size. The authors have reported that a V-jet nozzle configuration produces finer powders than the commonly used cone nozzle configuration using similar atomizing pressures. A schematic view of the nozzle configurations is shown in Figure 3.2. According to the authors, the relationship between the mean particle diameter and atomization pressure for the two nozzle configurations is given by the following equations:

$$D = 114P^{-0.58} \qquad \text{(for the cone configuration)} \qquad (1)$$

$$D = 68P^{-0.56} \qquad \text{(for the V configuration)} \qquad (2)$$

where D is the mean particle diameter, and P is the atomization pressure. The above equations were based on the results of the atomization of Fe–Ni alloy, carbon steel, high-speed tool steel, and pure nickel.

The process of high-pressure water atomization is an important development in the direction of producing atomized powders in the 10 μm range. The demand for these finer powders is increasing, especially due to the emergence of powder injection molding as an important fabrication route. Thus, it is expected that high-pressure water atomization could partially fill the gap between the supply and demand for reasonably inexpensive powders in the 10 μm particle size range. However, the powder shape could be a problem for these powders.

High-pressure water atomization can provide fine powders, as discussed earlier. The process also has the ability to produce a wide range of powder particle size, size distribution, and powder apparent density. With increasing

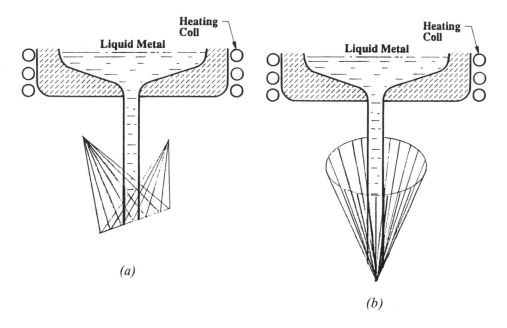

Figure 3.2: Schematic of water atomization nozzle configurations: (a) V-jet nozzle; and (b) commonly used conical nozzle.

atomization water pressure, the particle size tends to become finer and the particle size distribution also tends to become broader. This has been experimentally demonstrated by Kimura and coworkers in their experiments with water-atomized 316-L stainless-steel powders [3]. The atomizing water pressure, which was varied from 10 to 150 MPa, was found to yield finer average particle sizes with increasing water pressure. The cumulative size distribution of the powders with variation of the atomizing water pressure is shown in Figure 3.3. The same figure also shows the variation of the apparent density of the powders with varying powder particle sizes. According to the authors, they have been able to produce stainless-steel powders that are suitable for powder injection molding applications. By adopting a special atomizing system that provides advanced shape control, they have been able to increase the tap density of the powders significantly.

Little work has been carried out on the mechanism of the water-atomization process. It is suggested that the water jet is usually broken up into minute fragments before it actually impinges on the metal stream. Thus, tiny droplets of water impact the molten liquid stream, causing small appendages to appear on the molten liquid surface (like small waves) from which a metal droplet is

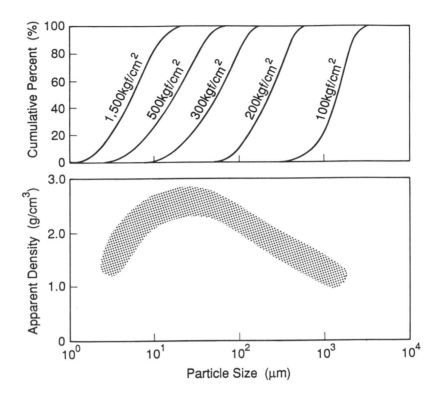

Figure 3.3: *Cumulative size distribution of water-atomized powders with different water atomization pressures and the variation in the apparent density with powder particle size for a stainless-steel powder. (Redrawn after Ref. 3.)*

ejected. The water droplet after striking the liquid film immediately evaporates. This is termed the "scrape" mechanism that operates during water-atomization events, and is schematically illustrated in Figure 3.4 [4]. Grandzol showed a simple relationship between the average particle size, D (μm), and the water velocity, V_w (m/s), which was later modified by Grandzol and Tallmadge [5] to yield.

$$D = \frac{S}{V_w \sin \alpha},$$ (3)

where α is the angle between the water jet axis and the molten metal stream axis (in degrees) and S is the normal velocity component. This equation predicts that it is the component of the water velocity normal to the metal stream that determines the final average particle size. However, more work

Figure 3.4: *The schematic steps involved in the "scrape" mechanism of water atomization.*

needs to be done in order to confirm the proposed equation and the postulated "scrape" mechanism.

Water atomization is extensively used to produce high-speed tool steel powders. However, ceramic contamination from the melt is a major concern in these materials. Thus, filtration by ceramic foam filters has been recently tried [6] with a sintered zirconia toughened alumina foam filter that could withstand molten metals up to 1953 K (1680°C). These filters were positioned within the atomizer tundishes. The tundishes were usually preheated to approximately 1273 K (1000°C) to prevent melt freeze-up. The tool steel powders produced by water atomization using the foam filters were very encouraging. The static strengths and the rolling contact fatigue lives were within the scatter bands for current aerospace bearing-quality wrought materials. SEM investigation of used filters revealed that they had indeed trapped ceramic inclusions that would otherwise have entered into the powder through the melt. Thus, the use of ceramic filters provides an opportunity to produce cleaner water-atomized powders.

Water atomization has also been used to produce prealloyed heat-treatable steel powders for the production of high-strength and dimensionally accurate P/M parts. The newly developed alloy steel known as Mannesmann steel, consists of prealloyed Fe–Ni–Mo powder with elemental additions of graphite and copper [7]. The alternate means of producing such steel components is by using diffusion-bonded steel powders that later homogenize during the sintering process. However, the material still has some local soft spots after heat treating, and these result in impaired properties. Also, the use of nickel, which does not have very high affinity for oxygen, results in powders that may not require a subsequent vacuum reduction, which is indeed a big economic factor.

In general, water atomization can produce powders that have an average particle size between 150 and 400 μm with an approximate cooling rate that can vary between 10^3 to 10^5 K/s. The particle shape is generally irregular and the powders usually have a wide size distribution. However, the recent developments in high-pressure water atomization have been successful in producing much finer powders.

Oil Atomization

Recently synthetic oils have been used as the fluid instead of water in order to combat the problems of high oxygen content in the water atomization process. Oil atomization has been found to be suitable for atomizing medium- and high-carbon tool steels. This process combines the advantages of a high quenching rate (faster than gas atomization) and low oxygen content (lower than that of water-atomized powders); however, for low-carbon steels, it becomes necessary to remove the carbon by a high-temperature wet hydrogen treatment [8,9]. The process of oil atomization is, however, suitable for medium- (1%C) and high-carbon (3%C) steel powders [10]. In these cases there is a higher tolerance to carbon and thus powders with an oxygen content of around 300 ppm can be used. Any changes in the carbon content due to atomization can be taken into account in the medium- and high-carbon steels but this is extremely difficult in case of low-carbon steels without accepting a coarser powder with more oxygen content.

In an investigation on oil atomization, ferrous-based alloy was melted in a furance and then poured through a preheated tundish into a bottom-tapped holding crucible leading to an atomizing nozzle. As the melt stream flows through, it is atomized by four high-pressure gas jets. The oil–powder mixture is allowed to settle for a few hours, after which the oil is drained and the powder is removed from the oil by secondary sedimentation. The excess oil removal can be carried out in a tank using compressed air.

The iron powder, after initial oil removal, still has 2 to 3 wt.% of oil, which must be removed without any residue and without any alterations in the microstructure and the oxygen level. The alternatives that are available include the use of solvent extraction or heating in air, inert gas, or reducing gas atmospheres. Among these alternatives, it has been found that a low-temperature heat treatment in protective atmosphere does the most efficient job of removing the oil without affecting the microstructure to any great extent. Heating the powder to a temperature sufficiently high to remove the oil but not high enough for diffusion events to occur is the best way of removing the oil. In iron containing low carbon, however, there is significant carbon pickup.

Thus, the process of oil atomization can be applied in the case of high-carbon steels, high-speed steels, bearing steels, etc. The process is also suitable for alloys containing high quantities of carbide-forming elements such as chromium and molybdenum. This process has been used to produce powders from high-speed steels having the compositions Fe–1.4%C–1.4%Cr and Fe–1.4%C–4.5%Cr–1.1%Mo–0.35%V (wt.%) [11]. The consolidated materials exhibit homogeneous and finely distributed carbides, which result in very good mechanical properties.

At this point the properties of the oil that is used for atomization should be mentioned. According to Kainer and Mordike [10], the oil must have several properties that are comparable to water. The viscosity of the oil must be low at room temperature, while the flash point should be high. The oil should be nontoxic in nature, have a low sulfur content, and should be recyclable. Therefore, oils used for hardening are quite suitable for use as the oil required for atomization.

This is indeed an interesting and economically viable process for producing high-carbon-content ferrous powders with the presence of carbide-forming elements. The high quench rate, which is equivalent to the quench rates in the water-atomization process, coupled with a low oxygen content makes this process quite attractive.

Gas Atomization

Gas atomization is the disintegration of an unconfined stream of liquid metal into fine particles by a focused jet or jets of atomizing gas. The design of the nozzle and the liquid-metal feed mechanism often varies, and the atomizing gas can be nitrogen, argon, helium, or air. The gas jet can be directed through a number of discrete openings or through an annular-ring type of opening. A representative version of each type of configuration is shown in Figure 3.5.

One of the basic variations in the gas atomization process is whether the liquid stream is disintegrated in a horizontal or vertical direction. The majority of gas atomization processes rely on the vertical direction of liquid metal flow. The horizontal gas atomizers are normally used for low-temperature metals and alloys. The molten metal is siphoned up through a feed tube and disintegrated by high-velocity gas. Vertical gas atomization is normally used for high-temperature materials, with inert gases being used as the atomization fluid. A schematic drawing of a horizontal gas atomizer is shown in Figure 3.6, while a schematic of the conventional vertical type of gas atomizer is shown in Figure 3.7.

The fine particles that are formed slightly after the point where the gas jet strikes the liquid rapidly lose heat and solidify in flight. The chamber size should be large enough that the particles produced can have free flight through the chamber and solidify prior to reaching the collection chamber wall. In this process, the recent drive is toward finer particle sizes (around the 10 μm range). Finer particles can be generated by higher gas flow rates, higher melt temperatures, and steep angles between the gas jet and the melt stream. The thermal conductivity of the atomizing gas, the metal surface energy, and the

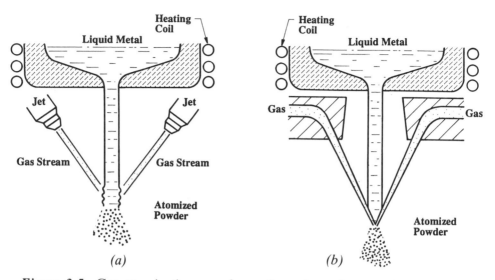

Figure 3.5: *Gas atomization nozzle configurations showing (a) twin-jet configuration; and (b) annular-ring configuration.*

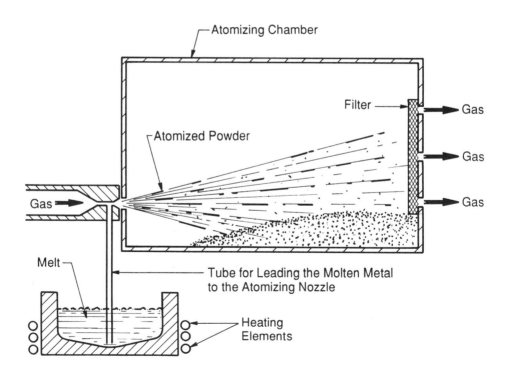

Figure 3.6: *Schematic of a horizontal gas atomizer.*

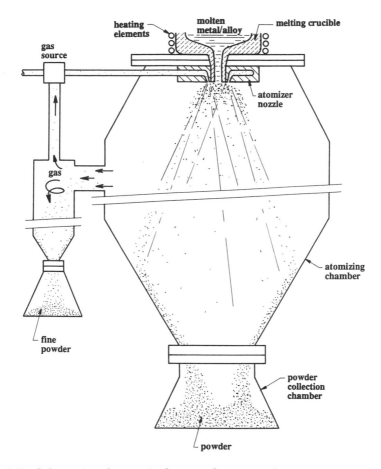

Figure 3.7: Schematic of a vertical type of gas atomizer.

atomizer nozzle design all play important roles in determining the final powder characteristics.

A "free-fall" configuration is often used in the conventional gas atomization technique. In this technique, the gas jet(s) strikes the metal stream at a point that is some distance away from the point where the liquid metal leaves the cruicible (i.e., flow becomes unconfined). Problems arise from the stream instability when very high gas velocity and sharp jet angles are used to try to produce finer particles. This conventional type of gas atomization can produce powders with a mean particle size usually greater than 50 μm.

For attaining smaller average particle sizes and rapid solidification rates, "close-coupled" (also known as confined or direct jet) gas atomization is the preferred route [12]. In this case the gas jet impinges on the stream of molten

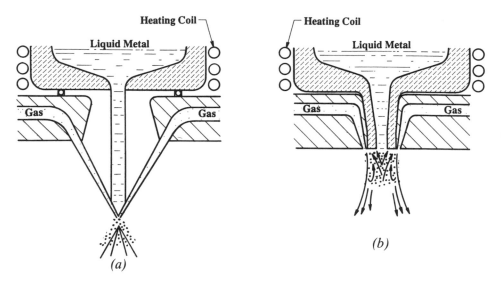

Figure 3.8: *Schematic of the two types of atomization nozzle design: (a) free-fall; (b) confined.*

metal at the point where it emerges from the melt guide tube. This results in the most efficient transfer of the gas kinetic energy to the liquid melt, which is the key to the formation of fine powders. Little of the gas energy is lost in turbulence as in the case of the open-die or free-fall arrangement. Also, a higher angle of impingement is possible, thus improving the fine particle yield. This form of gas atomization design is not without its share of technical problems. The major challenges are due to the proximity of the gas jet to the molten metal guide tube. Schematics of the two types of atomization nozzle design — "free-fall" and "confined" — are shown in Figures 3.8a and 3.8b, respectively.

High-pressure gas atomization (HPGA) was a natural evolution of the normal gas atomization process. The necessity to develop finer powders and to produce fast cooling rates resulted in the development the process. Lin et al. [13] have described some of the important parameters for the optimum performance of HPGA. In their work they found that use of 573 K (300°C) superheat, 4.5 MPa argon atomizing pressure, and a 2.5-mm-diameter pour tube was optimum for atomizing copper. HPGA of copper, amalgam, and a lead–tin alloy exhibited mean particle sizes of 25, 20, and 15 μm, respectively. Thus, high-pressure gas atomization is a possible way of obtaining fine, rapidly solidified powder. From the point of view of the large demand for finer powders, this is also a very important development.

The commonly accepted model for conventional gas atomization involves the formation of a stable sheet that later becomes wavy in nature. Cylindrical ligaments are torn off from the sheet and these in turn break up into number of small elongated droplets that then become spherical [14]. A schematic view of this is shown in Figure 3.9.

Bradley [15,16] developed a mathematical analysis of the stages in the formation of powders from the disintegration of the liquid melt. The detailed mathematical analysis has not been incorporated in this chapter, and readers are referred to Bradley's papers for the detailed analysis. Bradley also developed a graphic method of evaluating the droplet radius for different Mach numbers between 0.1 and 0.9. The universal curve of the Mach number versus a dimensionless parameter L is shown in Figure 3.10, where L is determined by

$$L = \frac{\lambda_{\max} \tau}{\rho_g U_s^2} \tag{4}$$

where

$\lambda_{\max}$ is the wavenumber of the fastest-growing amplitude [15];
τ is the liquid–gas interfacial energy;
U_s is the sonic velocity of the gas;
ρ_g is the density of the gas.

The diameter of the final droplet can be calculated from

$$d = \frac{2.95\tau}{L\rho_g U_s^2} \tag{5}$$

Knowing the Mach number $M = U/U_s$, the dimensionless parameter L can be determined from the universal curve depicted in Figure 3.10. The diameter of the particle can thus be predicted by substituting the value of L in equation (5). It should be pointed out that this model does not account for the gas impingement angle with the liquid and it assumes a stationary flat liquid surface and a liquid with finite depth. However, Bradley's model can be said to provide a reasonable agreement with experimental findings, though it provides no clue to the origin and magnitude of the particle size distribution [14].

Another form of mathematical correlation between the atomizing condition and the mean powder particle diameter has been developed by Lubanska [17]. The equation is

$$d_{av} = KD\left[\frac{\eta_m}{\eta_g W}\left(1+\frac{M}{A}\right)\right]^{1/2} \tag{6}$$

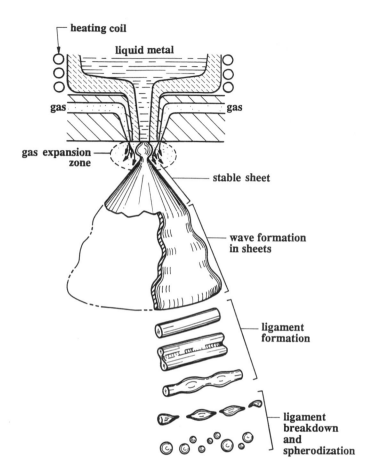

Figure 3.9: *Schematic of the stages by which the molten stream of metal is disintegrated into spherical powder particles during gas atomization.*

where

d_{av} is the is the mean particle size (μm);

K is a constant dependent on the conditions of spray ring and liquid stream, and is seen to vary between 40 and 50 for the variety of conditions investigated;

D is the diameter of the metal nozzle (mm);

M is the mass flow rate of the metal stream (kg/min);

η_m is the kinematic viscosity of metal (m²/s);

η_g is the kinematic viscosity of gas (m²/s);

A is the mass flow rate of the gas (kg/min);

W is the Weber number.

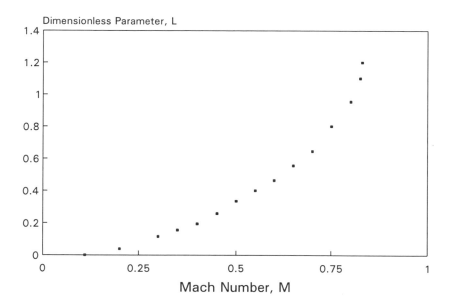

Figure 3.10: *Universal curve of Mach number versus a dimensionless parameter L. (Redrawn after Refs 15, 16.)*

The Weber number is dependent on the density and surface tension of the liquid metal, the velocity of the gas in the atomizing zone, and the melt stream diameter by the equation

$$W = \frac{V^2 D}{\rho \gamma} \tag{7}$$

where

γ is the liquid surface tension (dynes/cm);
ρ is the liquid density (g/cm^3);
V is the gas velocity at the impact of the gas jets with the metal stream (m/s).

It is difficult to determine accurately the gas velocity at the impact of the jet with the metal stream, which is determined from curves giving the velocity decay in sonic jets. A modified model developed by Coombs and coworkers [18] connects the average diameter of the powder particles to the atomization conditions according to

$$d_{av} = K_1 \frac{D^{1/2}(1 + M/A)^{1/2}}{(P + 1)} \tag{8}$$

where

K_1 is a constant that includes the gas and metal properties;
P is the gauge pressure in bar, and the rest of the symbols have the same connotation as in equation (6).

This modified model was experimentally verified on bronze, using three different melt stream diameters and two different atomizer nozzle cross-sectional areas. The correlation obtained by plotting the average particle size diameter versus $[D^{1/2} (1 + M/A)^{1/2}]/(P + 1)$ was quite good.

The above discussion provides a brief glimpse into the intricacies in the modeling of this process. Equally important is the development of the empirical parameter relationships between specific nozzles and the resultant particle size. Much more work needs to be carried out in the area of modeling of the atomization process.

With increasing demand for finer powders, the powder producers are adopting new techniques that could yield higher percentage of finer powders. Research and development effort in this area has therefore increased tremendously. The need for producing fine particle size with a narrow size distribution has resulted in the development of ultrasonic gas atomization. This form of atomization uses high-frequency supersonic gas jets as the atomizing medium.

The production of the ultrasonic waves is accomplished by accelerating high-pressure gas through a resonance cavity (like the Hartmann tube). The passage of the highly pressurized gas such as nitrogen, argon, or helium through the resonance cavities results in the attainment of ultrasonic frequencies of 80 to 100 kHz and supersonic gas velocities often reaching Mach 1.7 to 2.5 [19]. A schematic diagram of the ultrasonic gas atomization process and the nozzle is shown in Figure 3.11. A typical ultrasonic gas atomizer uses around 16 to 20 resonating cavities that direct the jets to a circle of small radius surrounding the liquid metal stream. Various investigators have used the ultrasonic gas atomization process to produce rapidly solidified powders of different metals and alloys [20–22].

Owing to the high velocity of the gas pulse, the mechanism for this form of atomization does not follow the usual ligamentation model that has been proposed for conventional gas atomization. Instead, the liquid metal stream behaves like an impacted low-shear-strength solid. More work is necessary in evaluating the mechanisms of the ultrasonic gas atomization process.

A novel form of convergent–divergent type of nozzle design developed by researchers of Banaras Hindu University has been successful in producing nitrogen gas velocity around Mach 2 using an atomizing gas pressure of 2 MPa [23]. The detailed description of the atomization unit and the special nozzle

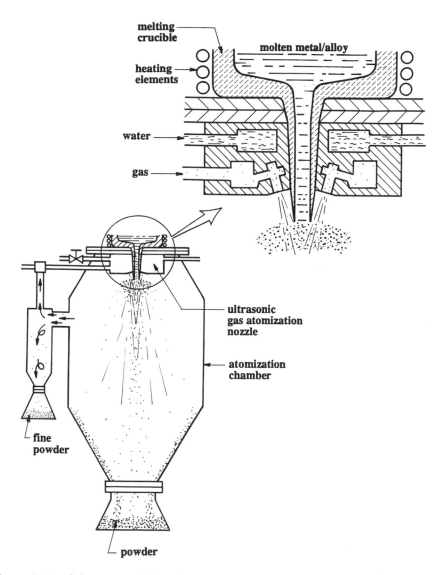

Figure 3.11: *Schematic of the ultrasonic gas atomization setup and an enlarged view of the nozzle design.*

construction is given elsewhere [24]. The process uses a refractory flow tube that is situated concentrically along the axis of a specially constructed annular convergent–divergent nozzle with an extended divergent section of large radius of curvature. This large radius of curvature in the divergent section induces a Coanda effect by which the fluid is deflected toward a curved solid surface and tends to follow the contour of the solid surface [25,26]. One end of the flow tube is inserted into the liquid metal container to supply the molten metal to the nozzle. When gas is passing through the nozzle, suction forces are created at the tip of the flow tube that cause the molten metal to be continually sucked up the tube to the tip. The atomization is accomplished by the gas jet meeting the liquid metal at the tip of the flow tube. According to the authors, further disintegration of the atomized droplets occurs due to the entrainment of the atomized particles in the high-velocity gas stream that is flowing along the annulus of the nozzle.

The authors have been able to process atomized powders with size ranges from below 25 μm to 180 μm. The general particle size distribution of the powders showed a bimodal trend. Very few satellites were observed on the particles, and the powder shape was essentially spherical in nature, with the coarse powders exhibiting an oblong appearance. The estimated cooling rate for this process is between 10^3 and 10^5 K/s.

The mechanism of atomization during this process is extremely complicated. It is suggested that the liquid melt at the tip of the flow tube is first stretched into a film, and then broken into tiny droplets by the action of the high-velocity gas impingement. These molten droplets are immediately entrained in the high-velocity gas film that is deflected toward the annulus of the nozzle. Secondary disintegration mechanisms include a form of stripping [27] or bag breakup. Depending on the gas velocity regime, both these secondary atomization mechanisms may be operative. A bimodal size distribution of the powders results in both cases [23].

Soluble Gas Atomization

In this process, the energy required to produce the disintegration of molten metal into powders is stored in the molten metal as a soluble gas (usually hydrogen and in some cases nitrogen). Approximately 1 to 3 ppm of the soluble gas is dissolved in the molten metal under pressure. The major energy for atomization is released due to the sudden change in the solubility of the gas when the material is cooled or released into a low-pressure chamber. One form of this process, commonly known as vacuum atomization, relies on the release of a stream of molten material supersaturated with the soluble gas into a vacuum chamber. The high velocity coupled with the desaturation of the soluble gas causes the melt to disintegrate into fine droplets, which subsequently solidify into powder particles.

In these processes, the molten metal is first saturated with the soluble gas by applying gas pressure (around 1 to 3 MPa) to the molten metal surface or even bubbling the desired gas through the melt. The pressurized melt is then forced into the atomizing chamber through an electrically heated ceramic transfer

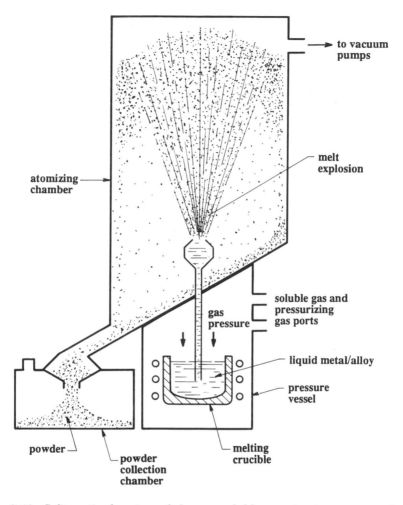

Figure 3.12: *Schematic drawing of the gas-soluble atomization process showing melt-explosion in a vacuum chamber.*

tube into a larger chamber that is under vacuum. The sudden release of the melt into this evacuated chamber causes the melt literally to explode into tiny fragments. A schematic diagram of this process is shown in Figure 3.12. Usually superalloys and steels are excellent candidate materials for this process. Materials like titanium that easily form hydrides have also been atomized by this process. The vacuum in the cooling chamber results in a much slower mode of heat transfer (radiation) from the particles and, thus, the solidification rates are quite slow.

A variation of the soluble gas atomization technique is the tandem atomization process, or TAP. This atomization process has been developed to meet the challenge of producing fine metal powders with a mean powder size less than 20 μm [28]. This process usually results in much higher yields of powders below 20 μm and at a lower cost compared to the general gas atomization processes where the finer powder yield varies between 2% and 15%.

The principle of the tandem atomization process (TAP) relies on the use of soluble gases to create a secondary melt explosion. To ensure that the molten metal takes into solution the soluble gases, gas pressure is applied to the molten bath. In certain cases, the gas is also bubbled through the melt in addition to the application of gas pressure to the molten bath. The purpose of this, as in the case of vacuum atomization, is to allow the metal to be saturated with the soluble gases. This is followed by standard inert gas atomization of the gas-saturated molten metal. As the atomized droplets cool, the soluble gas is rejected from the solidifying dendrites as the solubility decreases sharply when the transformation from liquid to solid occurs. This can easily be visualized from the variation in the solubility of hydrogen and nitrogen in iron with varying temperature as shown in Figure 3.13 [29]. This gas rejection causes the surrounding liquid to become super-saturated with the soluble gas until a gas bubble nucleation occurs. The formation of the gas bubble is immediately followed by the explosive growth of the bubble to cause a secondary explosion resulting in a second stage microatomization that leads to the formation of a very high yield of fine powders.

The possibility of gas entrapment within the powder and the formation of hydrides or nitrides (from reaction of the material with hydrogen or nitrogen) could be seen as some of the technical obstacles for the gas soluble atomization process. However, post powder production heat treatments, such as elevated temperature holding of the powder in vacuum, could eliminate or minimize the problem of retained soluble gases. This process is presently being scaled up to produce large quantities of powders. The facility includes a vacuum induction melting furnace with a capacity to melt approximately 90 kg of iron-based alloys. The melt chamber can be totally isolated from the atomization chamber, thus providing the capability of pressurizing the melt chamber with the soluble gas. It is postulated that the application of higher pressures of the soluble gas and other process modifications of a proprietary nature will result in average powder particle sizes of 1 μm. This is a process with good potential for providing ultrafine powders at an economic price and could partially fulfill the tremendous appetite of the metal injection molding communities for fine atomized powders. However, the control of internal porosity within the powders poses a tremendous challenge.

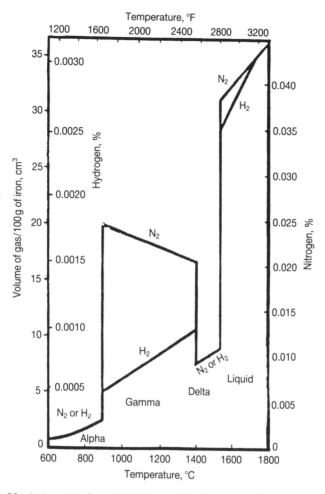

Figure 3.13: *Variation in the solubility of hydrogen and nitrogen in iron with varying temperature. (Reprinted with permission from Ref. 29.)*

Vertical–Horizontal (VH) Atomization

Another form of gas atomization technique, referred to as vertical–horizontal atomization (VH-atomization) has been introduced by Anval Nyby Powder AB (ANPAB), a subsidiary of Valinox SA, Paris, France. This process results in lower gas consumption, requires lower gas pressures, and is capable of producing fine powders [30,31]. Three independent gas pressures control the atomization process. They are furnace gas pressure, froth gas pressure in the vertical ceramic tube, and the atomizing gas pressure.

In this process, a furnace on a platform is raised toward a ceramic tube through which bubbling gases are added. This creates a froth of gas and steel, which enters the horizontal part of the nozzle. When the froth enters the horizontal atomizing area, the third gas tank applies the atomizing gas pressure, which disintegrates the alloy "froth" into fine powders. The metal particles do not accelerate to the same degree as the gas, causing a "drag force" that causes further disintegration. Thus, the three pressures that determine the process efficiency are the furnace gas pressure, the froth gas pressure in the vertical ceramic tube, and the atomizing gas pressure.

A number of nickel-based alloy and steel powders have been produced by this process. A typical example of the processing parameters for the atomization of a nickel-based superalloy, Nimonic 80A, as outlined by Aslund and Tingskog [30] is as follows: a 1500 kg melt was atomized using a maximum gas pressure of 0.7 MPa (7 bar); the metal flow rate was 1.5 kg/s (200 lb/min) and the gas consumption was 0.07 Nm³/s (250 Nm³/h). The powder was collected in argon and had an oxygen content of 95 ppm. The average grain size of the powder was 54 μm and it was spherical in shape.

Liquefied Gas Atomization

A new process of atomizing metals and alloys with the help of liquid gas has been developed by Greishiem of Germany in cooperation with Leybold AG and the University of Erlangen-Nürnberg. The process is said to produce quenching rates at least one order of magnitude higher than those obtained by gas atomization. This process combines the advantages of fast cooling rates, lower gas consumption, a close powder particle size distribution, and the production of clean powders due to the inert gas atmosphere.

For logistic reasons, industrial gases are transported and stored as cryogenic liquids. Thus, the liquid gas has an inherent "cold" associated with it, which is never utilized when powders are produced by gas atomization techniques. It was conceived that the use of cryogenic liquid gas for atomization would result in a very high quenching rate, as can be seen from equations (9) and (10) [32]:

$$Q_1 = M\left[\int C_{pM}\,dT + H_1 + \int C_{pS}\,dT\right] \tag{9}$$

$$Q_2 = m[C_{pL}(T_u - T_1) + H_2 + C_{pG}(T_2 - T_u)] \tag{10}$$

where

Q_1 is the quantity of heat to be removed;
Q_2 is the amount of heat removed by the gas, and $Q_1 = Q_2$.

The other notations are outlined below.

C_{pM} = specific heat of the melt;
C_{pS} = specific heat of the solid metal;
C_{pG} = specific heat of the liquid gas;
C_{pL} = specific heat of the gas;
H_1 = enthalpy of solidification;
H_2 = enthalpy of vaporization;
T_u = transition temperature of liquid to gas;
T_1 = temperature of the liquid gas;
T_2 = temperature of the gas;
m = mass of the gas;
M = mass of the melt.

With liquid gas the enthalpy of vaporization can be utilized and the mass m is much higher for a liquid than for a gas. All this translates into increased production capacity and reduced operating costs along with finer powders that have undergone very rapid quenching.

A schematic of the liquid gas atomization unit and its nozzle arrangement is shown in Figures 3.14a and 3.14b [32], respectively. The nozzles can produce either a round or a flat jet. High-pressure lines are used to feed the cryogenic liquid gas to the nozzles, which are placed directly under the melt flow stream. To attain the high working pressures of 30 to 60 MPa, the liquid gas pressure is raised by a piston pump. With the use of additional cooling by heat exchange, a slim liquid gas jet is maintained over a length of 0.1 to 0.3 m. In case of liquid nitrogen, pressures up to 65 MPa can be reached (with precooling), while the maximum pressure that can be reached with argon is around 30 MPa.

Powders of lead–tin and copper–phosphorus have been produced by this atomizer. The powders are spherical in nature and, due to the fast cooling rates, the powder particles do not stick to each other, thus eliminating satellite formation. The cooling rate, which can be measured from the secondary dendritic arm spacing, was used to calculate the quenching rate of the copper–phosphorus powder. It was found that particles of size 40 μm were cooled at a rate of 10^6 K/s. The high cooling rate is obtained by the production of finer particles as well as the higher heat transfer coefficient. The final powder particles do not have an oxide skin and thus do not require the post-reduction annealing of the powders.

Gas–Water Atomization

The need for finer powder particle size has led to the development of two-step atomization [33,34]. The process relies on the atomization of the flowing metal

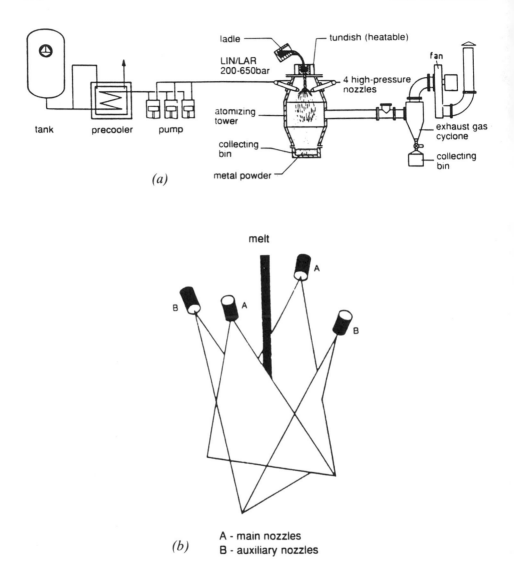

(a)

(b)   A - main nozzles
          B - auxiliary nozzles

Figure 3.14: *(a) Flow chart of the liquid gas atomization process; (b) schematic nozzle arrangement for liquid gas atomization. (Reprinted with permission from Ref. 32.)*

stream by two consecutive fluid media, namely gas and water. A flowing stream of liquid metal is first impacted by a high-pressure gas jet as in the case of normal gas atomization. Within a short flight distance, the disintegrated particles are subjected to impact by a high-pressure water jet. The resultant powder is finer than the powder obtained by either gas or water atomization

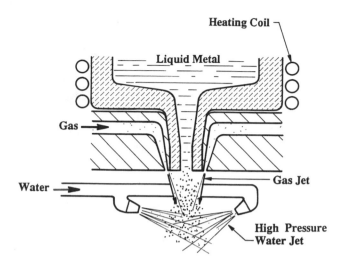

Figure 3.15: Schematic arrangement of the two step gas–water atomization process. (After Ref. 34.)

techniques. A schematic arrangement of the two-step atomization process as reported in Metal Powder Report [34] is shown in Figure 3.15 (after the work of Stock).

In their experiments, the gas pressure was varied from 0 to 3 MPa and water pressure from 0 to 15 MPa. Keeping the water pressure constant, a 0.5 MPa gas pressure results in a 40% decrease in the particle size. The water pressure does not have such a significant effect as that of gas pressure.

The principle underlying the process has also been outlined in a paper presented by Stock et al. [33]. The initial impingement of the high-pressure gas jet on the liquid stream causes its disintegration into very fine droplets. The droplets, still molten, travel a considerable distance before they solidify to form the powder. During this long flight, collision with other droplets results in the formation of coarse powders. Secondary impingment of the molten droplets with the water jet causes a rapid extraction of heat and freezing of the fine droplets before they undergo a large number of collisions to form the coarser particles.

From the above discussion it can easily be visualized that one of the key parameters that will influence the two-stage atomization process is the distance of separation between the gas and the water jets. The closer the two jets, the finer will be the powders as they are frozen in their flight path much more quickly. Other general single-fluid atomization parameters such as gas pressure, superheat, heat extraction capacity of the gas used for atomization, etc., will also influence the final powder particle size. One of the interesting aspects

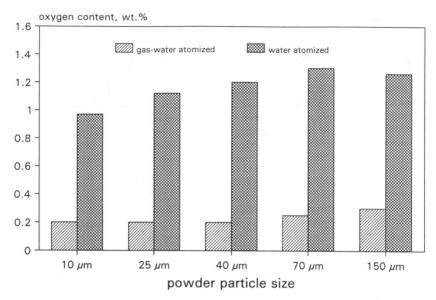

Figure 3.16: *Oxygen content of powders obtained by the gas–water atomization process compared to water-atomized powders.*

of this process is the particle shape. Gas atomization generally produces spherical particles, while water atomization yields highly irregular-shaped powders with high surface oxide contamination. As the proportion of spherical particles increases, the compressibility of the powder decreases. The two-stage atomization process produces powders that have particle shape that is in between the water- and gas-atomized powders. Also, the oxygen content of the powders obtained by this new two-step atomization process is much lower than that of water-atomized powders, as seen in Figure 3.16.

The two-step atomization process has the potential of providing fine and clean powders where the powder shape can be controlled to some degree by the proximity of the two fluid jets. The control of the particle shape and size provides an opportunity to produce powders that could be ideal for powder injection molding.

Ultrasonic Atomization

Ultrasonic atomization is regarded as a powder processing technique with the potential for inexpensive powder production capability. The ultrasonic atomization process is different from the ultrasonic gas atomization discussed earlier and the distinction should be carefully noted. In the present process, very little gas is consumed during atomization and the classic impingement of the liquid stream with a high-pressure gas jet does not occur. There are two types of ultrasonic techniques for atomization: capillary wave atomization and the standing wave technique [35]. For specific applications, this is an alternate powder production technique with large potential as it does not require a large chamber and has very low gas consumption.

The principle of the capillary wave atomization process relies on a thin layer of liquid metal evenly covering the surface of a vertically vibrating (with respect to its surface plane) resonator, forming a stationary capillary wave when the elongation amplitude exceeds a certain limiting value. The schematic design of the process and the principle of capillary wave atomization are shown in Figure 3.17.

With very high velocity amplitude, small droplets are expelled from the peaks of the cones as shown in Figure 3.17. The diameter of the particle is often related to the capillary wave length as shown in the figure. The process involves the delivery of molten liquid through a nozzle to the surface of a resonator that resembles a hollow cone vibrated by a central stem (looking similar to an umbrella). The resonator material should be at a temperature at which the molten liquid does not solidify, and also the resonator material should not react with the liquid metal. The liquid metal forms a thin layer on the resonator, which is vibrated at high frequencies. Once the proper amplitude of oscillation is reached, small droplets are ejected, and they solidify to form spherical metal powders. The particle diameter decreases as the resonator frequency is increased. The high efficiency of the process results in extremely low energy consumption and very little gas usage. The low speed of ejection of the liquid droplets allows ample time for the material to solidify, resulting in the possible use of small, compact atomizing chambers.

The technique of standing wave atomization relies on the static and dynamic forces and gas currents in a standing ultrasonic wave. The atomization occurs at the velocity antinode where the forces act like a gas jet on the boundary of a liquid. However, in this case, no gas is consumed. A cone with a central diaphragm is formed due to the Bernoulli forces, and shear forces of the dynamic gas oscillation disrupt the diaphragm. The principle of standing wave atomization and a schematic design of the process are shown in Figures 3.18a and 3.18b, respectively.

A sound-radiating transmitter and a coaxially aligned reflector placed at a resonating distance create an intense standing ultrasonic wave. Liquid metal is introduced through a nozzle into the pressure nodes of the horizontal standing wave. The shear force of the dynamic gas oscillation results in the atomization of the liquid metal. The powder size decreases with increased gas pressure due to the increased sound intensity. The flight distance of the molten droplet is very small when the chamber gas pressure is high, resulting in a small, compact

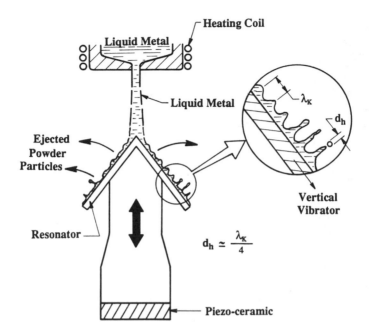

Figure 3.17: *Schematic design and the principle of the capillary wave atomization process.*

atomization chamber. The advantage of this process over the capillary wave atomization technique lies in the ability to atomize materials without the contact that is required by the capillary wave atomization process. However, the problem of guidance of the molten stream of metal and the design requirement to prevent nozzle freezeup during atomization prevents the large scale commercialization of the process at present.

Vibrating Electrode Atomization

There are alternate means of providing the energy that will subsequently cause melt disintegration: one is to use vibrational energy to cause the melt breakup. This has been used in the process of vibrating electrode atomization (VEA).

Vibrating electrode atomization is a variation of the rotating electrode process that will be discussed in a later part of this chapter. In the rotating electrode process powder particles are produced by the action of the centrifugal forces, while in this process the vibration of the consumable electrode itself causes particle formation.

The material to be atomized is first formed into an electrode. This consumable electrode forms a resonating rod, the free end of which forms an

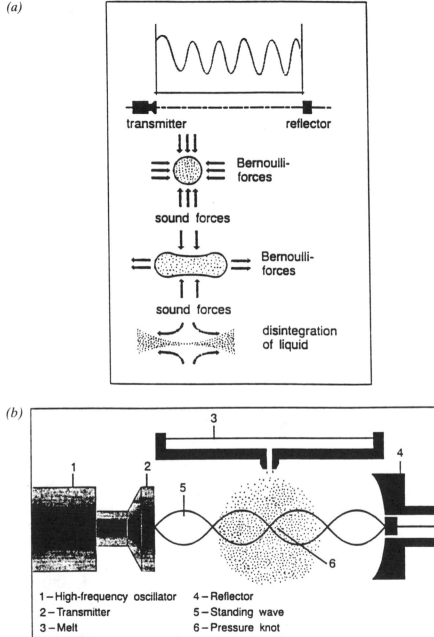

Figure 3.18: (a) *Principle of standing wave atomization.* (b) *Schematic design of the process by standing-wave atomization. (Reprinted with permission from Ref. 35.)*

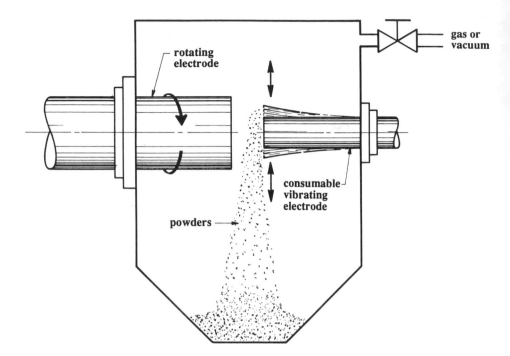

Figure 3.19: Schematic of the vibrating electrode process.

arc with a water-cooled copper rod. The water-cooled copper rod rotates slowly and the arc that it forms with the vibrating end of the consumable rod causes local melting of the front end of the consumable electrode. Spherical particles are formed by this process and a very narrow size distribution can be obtained by vibrating the rod at its resonant frequency. The powder size can be controlled by the length and diameter of the vibrating rod.

The process has not gained much popularity, partly due to the prior commercialization of the rotating electrode process. Clean, low-oxygen-containing powders can be produced by this process [36]. A schematic of the vibrating electrode process is shown in Figure 3.19.

An interesting variation on the general theme of vibrating electrode atomization has recently been described by Kito et al. [37]. They have developed a new atomization technique for the production of spherical powders. The process utilizes the simultaneous application of an ultrasonic oscillator to provide the mechanical energy for disintegration and a plasma arc for the thermal energy.

The metal or alloy that will be atomized is prepared in the form of a cylindrical rod having a diameter of approximately 1×10^{-2} m and a length of approximately 5×10^{-2} m to 7×10^{-2} m depending on the material. The top area of the cylindrical rod serves as the anode of the arc electrode that is established between the cylindrical rod and the internal electrode of a plasma torch. The cylindrical rod is mechanically attached to a vibrating horn connected through the driving unit to an ultrasonic oscillator. The frequency of oscillation is 15 kHz, the power is 600 W, and the amplitude of the oscillation is rated as 20 μm along the axis direction of the cylindrical anode. The arc current is supplied through a 1000 kV DC power supply. The vibration unit and the horn are electrically isolated with an insulator. Helium gas, supplied through the plasma torch at a high flow rate of 16×10^6 m^3/s, serves as a rapid cooling medium and also prevents oxidation. The oscillating part, which is made up of the anode, vibration unit, horn, and insulator, is a complete resonant structure. Thus, the anode length has a direct bearing on the amplitude of oscillation and the powder production rate. Figure 3.20 shows a schematic diagram of this process [37].

Once the arc between the anode and the cathode has been established, the extreme top part of the cylindrical anode is molten. The rapid vibration of the anode itself causes molten droplets to be ejected at a low velocity of around 3 to 5 m/s. The ejected particles are rapidly cooled by the flowing helium gas at calculated cooling rates exceeding 10^5 K/s. Calculations indicate that an ejected iron particle will be cooled from its melting point to room temperature within 0.01 s, thus, requiring 0.05 m of total flight distance before solidification.

To model the process, it can be assumed that the disintegration of the liquid pool formed on top of the electrode will be atomized into smaller droplets when the vibrational forces due to atomization are equivalent to the surface tension of the liquid. The calculated particle diameter, D, can be related to the various ultrasonic vibration conditions and the surface tension and density of the liquid material by:

$$D = \left[\frac{6\gamma}{\rho\lambda\omega^2} \right]^{1/2} \tag{11}$$

where

D is the calculated particle diameter;
ω is the angular frequency of vibration;
λ is the amplitude of the ultrasonic oscillation;
ρ is the density of the molten alloy;
γ is the surface tension of the liquid alloy.

Calculations based on this equation reveal particle sizes of 76, 93, and 106 μm for copper, iron, and titanium, respectively. These calculated average particle diameters are in good agreement with the experimental results on the

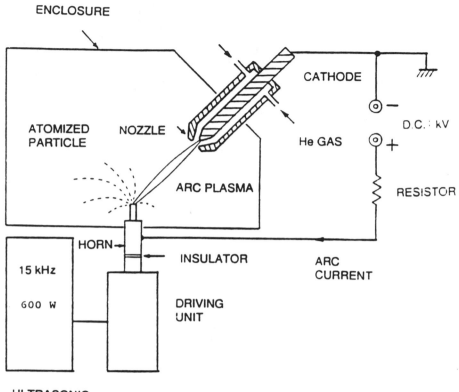

ENCLOSURE

CATHODE

ATOMIZED   NOZZLE
PARTICLE

D.C.: kV

He GAS

ARC PLASMA

RESISTOR

HORN

15 kHz   INSULATOR   ARC
                                     CURRENT

600 W   DRIVING
                   UNIT

ULTRASONIC
GENERATOR

Figure 3.20: Schematic diagram of high-frequency (ultrasonic) vibrating arc electrode atomization. (Reprinted with permission from Ref. 37.)

atomization of these materials by this novel process. Figure 3.21 shows photomicrographs of the three powders produced by this process [37]. Note the spherical morphology of the powders and the approximate particle size of the powders. The typical operating conditions of this process as outlined by the authors are given below [37].

Distance between anode and cathode	50 mm
Inert gas flow rate (helium)	100 liter/min
Anode bar diameter	10 mm
Anode bar length	50 to 67 mm
Arc current	10 to 40 A
Operation time	30 s
Frequency of ultrasonic generator	15 kHz

| Power of ultrasonic generator | 600 W |
| Amplitude of ultrasonic generator | 20 μm |

This process provides several advantages over the conventional atomization processes and even the relatively new processes such as the rotating electrode processes. The advantages as summarized by the authors are as follows:

1. Due to the use of a plasma arc as the heating source, practically all high-temperature metals and alloys can be atomized by this process.
2. The probability of gas-induced porosity is reduced.
3. The lower ejection speed of the particles and the rapid cooling rates reduce the requirement for the large cooling area that is required for conventional atomization processes. The cooling cavity is smaller even than in the rotating electrode processes where the ejected velocity of the particles is at least an order of magnitude larger.
4. The device works without any moving components, thus providing operational safety.
5. As the process does not require any crucible, the powders formed can be extremely pure, with the purity level being dictated by the purity of the anode. Thus, processing of reactive powders with high melting points is attainable.

The process has been verified at laboratory scale. More research is needed to scale up this process for fabricating production quantities of powders. However, the advantages of the process will provide a great deal of impetus for further development.

Melt Drop Atomization

When a crucible containing molten metal is vibrated while the molten metal is flowing down due to the force of gravity, the melt disintegrates to form a powder. In the melt drop technique, molten metal flows out through number of equally sized holes in a nozzle that is at the bottom of the crucible containing the melt. The chamber into which the molten metal enters can either be filled with inert gas or be under vacuum. The crucible is vibrated by a variable-speed direct-current motor at frequencies that vary from 7 to 25 cycles per second, and the amplitude of vibration is around 1 cm. If the orifices in the nozzle through which the liquid metal exits are small, additional gas pressure is needed above the liquid metal surface.

This process produces coarse powders and is capable of producing millimeter sized granules or shots. This technique has been used to produce coarse powders of aluminum, beryllium, copper, lead, and superalloys.

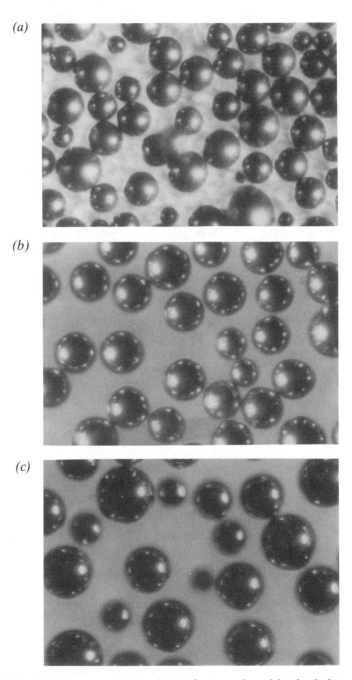

Figure 3.21: *Photomicrographs of powders produced by high-frequency vibrating arc electrode atomization: (a) copper; (b) iron; (c) titanium. (Reprinted with permission from Ref. 37.)*

Electrode Induction Melting with Inert Gas Atomization (EIGA)

Inert gas atomization is one of the leading processes for the production of high-temperature metal powders for use in high-strength high-performance applications. Reactive and refractory metals, intermetallic compounds and superalloys are examples of such high-temperature high-performance materials that require super-clean, low-oxygen-containing rapidly solidified powders. Inert gas atomization is the most efficient and economical way for meeting the stringent requirements of these powders and also producing them in large production quantities. However, one of the major problems encountered during the production of such powders is ceramic impurity pickup due to the ceramic-lined crucible in which the material is melted prior to inert gas atomization. Ceramic inclusions in these high-performance materials are a major cause of their failure during operating conditions. The industry has adopted various ingenious means to combat this problem. Processes have been developed in which the lighter inclusions are allowed to float through the melt toward the surface before the melt is poured though a "teapot spout," and ceramic reticulated foam filters have been used to sieve out the ceramic inclusions from the molten metal.

Obviously the best way to handle the ceramic inclusion problem is to eradicate the root of the problem, which is the ceramic crucible. This then leads to the concept of ceramic-crucible-free or even crucible-free melting coupled with inert gas atomization. This concept has been utilized by the researchers of Leybold AG, Germany, to develop two different techniques known as "electrode induction melting combined with inert gas atomization," or EIGA, and "plasma melting in combination with inert gas atomization," or PIGA [38–40]. The PIGA process will be discussed in detail in a later section of this chapter dealing with plasma atomization.

The EIGA process is essentially an induction drip-melting atomization process that utilizes a rotating electrode that gradually moves down through a specially designed induction melting coil. A schematic diagram of the process is depicted in Figure 3.22. The desired material is formed into an electrode and does not require a ceramic crucible. Continuous melting can be produced by melting from the surface to the center of the electrode. This is dictated by the generator frequency, which has to be as high as 200 kHz to attain the desired energy penetration on the surface of the electrode. The arrangement of the induction coil is also of great importance. The special arrangement provides the desired superheat to the melt, which flows down the front end section and is fed directly into the atomizing nozzle. The atomization is carried out by a

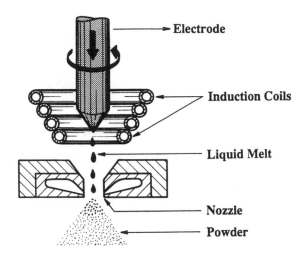

Figure 3.22: Schematic diagram of the EIGA Process.

high-velocity inert gas stream. Automation of the process leads to an efficient atomization process that has been utilized to successfully produce powders of pure titanium, titanium aluminides, Ti–6Al–4V alloy, superalloy, and FeGd. A comparison of the particle size distribution of some titanium-based alloy powders produced by the EIGA process (indicated as "Leybold AG Process") and two other processes, PREP (plasma rotating electrode process) and a normal inert gas atomization process, is shown in Figure 3.23 [38]. One of the disadvantages associated with this process is the need to prefabricate electrodes from the desired material; and, except for degassing, no additional refining steps can be used.

This new technique is quite suitable for producing ceramic-free powders of high-performance materials. The commercial use of this technique will enhance the use of reactive, high-temperature and high-performance materials.

Roller Atomization

This atomization process provides an excellent method for producing flaky powders of a large number of materials. Flakes of aluminum, copper, bronze, tin, etc., have been produced by this method. The process has, however, not been utilized to its full extent.

The technique in essence consists of two inwardly rotating rolls through which the molten metal stream is passed [41,42]. The molten metal cavitates within the roller, which causes perforations and streamers. High roller speeds

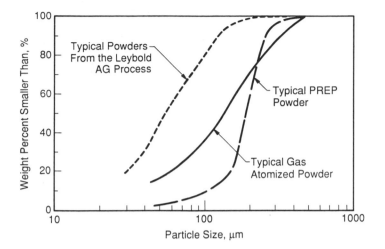

Figure 3.23: *Comparison of the particle size distribution of some titanium-based alloy powders produced by the EIGA process (indicated as Leybold AG Process) and two other processes — PREP and a normal inert gas atomization process.*

cause the liquid metal to flow around the cavitation points and form thin threads that break down to form the desired powder, which is cooled in water. A typical solidification rate is around 10^2 K/s or higher.

The powders produced by this method are relatively coarse with an average powder particle size around 200 μm. The typical processing conditions of this process are outlined below [43] and a schematic drawing illustrating the roller atomization concept is depicted in Figure 3.24.

Roll gap	60 μm
Roll rotational speed	120 000 rpm
Stream diameter	2.4 mm
Superheat	373 K (100°C)

Electrohydrodynamic Atomization

Electrohydrodynamic (EHD) atomization is a process by which a very fine liquid spray is produced in a vacuum atmosphere [44–46]. The instrument in which this electrostatic spraying is carried out is known as a micro-particle processor. A schematic diagram of the EHD spray process for producing both powders and coatings is depicted in Figure 3.25 [46]. The process can generate

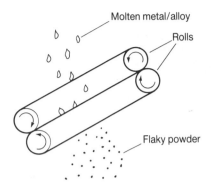

Figure 3.24: Schematic drawing of roller atomization.

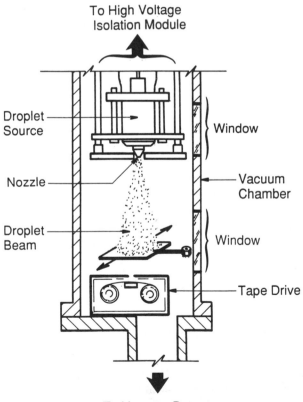

Figure 3.25: Schematic of electrohydrodynamic spray atomization.

microcrystalline to amorphous powders. Particle sizes produced by this process are in the submicrometer range and cooling rates above 10^7 K/s have been achieved. Powders of 3 to 100 μm have also been produced by this process.

The fine powder is produced by the ejection of charged molten droplets from a conductive molten liquid surface that is exposed to intense electric fields. The large electric field produces a continuous beam of fine droplets that solidify in flight as fine powders. The atomization event is carried out in a vacuum atmosphere of 0.0013 MPa (10^{-5} torr). The process allows the possibility of producing various carbides, nitrides, or borides by ejecting the molten droplets through gaseous atmospheres where they can react to form the desired compound.

Powders and coatings of various material such as Sn, Cu, Al, Si, and alloys of aluminum, have been produced by this process [44,45]. As an example, Al–1%Mn alloy powder having an average particle size of 0.01 μm (10 nm), with the smallest particle being below 0.003 μm (3 nm), have been produced by this process. The machine can typically produce 8 g of aluminum alloy in approximately 144×10^2 s (4 hours). The process is suitable for producing very specialized powders for research and development purposes [46]. One very interesting application of powders produced by this process is in the area of standards for analytical electron microscopy (private communication by H. Fraser with the authors of Ref. 46). Also, the low-temperature submicrometer alloy powders could be melted and resolidified in the electron microscope, whereby sophisticated in-situ studies can be made on the formation and reversion of metastable phases, solidification, coarsening, and other thermally activated transformations. Although the scaling up of this process to produce production quantities of powders will be difficult, the process has a niche in the sophisticated fundamental research area where its role is expected to remain of vital importance.

Centrifugal and Rotary Atomization

A special class of atomization technique that utilizes the breakup of a thin liquid film by centrifugal forces instead of the conventional disintegration of the molten stream of metal by fluid impingement is generally termed "centrifugal atomization." In some atomization processes a molten stream of metal is allowed to fall on a rapidly spinning disk rotating at high velocities. The molten metal stream is disintegrated on contact with the rotating disk as opposed to its breakup by high-velocity fluids. These processes are generally known as "rotary atomization." Myriad variations based on these general concepts have been utilized to produce powders of various metals and alloys. Prealloyed powders of aluminum, zirconium, titanium, iron, and various

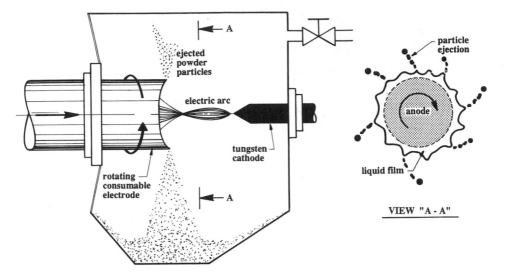

Figure 3.26. *Schematic of the rotating electrode process.*

superalloys have been produced on a commercial scale by some form of centrifugal or rotary atomization process.

One of the commonly used commercial processes that is based on this concept is the "rotating electrode process" better known as REP. A schematic diagram of this technique is shown in Figure 3.26. Material of the desired powder composition is fabricated into the form of a rod electrode that serves as the anode. This consumable rod electrode is usually rotated at velocities of 250 revolutions per second and gradually moved in as the electrode is consumed. These motions are accomplished by external drives. The melting of the electrode at one end is accomplished by a stationary tungsten electrode. This can, however, result in small amounts of tungsten contamination of the powders. The process is usually carried out under some form of protective atmosphere in order to prevent the oxidation of the powders.

Powders produced by the rotating electrode process are generally spherical in nature, and have a coarse particle size compared to gas-atomized powders. A typical powder produced by the rotating electrode process has an average powder particle size of 150 to 250 μm, and the particle size distribution is usually narrow. Thus, the cooling rates that can be obtained by this process are low and the degree of superheat that can be imparted is also limited. The process has been used extensively to produce powders of reactive and refractory materials.

Champagne and Angers [47–49] have discussed the physics of droplet formation during the centrifugal atomization process. They have identified three distinct mechanisms by which droplet formation can occur. These are schematically illustrated in Figure 3.27 and are denoted as

Mechanism 1   Direct droplet formation (DDF)
Mechanism 2   Ligament disintegration (LD)
Mechanism 3   Film disintegration (FD)

Champagne and Angers divided the parameters controlling the atomization mechanisms into two broad groups, namely, material parameters and operating parameters. The two parameters that usually control the operating mechanisms discussed above include a number of parameters as outlined below.

Operating parameters.
ω = angular velocity of the rotating electrode (rad/s)
D = diameter of the rotating electrode (m)
Q = liquid supply rate or the melting rate of the electrode (m^3/s)

Material parameters.
ρ_L = density of the liquid metal
γ = surface tension of the liquid metal (N/m)
η = liquid metal viscosity (Pa.s)

It was determined that the ratio of two parameters, the first of which incorporates the process variables and the second the material variables, determine the transition from one mechanism to the other. The ratio that governs this transition from one mechanism to the other is

$$\frac{Q\omega^{0.6}/D^{0.68}}{\gamma^{0.88}/\eta^{0.17}\rho_L^{0.71}}$$

In this ratio, the change of mode from mechanism 1 to mechanism 2 (i.e., from DDF to LD) occurs at a ratio of 0.07, and from mechanism 2 to mechanism 3 (i.e., from LD to FD) at a ratio of 1.33. It is obvious from the form of the ratio that an increase in the angular velocity and liquid supply rate and lowered diameter of the rotating electrode will tend to push the transition from mechanism 1 towards mechanism 2, and from mechanism 2 to mechanism 3. This model has been experimentally verified on number of metals. Champagne and Angers have also provided a diagram that predicts the various domains of existence of the three modes of centrifugal atomization. The diagram was created by plotting the denominator of the above ratio, which

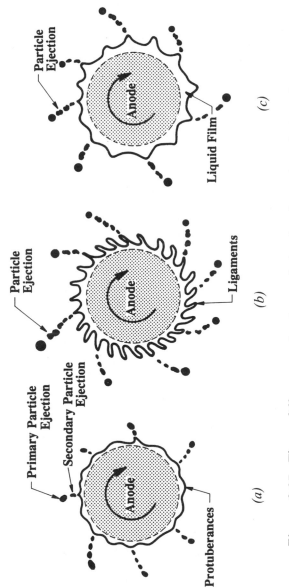

Figure 3.27: Three different mechanisms of droplet formation during the centrifugal atomization process according to Champagne and Angers: (a) direct droplet formation; (b) ligament disintegration; (c) liquid film disintegration. (After Refs. 47–49).

includes all the material parameters, as the X-axis and the numerator of the ratio, which includes all the operating parameters, as the Y-axis. Thus, the X-axis denoted by $[\gamma^{0.88}/(\eta^{0.17}\rho_L^{0.71})]$ and the Y-axis denoted by $[Q\omega^{0.6}/D^{0.68}]$ are plotted, and the three zones showing direct droplet formation, ligament disintegration, and film disintegration are clearly distinguished. This figure provides the operator the capability to determine the prevailing atomization mode depending on the chosen operating parameters.

The shape, size, and size distribution of the powder are also affected by the mechanism of droplet formation. In the first mechanism of direct drop formation, shown in Figure 3.27a, the supply of liquid to the edge of the electrode is small. The liquid forms small surface protuberances, and the particle ejection is controlled by two opposing mechanisms, namely, centrifugal force causing instability and the surface energy. If particle formation is dictated by the direct drop formation mechanism, two droplets, one large from the outermost end and one small from the inner neck, are ejected from the surface protuberances. Thus, a distinct bimodal powder size distribution is generally obtained. With increasing liquid supply to the rotating electrode edge, the larger amplitude of the protuberances causes a breakup of the elongated ligaments. The trend is toward a more tear-drop type of powder shape. The mean powder particle size is usually increased by an increase in the melt rate, which is generally in the order of 10^{-7} m^3/s. An empirical formula developed by Champagne and Angers (equation 12), can predict the mean volume–surface diameter, d_{vs}.

$$d_{vs} = 4.63 \times 10^6 \frac{\gamma^{0.43} Q^{0.12}}{\rho_L^{0.43} D^{0.64} \omega^{0.98}} \tag{12}$$

where the parameters have the same meaning as discussed above.

The mechanisms of disintegration during centrifugal atomization have also been extensively studied by Halada and coworkers [50], who investigated the mechanisms both theoretically and empirically. From their work, the authors have produced a centrifugal atomization (CA) diagram that shows the state of atomization for liquid properties and atomizing conditions. The CA diagram is based on a complex relationship between Q, a dimensionless flow rate parameter, the Weber number W, and the Reynolds number Re. The Weber and Reynolds numbers for rotation can be calculated from:

$$W = \rho\omega^2 R^3/\gamma \tag{13}$$

$$Re = R^2 \omega\rho/\eta \tag{14}$$

where

ρ is the density of the liquid (kg/m^3);
ω is the angular velocity (rad/s);
γ is the surface tension of the liquid (N/m);
η is the viscosity of liquid (Pa.s);
R is the radius of the rotating disk.

The authors concluded that liquid metal atomization was primarily through the mechanism of direct drop formation. According to the theoretical calculations by Halada and coworkers [50], the diameter d_p of the particle formed when the direct drop mechanism is operative is given by

$$d_p = 3.2 \ R \ W^{-1/2} \tag{15}$$

where R and W are the radius of the rotating disk and the Weber number, respectively.

It can be predicted from the above equation that an increase in the rotation speed will result in particles of smaller size. This equation also predicts that the melt flow rate will not be a factor influencing the particle size.

A variation of the rotating electrode process is the plasma rotating electrode process, or PREP. In this process the end of the rotating consumable electrode is melted by a plasma gun and the powder is formed by the ejection of the molten material by centrifugal forces. This process will be discussed in detail in a later part of this chapter on plasma processing. Another variant of this process is the laser spin atomization process in which a laser beam is used to melt the front face of a consumable rotating electrode spinning at 8000 to 35 000 rpm. A 5.2-kW carbon dioxide laser beam has been used to form the melt from which particles are rapidly ejected by centrifugal forces. The ejected particles are rapidly cooled by jets of helium gas and the particle shape is mostly spherical.

Different forms of centrifugal atomization (sometimes referred to as rotary atomization) rely on an external source of molten material. A number of these processes have been discussed by German [51] and are illustrated in Figure 3.28. In the majority of cases a molten stream of metal is allowed to fall on a chilled rotating member, which can be in the form of a wheel, cup, or mesh. The centrifugal force results in the droplet formation. Other variations include melt extraction, where a rotating wheel is used to produce powders or ribbons (Figure 3-28c). The spinning wheel is made to touch the surface of a molten pool of metal or alloy, and the high centrifugal force causes the melt extraction. Coarse powders can be produced by intentionally texturing the melt

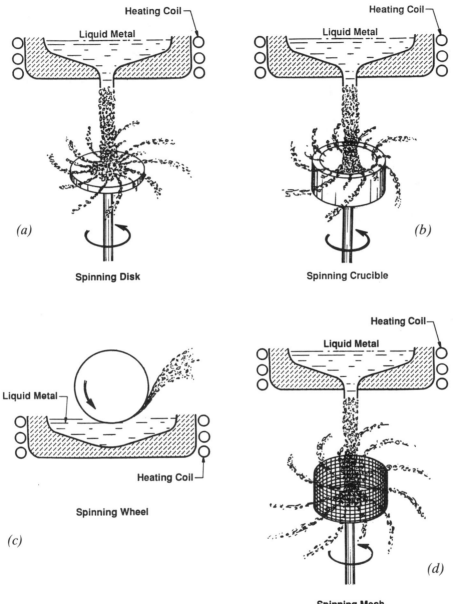

Figure 3.28: *Different forms of centrifugal atomization: (a) rotating disk; (b) rotating cup; (c) melt extraction; (d) rotating screen. (Redrawn after Ref. 51.)*

extraction wheel. One modification of the process directs a molten stream of liquid metal on to a rotating disk that atomizes the material into numerous droplets by centrifugal forces. The droplets are then allowed to fly into a water vortex barrier, which rapidly solidifies the powder particles. Modifications of some of these processes can be used to produce rapidly solidified powders and also amorphous powders. Numerous variations based on the above concept have been developed by various companies and research organizations. This chapter will only deal with a few of those processes.

Before discussing the individual processes, it will be useful to consider the equations that govern the particle sizes and the cooling rates for centrifugal atomization processes where an unconfined liquid metal stream is allowed to impinge on a rotating disk. The diameter of the particles that are produced by this form of centrifugal atomization is connected to the various atomizing conditions by [52]

$$d = \frac{1}{2\pi N}\left(\frac{6\gamma}{\rho R}\right)^{1/2} \tag{16}$$

where

d is the particle diameter (mm);
N is the speed of the rotating disk (rpm);
γ is the surface tension of the liquid metal (N/m);
ρ is the density of the liquid metal (kg/m^3);
R is the radius of the disk (mm).

For centrifugal atomization processes [53], the cooling rate of the particles that can be roughly measured from the secondary dendritic arm spacing in the microstructure can be connected to various processing parameters by

$$S_d = \frac{52.7 d^{0.40}(\rho C_p)^{0.29} h_f^{0.17}}{[K(T_p - T_g)]^{0.29}(\rho_g DR)^{0.17}} \tag{17}$$

where

S_d = secondary dendritic arm spacing (μm);
ρC_p = thermal capacity of liquid per unit volume (cal/cm^3);
h_f = viscosity of the cooling gas (g/(cm s));
K = thermal conductivity of the cooling gas (cal/(cm s K));
T_p = temperature of particle (K);
T_g = temperature of the wall of the chamber (K);

ρ_g = density of the cooling gas (g/cm^3);
D = diameter of the rotating disk (cm);
R = speed of revolution of the rotating disk (rpm);
d = particle diameter (cm).

From equations (16) and (17), it can be generally surmised that smaller particle sizes and faster cooling rates will be favored by faster speeds of the spinning disk, a lowered surface tension of the liquid metal (which is favored by higher melt temperatures), larger radius of the spinning disk, and higher thermal conductivity of the cooling gases. With this theoretical background, readers should have a better appreciation of some of the centrifugal atomization processes that are discussed below. For example, the role of forced convection cooling by helium gas in providing powders with faster cooling rates can be understood qualitatively from the above equations.

One form of rotary atomization process known as the electron beam rotating disk (EBRD) process has been developed in order to produce high-quality powders. The starting material in the form of a slowly rotating bar is melted so that the molten metal falls into a rotating copper crucible. One of the electron beams is focused on the edge of the rotating crucible, which is water quenched. Molten metal particles are thrown off the rim and deflected downwards by a copper shield that is water cooled.

The process does not produce ceramic inclusions due to the cold crucible used for atomizing. The powder size produced by this process varies from 30 to 50 μm and the particle shape is spherical in nature. A schematic of the process as shown in the paper by Aller and Losada [54] is shown in Figure 3.29.

One of the most successful rotary atomization techniques was developed by Pratt and Whitney to produce rapidly solidified powders. They use the principle of forced convection cooling of molten particles [55]. In these investigations, a vacuum induction melting furnace with a capacity of 46 kg was used to supply the molten metal, which was allowed to fall as an unconfined stream on to a rapidly rotating disk spinning at approximately 24 000 rpm. The molten stream of metal or alloy is disintegrated on impact by the rapidly spinning wheel and the particles formed are accelerated outward from the central source into a high-conductivity helium gas quenching medium. The cooling rate is usually around 10^5 to 10^6 K/s, which is well within the rapid solidification range. The powders are spherical in nature and are free of satellites and entrapped gas pores. Approximately 70% of the particles are in the range 10 to 100 μm.

A new process of atomization, known as rotating water film atomization, which is in principle similar to the atomization approach in which a molten stream of metal is allowed to free-fall on to a fast spinning metal disk (see

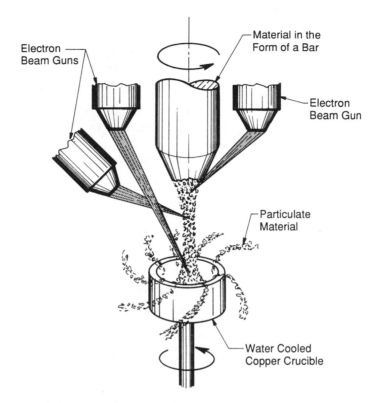

Figure 3.29: *Schematic diagram of the electron beam rotating disk (EBRD) process. (Reprinted with permission from Ref. 54.)*

Figure 3.28a), has been described by Ramon and coworkers [56]. Atomization occurs as the melt stream strikes the rotating disk and disintegrates due to the centrifugal forces. The new process provides an interesting twist to this approach. The rotating wheel has a thin layer of water on top of it, and the metal actually hits the layer of water and is atomized. This results in rapid cooling, though the droplets ejected are not fully solidified on impact.

The powders produced by this process have elongated shape. This is probably due to the rapid cooling, which does not allow the powder particles sufficient time to spherodize. The powder particle size is also quite fine.

The process involves melting and ejecting the molten metal or alloy through the bottom of the crucible in a thin stream. Often an argon overpressure is used on the surface of the melt to pressurize the molten stream through the orifice. The molten metal impinges on a fast-spinning aluminum wheel and the tangential speed at the point of impact is around 140 m/s. The impact at this velocity disintegrates the liquid into numerous tiny droplets that are hurled away from the spinning disk. The rotating disk has a specially prepared surface such that a stable film of water can be maintained during atomization. The water serves the dual purpose of cooling the rotating wheel and also providing rapid cooling for the atomized powders.

This process has been used to atomize aluminum–chromium alloys as model materials [57]. The atomization unit could atomize around 0.5 kg of the aluminum alloy per run. The total unit layout measured approximately $1.2 \times 1.5 \times 0.5$ m. The particles formed had a thick oxide layer and the relative particle size distribution was 20% below 56 μm, 14% between 56 and 100 μm, 30% between 100 and 200 μm, and 36% between 200 and 400 μm. The cooling efficiency is similar to that obtained by helium gas atomization and faster than that obtained by argon or nitrogen gas atomization.

One of the major problems of this process has been the velocity gradient in the gas layer entrained just above the rapidly spinning wheel. The gas layer immediately adjacent to the wheel is also rotating at a high velocity. This velocity rapidly decreases as the distance from the wheel increases. The molten metal stream is virtually reflected back toward the crucible bottom. This often results in the coating of the crucible surface and the clogging up of the nozzle. Presently around 15% to 20% of the metal is reflected back and, thus, constitutes a major problem.

Another process based on atomization in the presence of a secondary liquid film has been developed by Battle Research Laboratory, Ohio. This process utilizes the atomization of a molten stream of liquid inside a rotating cup that also contains the coolant liquid film. The atomization and the cooling of the particles is done in the presence of the cooling liquid. The high-density metal particles disrupt the vapor of the cooling liquid envelope and thus improve the heat transfer by the direct contact of the particles with the liquid. A variety of particle shapes of materials such as tin, copper, nickel, and superalloys have been produced by this process. A glassy alloy of Fe–40Ni–14P–6B (at.%) was also produced by this process. This indicated that cooling rates were in the range of 10^6 K/s.

Centrifugal shot casting, as the name implies, is another variation of the centrifugal atomization process. This process can produce powders from a number of reactive and refractory metals having melting points as high as 3775 K (3500°C). The process is carried out under a controlled atmosphere to prevent any oxidation of the material. Since the inert gas is used only as a protective atmosphere and not for supplying the energy to disintegrate the molten metal, the volume of inert gas consumed is very low. The shape of the powders produced by this process can be spherical or flaky in nature. The development of this process by Harwell, one of the UK Atomic Energy

Authority's development laboratories, has been described in *Metal Powder Report* [58].

In the centrifugal shot casting process the material to be atomized is present in the form of a skull on a water-cooled crucible. An arc is struck between the water-cooled crucible and a stationary electrode. The crucible is rotated, while the heat due to the arc melts a part of the skull material. Due to centrifugal forces created by the rotating crucible, the molten material rises up the side wall until it reaches the lip of the crucible. At that point, the material flowing over the lip of the crucible breaks up and forms the desired powders. For conductive metallic materials, the stationary electrode is made up of the metal or alloy which will be atomized. During the process the electrode also melts and drops of metal drip into the rotating crucible. A part of the skull remains intact and prevents direct contact of the molten material with the crucible, thereby producing a clean, contamination-free powder under inert gas cover. Thus, this process also utilizes the "cold crucible" concept for producing clean powders of reactive and refractory metals and alloys. In the case of a nonmetallic ceramic material, a nonconsumable hollow electrode is used. The material to be atomized is fed down through the annular hole in the center of the electrode, through the arc plasma, into the rotating crucible.

A large variety of powders have been produced by this process. Stainless steels, high-strength steels, low-alloyed steels, refractory metals such as titanium and niobium, and tool steels are some of the materials that have been produced in powder form via this centrifugal process. Materials that are reactive and oxidizing in nature, and materials for which contamination from the crucible is a major problem, are those that are most suitable for atomization by this process. The powder size can vary from 150 to 1000 μm, with the main controlling parameter being the rotational speed of the crucible. The oxygen content of the powders can be maintained to the level of the original feedstock material, which is the electrode and the skull. The process has good potential for scaleup with the possibility of being developed into a continuous process. It could compete to some extent with the PIGA process, which will be discussed later in this chapter.

Another form of centrifugal atomization process, developed by Dow Corning Corporation, has been used to produce a variety of rapidly solidified aluminum alloy powders such as Al–5Cr–2Zr and Al–Si alloys. The same process has been used by Alcan, who also carried out extensive evaluation of the powders produced by this process [59]. There are a couple of unique features of this process that are worth discussing. The process uses an organic liquid, which acts as the coolant, and there is the provision of vanes on the spinning rotor that impact the airborne molten droplets and break them into small

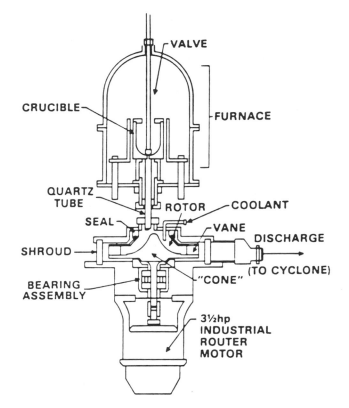

Figure 3.30: Schematic diagram of the rotary atomization equipment with rotating vanes and an orifice for supplying the organic cooling fluids. (Reprinted with permission from Ref. 59.)

particles. A schematic diagram of this atomizing unit is shown in Figure 3.30, which shows some of the key features of the equipment. The patented process [60,61] allows a stream of liquid metal to fall on a spinning rotor. The laboratory atomizer unit has an approximate diameter of 22 cm and height of 3.6 cm, and is equipped with eight steel vanes located at the periphery. During atomization, the rotor speed is around 12 000 rpm. The stream of liquid metal comes into contact with the side of the cone (on the rotor) and is broken into coarse particles that are ejected away from the center at low angular velocities. Assuming that the rotor rotates at speeds around 10 000 to 12 000 rpm, the approximate velocity of the vanes is around 100 m/s. The coarse millimeter-sized droplets that are ejected after striking the cone are subsequently impacted by the rotating vanes. This impact of the still-molten particles results in the mechanical atomization of the droplets into small particles or sheetlike pieces that travel in numerous directions. Most of the material forms a liquid film on the leading face of the vane. These small liquid films are accelerated to the velocity of the vane as they slide on the vanes, and are ejected from the edge of the vane. During this process, ligaments form from the sheets, and the ligaments then form tiny droplets or teardrop-shaped powder particles. Some of the particles, after impact with the vanes, solidify and adhere to the vane. However, they are soon ejected due to the centrifugal forces. The coolant, absolute methanol, is introduced into the system at a rate of $30 \, m^3/s$. The coolant serves the twin purpose of keeping the vanes and the rotor cooled while it also forms a dense atmosphere during the atomization process, which leads to rapid cooling of the particles by very fast heat extraction. Sometimes the ligaments formed are already solidified and can be found in the form of film-shaped particles. Some degree of agglomeration also occurs due to the secondary collisions between small particles. The cooling rate of the particles finer than 45 μm was between 10^4 and 10^5 K/s.

Plasma Atomization

Plasma has been used extensively in forming powders of various materials. The plasma process may be divided into two broad categories. In one case, plasma is used to form a melt pool that is then atomized by some conventional process. In this case, the plasma essentially serves as the source of thermal energy. In the second case, the plasma is used to form the molten particles from either agglomerates or other starting powders and can be atomized by impacting the molten droplets traveling at very high velocities against a moving or stationary substrate material.

One of the uses of plasma atomization is described in the plasma rotating electrode process, popularly known as the PREP process. This is a special form of centrifugal process and is very similar to the concept of the rotating electrode process, whose schematic diagram was shown in Figure 3.26. In the case of PREP, the electrode to be atomized is also consumable in nature and is usually rotated at velocities around 15 000 rpm. The melting of the electrode at one end is carried out by establishing a plasma arc instead of an electric arc as shown in Figure 3.26. As the electrode is consumed, it is also gradually moved in. The process is usually carried out under protective atmosphere to prevent oxidation of the powders. The powders produced by PREP are usually coarse and spherical with average particle sizes usually in the range of 100 to 500 μm. The atomization occurs when the centrifugal forces acting on the edge of the rotating melt pool formed on the surface of the consumable electrode overcome the surface tension effects of the melt. This process has been used to process various kinds of powders that include powders of titanium and titanium alloys, superalloys such as IN-738, Rene-95, Coast-64, and various cobalt based alloys. Since this process does not have provision for forced convective cooling as in the process developed by Pratt and Whitney, the cooling rates are usually slower.

Plasma melting in combination with inert gas atomization, or simply the PIGA process, uses a plasma torch to form a melt of the desired material inside a water-cooled copper crucible. In this process a thin skull of the desired material is formed and prevents the contamination of the actual material through the crucible. The molten material in the water-cooled copper crucible is guided into the gas nozzle where it is atomized. The new guiding system provides additional heating of the molten material if desired.

The main features of the PIGA unit are illustrated schematically in Figure 3.31. The process obviously has the advantage of utilizing scrap, granules, bars, etc., as feedstock material, and also additional alloying can be accomplished during melting. The new high-power inert gas plasma torch, which operates at high pressures, is also suitable for the handling of reactive metal alloys containing volatile constituents. The disadvantages include the use of very high-power plasma torches exhibiting a high degree of erosion of the electrode, which can contaminate the metal powder.

The plasma melt and rapid solidification (PMRS) process goes to prove the versatility of the plasma process in producing rapidly solidified powders [62]. In this case, agglomerated powders serve as the plasma melt feedstock, which is then melted by the plasma and allowed to rapidly solidify. The major difference between this process the plasma processes described earlier is the fact that in this case the melting is containerless and the powder is produced directly as it exits the plasma gun without the use of a secondary atomizing energy input. It will be seen later how the plasma gun itself can provide the thermal energy for melting the particles and also the kinetic energy for disintegrating the molten droplets into smaller particles. Compared to this, in the earlier cases, the plasma was used to simply serve as a source for thermal energy that was used to create a melt pool, which was then atomized by applying another energy source. Examples of that are the PIGA process, where the plasma is used to form a melt pool inside a container that is then atomized by inert gas, or the PREP process, where the plasma forms a containerless melt pool on one surface of a rotating electrode and the melt pool is then atomized by centrifugal forces.

One of the keys to successful plasma melt and rapid solidification process is the process of agglomeration of the powder particles that will be melted during the plasma melting step. In general, finer starting powder for producing the agglomerates is always desirable. If an alloy powder is to be made from individual elemental powders, it is essential to have the individual elemental powders as fine as possible to attain good chemical homogeneity in the

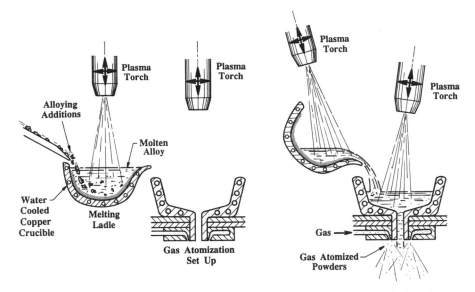

Figure 3.31: *Schematic illustration of the main features of a PIGA unit.*

individual agglomerates that go to form the individual alloy powder particles. Smaller constituent particles have been observed to produce more spherical and uniform agglomerates, which allows better flowability and thus higher feed rate into the plasma jet. Also the finer particles result in a stronger agglomerate, which results in much lower premature agglomerate breakups. Thus, it can be appreciated that attainment of the correct agglomerate particle size to constituent particle size ratio is very important in obtaining statistically correct chemical homogeneity and high flowability of the powders. Investigations carried out by Johnson and coworkers [62] have established an empirical approach to determining the correct agglomerate/constituent particle diameter ratio. The agglomerate diameter has to be at least 5 times greater than the constituent particle diameter for a good agglomerate with desirable attributes.

There are a number of ways by which the agglomerates may be formed. These methods include tray drying (for very small quantities of powders), fluidized-bed agglomeration, and spray drying of constituent powder mixtures. Detailed description of the individual agglomeration processes may achieved by considering a real system of tungsten heavy alloy, with a composition of 93W–4.9Ni–2.1Fe powder [62]. The tray drying approach produces small quantities of agglomerates (less than 5 kg), mainly for research and development purposes. In this case, fine elemental powders of tungsten, nickel, and iron, with average particle sizes less than 5 μm, are used as the starting material. The elemental powders blended in the proper ratio are mixed with a desired solvent, which can be either water- or hydrocarbon-based. The solvents normally have a dissolved binder such as poly(vinyl alcohol) or poly(ethylene glycol). The mixture of the elemental powder constituents in the solvent forms a slurry that is agitated to disperse the materials uniformly within the binder. The agitation, along with some heating of the slurry, results in the removal of the solvent material, leaving behind agglomerates formed from uniform mixture of the elemental powder particles that are held together by the binder. Agglomerates are screened to the correct size fraction, with the over- and undersized agglomerates being recycled into the slurry stage.

The fluidized bed type of agglomerator can be used to form pilot quantities, around 45 to 90 kg, of agglomerates. The setup consists of a fluidized bed where the powder mixture is fluidized while at the same time a binder solution is injected in a controlled manner. The powder flow helps in the process of nucleation and growth of the agglomerates. Controlled air flow in the system helps to eliminate the residual amount of solvent and produce dry agglomerates that are held together by the residual binder.

The process of spray drying, which is extensively used in the chemical and pharmaceutical industries, can also be used to form the agglomerates with fine powders. Details of the spray drying process are given in the *Metals Handbook* [63]. This process is suitable for producing large quantities of agglomerates (more than 90 kg). In this case, a powder–solvent (containing the binder) slurry is atomized into a chamber where the individual atomized particles, containing a number of elemental particles forming the heavy alloy composition, are met by a stream of hot gases. The hot gas rapidly removes the solvent and produces strong agglomerates that can be collected from the bottom of the chamber.

The agglomerates that are produced by the first three agglomeration techniques still contain the binders, which need to be removed before the agglomerates can be fed into the plasma flame. The process is known as dewaxing and sintering and is carried out by passing the agglomerates through a controlled-atmosphere furnace where a protective atmosphere of argon, nitrogen, hydrogen, or vacuum is used to remove the binders. The details of this form of dewaxing process are somewhat akin to the debinding step in particulate injection molding, with the major difference being that the size of the agglomerates is much less than the general thickness of the injection-molded parts. Proper dewaxing should remove almost all the binders without leaving behind any residue. The agglomerates with the binder removed are extremely fragile and can easily break up during the subsequent handling stages such as screening and feeding of the agglomerates into the plasma chamber. Thus, these agglomerates are further heated to allow interparticle bond or neck formation, which will render stronger agglomerates that can be handled easily. Too low a sintering temperature or too short a time will not

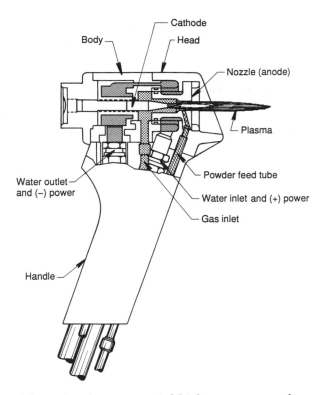

Figure 3.32.: Schematic of a commercial high-temperature plasma gun showing its key features.

produce sufficiently strong agglomerates. Too long a sintering time or too high a sintering temperature could allow bonds to form between the agglomerates themselves, resulting in large sheets formed from agglomerated particles that are stuck together. Such sintered material is obviously unsuitable for serving as the feedstock for plasma processing. The optimum sintering cycle is determined experimentally, and results in the formation of free-flowing powders with good chemical homogeneity. A well-sintered material has a density around 50% to 75% of theoretical.

The above discussion primarily covers the steps required to form the feed material for the actual plasma atomization process. The process of plasma melting is essentially the same for all the agglomerates that have been discussed.

All these containerless plasma atomization processes use a plasma gun to carry out the melting and atomization. A schematic of a commercial high-temperature plasma gun showing its key features can be seen in Figure 3.32 (after ref. 71). For this sort of operation, the plasma is simply a source of a confined very high-temperature flame that is created by passing a gas through an electrical arc that is sustained between a thoriated tungsten cathode and a water-cooled copper anode. There are ports through which the agglomerate may be injected. The agglomerates, which are entrained in an inert gas carrier and carried into the plasma jet, are then melted by the plasma and ejected through the front nozzle at high velocities. The gas used for the plasma can vary from a single gas such as nitrogen, argon, helium, or hydrogen, to a mixture of any of the above gases. The gas velocities in commercial plasma guns can be as high as Mach 2. Plasma exit velocities can be in the range of 0.8 to 3.0 times the sound velocity. The core temperature within the plasma can exceed 10 000 K and often lie between 10 000 to 50 000 K [64,65].

The actual process is fundamentally uncomplicated. The electrical arc struck between the two electrodes energizes the gas, which results in a concentrated flame of around 12.7 mm diameter and approximately 50 to 75 mm long, with exceedingly high core temperatures. The high temperatures can melt any material that is injected into the plasma. The powder feedstock, which is usually presized to a range of 50 to 150 μm, is injected into the plasma flame, where the particles are melted and accelerated to nearly sonic velocities by the plasma gas. However, as with all processes, there are a great many subtleties associated with the use of plasma melt processing.

The requirement of finer powders has also affected the plasma powder processing industry. As a result, a process termed "microatomization" of plasma melted material was developed. Initial work on microatomization was carried out by Cheney [66] and then by Cheney and Pierce [67–69]. The concept was later extended for various materials by Johnson and coworkers [70], Cheney and co-workers [71–73], and Paliwal and Holland [74]. In this process, the key step is the disintegration of the plasma-melted droplets by their impact on a substrate material. The substrate may be cold or hot, stationary or rotating. The process concept is similar to the well-documented work of Edgerton and Killion [75], who demonstrated, through the use of high-speed photography, the disintegration of a drop of milk splashed on to a hard surface. From the original droplet that impinges on the hard surface, a large number of disintegrated tiny droplets are produced. An illustration showing a liquid droplet disintegrating into tiny microatomized particles is shown in Figure 3.33. The theory is that any liquid, including molten metals, ceramics, intermetallic compounds, etc., will be microatomized into tiny fragments if it is impacted with sufficiently high velocities against a solid substrate. Essentially the kinetic energy of the molten droplet itself is used to overcome its surface energy requirements for the production of fine powders. One of the obvious preconditions will be that the liquid should not stick to the substrate material. The situation was ripe for the development of the plasma

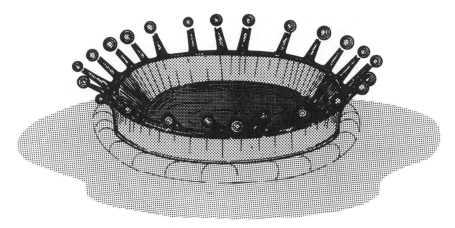

Figure 3.33: *Schematic of a liquid droplet disintegrating into tiny microatomized particles. (Redrawn after Ref. 75.)*

microatomization process. The plasma melt feedstock was readily available in terms of other coarse powders or from agglomerates of individual elemental powders as described earlier; and commercial plasma equipment was already in use for variety of applications such as plasma spray coating and melting of steels. Thus the two main ingredients, appropriate feed material and the plasma gun, were readily available. The plasma gun exit velocity is very high and would thus allow the molten material to strike the substrate at a very high velocity and produce efficient microatomization. It was a matter of putting the parts together to produce plasma microatomized powders.

The process of microatomization must have a mechanism to rapidly heat the powders to form the molten droplets; it must be able to accelerate the molten droplets at a high velocity toward a substrate material; and the droplets must impinge on a substrate material to which it should not stick. RF or DC plasma guns can be used as the melting and high velocity generating device, with the DC gun being preferred due to its ability to achieve higher gas velocities. Two different events can occur when the molten droplet strikes the substrate material. It can either spread and solidify on the surface, producing a plasma-sprayed coating, or it can be microatomized to form fine powders. For coating, adhesion of the droplet to the surface is preferred, and thus the substrate is prepared appropriately by increasing the surface roughness by various means. Having a warm substrate also tends to promote adhesion of the droplets to the substrate material [65,76]. Thus, it seems that the use of a cold and very smooth substrate that does not react with the molten liquid would provide the best microatomization conditions. It is better to rotate the cold substrate so that new areas of the substrate are continually exposed to the

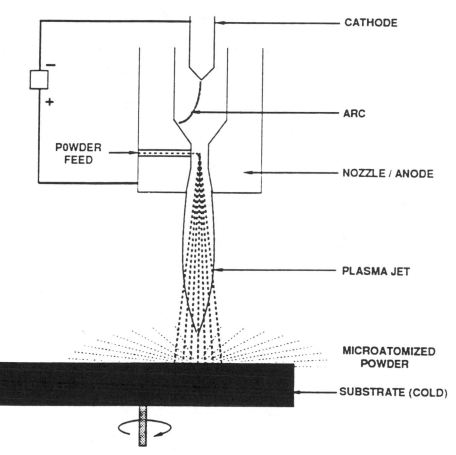

Figure 3.34: *A schematic diagram of the cold substrate plasma microatomization setup. (Reprinted with permission from Ref. 74.)*

impinging molten particles. This will minimize the tendency of the particles to stick to the substrate, as allowing the particles to strike one area continually will result in local heating of that zone and increased adhesion. To keep the substrate cool, it is necessary to have the plasma flame at sufficient distance from the substrate. This again means that the molten particles have to travel through a colder zone, which increases their chance of being solidified before impinging the substrate. A schematic diagram of the cold substrate plasma microatomization setup is shown in Figure 3.34 [74].

In experiments on the cold substrate plasma microatomization process, the plasma torch was mounted on a chamber that contained a cold wheel rotating at 300 rpm [71]. Powder feeds of size finer than 75 μm (-200 mesh) were fed at a rate of 5×10^{-4} to 7.5×10^{-4} kg/s (4 to 6 lb/h). The experiments demons-

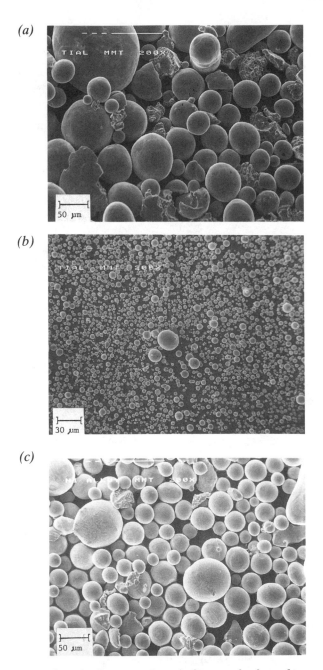

Figure 3.35: Metal and alloy powders before and after plasma microatomiza-
tion: (a) TiAl powder before plasma microatomization; (b) TiAl powder after
plasma microatomization; (c) Ni-based superalloy powder before plasma
microatomization;

(d)

(e)

(f)

Figure 3.35 contd: *(d) Ni-based superalloy powder after plasma microatomization; (e) stainless-steel powder before plasma microatomization; (f) stainless-steel powder after plasma microatomization. (SEM photomicrographs courtesy R.F. Cheney Micro Materials Technology, Inc.)*

(a)

(b)

Figure 3.36: Ceramic powders before and after plasma microatomization: (a) alumina powder before plasma microatomization; (b) alumina powder after plasma microatomization;

trated that the velocity of the droplets just before impact with the substrate was the main criteria that determined the average particle size of the final product. The chemistry of the final product is dependent on the quality of the starting feedstock. However, there is some degree of oxygen pickup that is dependent on the reactive nature of the powder. According to the authors, an oxygen pickup of 1500 ppm is expected for ultrafine highly reactive powders, which can be reduced drastically by simple design changes.

Some of the metal and alloy powders before and after microatomization are shown in Figures 3.35a through 3.35f, which represents powders of TiAl, Ni-based superalloy, and stainless steel. The process is equally capable of microatomizing ceramic materials into fine powders, as shown by the SEM

(c)

(d)

Figure 3.36 contd: *(c) calcio-stabilized zirconia powder before plasma microatomization; (d) calcio-stabilized zirconia powder after plasma microatomization. (SEM photomicrograph courtesy R.F. Cheney Micro Materials Technology, Inc.)*

photomicrographs of calcio-stabilized zirconia and alumina before and after microatomization as shown in Figures 3.36a through 3.36d. The particle size distributions of some of the powders (as measured by Microtrac) are shown in Figure 3.37. The mean particle sizes of the TiAl, Ni-based superalloy, 316 stainless steel, alumina, and calcio-stabilized zirconia are 18, 15, 17, 15, and 5 μm, respectively.

It has been found that partially solidified or resolidified particles bounce off the cold substrate, resulting in a poor yield of microatomized fine particles. Nevertheless, adhesion of molten fragments or partially solidified particles does occur on the substrate and, if their build up is allowed to continue,

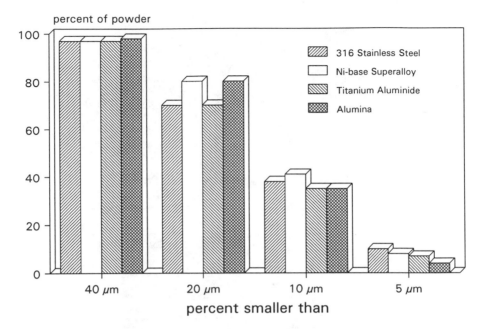

Figure 3.37: *Figure showing the weight percentage of powders below 40, 20, 10, and 5 μm (as measured by Microtrac) for plasma-microatomized 316 stainless steel, nickel-based superalloy, titanium superalloy, and alumina. (Data from Ref. 71.)*

microatomization efficiency plummets as a result of more particles sticking on the rough surface. It is thus important to continually remove the adhered material from the substrate in the cold substrate microatomization process.

To combat this problem, an alternate hot substrate microatomization route has been developed [74,77]. In this approach, as the name suggests, the substrate material is kept hot. According to the previous discussion, the hot surface should tend to promote adhesion between the substrate and particles. This is prevented by keeping the substrate at a temperature higher than the melting point of the material that is being microatomized. The plasma gun can be operated so as to serve the dual purpose of melting and forcing the particles at high velocity toward the substrate and also to heat substrate material to the desired temperature. Some of the unmelted particles will stick to the surface but, due to the temperature of the substrate, will soon melt and be blown off by the plasma jet. Thus, the problem of buildup of deposited particles that occurs in case of cold substrate atomization is obviated. Another advantage of the hot substrate method is the fact that the substrate can be stationary, since any deposited material is melted and blown away by the plasma jet. A

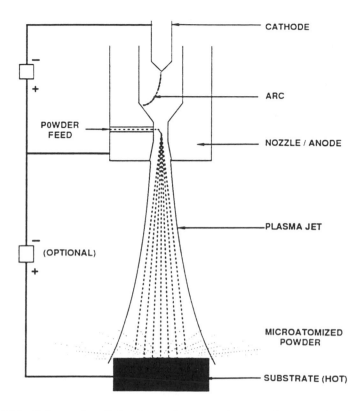

Figure 3.38: *Schematic of the hot substrate plasma microatomization process.* *(Reprinted with permission from Ref. 74.)*

schematic view of the hot substrate plasma microatomization process is shown in Figure 3.38.

It has been found that the operation of the plasma process under low pressure environment results in higher plasma jet velocities and a more diffuse plasma jet, conditions that are extremely suitable for the hot substrate microatomization process. Copper has been microatomized in experiments on hot substrate microatomization. The substrate material was made from 2 wt.% thoria-dispersed tungsten, which could withstand high temperatures and did not react with copper. Two types of plasma gun were used. The nozzle throat diameter was 6 mm in one case and 8 to 12 mm in the other. The gas flow rate was between 1.7×10^{-3} and 3.4×10^{-3} m^3/s (100 to 200 liter/min). The plasma guns were operated in a vacuum chamber where the pressure was varied between 1000 and 2000 Pa. The impurity contents in the powders were found to be 150 ppm carbon, 1300 ppm oxygen and 50 ppm nitrogen. The average powder particle sizes were 3.8 and 6.7 μm.

It can be concluded that plasma atomization provides a unique opportunity for producing fine, spherical, rapidly solidified powders of practically any material. The process offers the advantage of containerless melting and thus reduces the chances of contamination from the crucible. The process offers great energy savings as the heating is extremely focused and does not go to heat the furnace walls and structure. There is practically no limitation on the melting point of the material in the cold substrate microatomization process. The process can be used to produce novel alloys and composites as the plasma feedstock can be agglomerates made up of any combination of elements. The process can be somewhat scaled up to provide powders that would be ideally suitable for particulate injection molding. One of the problems that has to be dealt with is the relatively high levels of oxygen pick up, especially in the case of fine reactive powders. However, that is definitely not insurmountable, as the proper sealing of the system and the use of a clean protective atmosphere can significantly reduce the problem.

Spray Forming

This section deals with a technique that bypasses the powder production step and directly forms an almost fully consolidated part. The process, which starts with the melt atomization step, is commonly known as the "spray deposition" or the "spray forming" technique. This recent powder consolidation technique is rapidly becoming a commercial reality. The process, which forms the desired shapes by directly using the atomized powder droplets before they are solidified, was pioneered by Singer [78,79]. It was later commercialized by Osprey Metals Ltd. and is popularly known as the Osprey Process [80]. Developments in this area have been so rapid that it would be an impossible task to cover the subtle details of this process. Instead, this section will briefly cover the basic principle of spray forming, and touch on some of the developments in this area.

In the process of spray forming, a molten stream of liquid metal or alloy is disintegrated into fine droplets in the form of a spray. The process of disintegration or atomization of the melt is usually accomplished by impinging the liquid metal stream by high-velocity inert gases. This part of the process is exactly similar to the process of atomization discussed earlier. During the usual process of atomization, the atomized particles are allowed sufficient free flight to cool and solidify particles before they impinge on any solid substrate, which in the case of atomization is the chamber walls. However, the principle in the "spray deposition" process is to allow the atomized powders to be accelerated

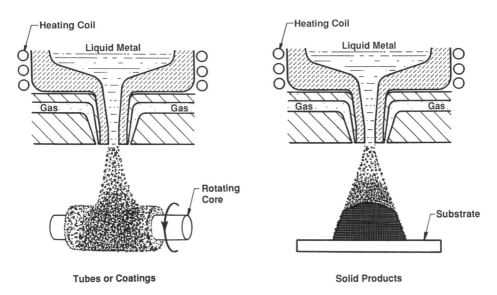

Tubes or Coatings   Solid Products

Figure 3.39: *Schematic of the spray deposition process and the substrate motions that have been used to create different shapes.*

toward a solid substrate material and impinge on the substrate while the droplets are not completely solidified. This results in a flattening and consolidation of the particles and the layer thickness of the spray deposit can be gradually built up to form a thick deposit. The process thus combines several advantages of metal-shaping techniques such as processing of near net shapes in one melt-heat operation, retention of the fine-scale microstructural characteristics of powder processed materials, reduction in the oxygen content of the final material as the material is never exposed to the atmosphere before consolidation, and also the benefits of rapid solidification. By moving the substrate in different ways, various shapes such as solid disk preforms or even thin- or thick-walled tube types of specimen can be fabricated [81]. A schematic of the spray deposition process and the substrate motions necessary to create a solid and a tubular product is shown in Figure 3.39. Efforts to model and optimize the spray deposition process are being made by several research groups.

Various materials have recently been processed by the Osprey Process. A liquid dynamic compaction process has been used by Lavernia et al. [82] to consolidate an aluminum alloy to 95% density followed by subsequent hot extrusion of the spray-deposited material to full density. Nickel-based super-alloys spray deposited by Bricknell [83] have exhibited excellent mechanical properties, probably due to the very low oxygen content of the material.

Tubular products of another nickel-based alloy, Alloy 625, were investigated by Morgan et al. [84]. They concluded that the porosity was most sensitive to variations in the processing parameters, and the greatest porosity is observed adjacent to the inner diameter of the preform. The porosity could be controlled to less than 1%. Recently a bimetallic tubular structure of Alloy 625 and copper has been fabricated by spray deposition technique. Generally the bonding during the spray deposition process is produced by preheating the substrate to a temperature slightly below that of the absolute solidus temperature of the sprayed material. Spraying of Alloy 625 on copper would therefore require the copper substrate to be heated to an unacceptable temperature. An alternative approach was the use of an inner layer of brazing material with a lower melting point that could create bonding between the substrate and the sprayed deposit. The results obtained from a preliminary study with a layer of aluminum proved to be encouraging [85]. Investigations are under way to spray deposit high-speed steels, CuZr, titanium, and a variety of other conventional and exotic materials. The process is capable of producing fairly large tubes and flat disks within a short time. Gillen et al. [81] have reported that Sandvik Steel in Sweden have been commercially processing 300 kg stainless-steel pipes that could be spray deposited in 180 s (3 minutes). This process is clearly one of the exciting new P/M fabrication techniques with a great commercial potential.

A new form of thermal spraying technique has been developed in Germany to produce metal matrix composites [86]. The process involves the use of a special computer-controlled fiber-wrapping unit that can produce fiber wrapping with a high precision of spacing to produce a fiber mat around a polished support. Powder of the matrix material is then plasma sprayed over the wrapped fiber to produce a layer of the desired thickness. Once the desired sprayed thickness is attained, the fiber–matrix monotape is created by cutting the spray-deposited composite layer. Thus, the width of the monotape is equivalent to the perimeter of the sample while the length is controlled by the length of the support over which the fiber is wrapped. To produce a multilayered composite material, the plasma-sprayed material surface is first ground and then the fiber is wound on top of the ground surface and resprayed. It has been observed that if the as-sprayed layer is not ground before the wrapping of the next layer of the fibers, porosity is retained between the fiber and the matrix. This occurs primarily due to the rough surface of the as-sprayed deposit. Stainless-steel fibers have been coated with a variety of heat-resistant Ni–Cr–Al-based matrix materials. According to the mechanical test data obtained, it was concluded that it is possible to produce composites with twice the tensile strength of the as-sprayed monolithic matrix material.

The Future

The new directions in atomization that are presently emerging have a clear bias toward the production of finer and cleaner powders with homogeneous and rapidly solidified microstructures. Another area for future development is the rapid quality control of atomized powders. There is an urgent need for rapid on-line characterization of the atomized powders and also the incorporation of sensors for feedback control of gas atomization parameters. Thus, there is a tremendous challenge for producing "intelligent sensors" for the atomization processes of the future. Jiang and coworkers has provided an excellent overview of this rather complex area of intelligent processing for atomization [87].

The production of cleaner powders with low oxygen or ceramic inclusion contents is becoming extremely important. The development of the "cold crucible" concept, which holds the possibility of eliminating the ceramic inclusion problem, is expected to play an important role in this respect. The EIGA process described earlier also provides a source for melts that do not rely on crucibles. Yet another technique that has been applied to produce clean castings is now being applied in the atomization process: filtering out the ceramic inclusions by passing the melt through a ceramic filter. A recent report published in the Technical Spotlight section of *Advanced Materials and Processes* [88] describes an intriguing levitation-melting method developed by Inductotherm Corp., Rancocas, NJ. The feasibility of this process has been demonstrated on Al–Li alloys. Work on ferrous alloys, titanium alloys and superalloys is in progress. The key to this process is an induction coil that simultaneously serves the purposes of heating, stirring, and confining the melt. The induced current interacts with its associated magnetic field to produce an electromagnetic field whose rotational component stirs the melt and whose irrotational part creates nonshearing magnetic pressure on the boundaries of the melt. The magnetic pressure pushing against the free outside surface of the melt acts as a "deformable" crucible held by an invisible magnetic wall. This levitation-melting method has until now been used only for casting processes. However, with suitable adaptation it is possible to utilize this unique concept for producing clean powders of reactive metals and alloys. The stirring action of this process could help in forming a homogeneous alloy within the melt. This process should be carefully looked into by companies involved in atomization of reactive metals and alloys. This development could support the drive toward cleaner atomized powders.

The need for producing rapidly solidified powders has resulted in the development of new techniques such as liquid gas atomization, two-step gas–water atomization, plasma microatomization, and the secondary quenching

Table 3.1. Particle shape, average size, and size distribution obtained by various atomization techniques

Technique	Particle shape	Typical average particle size, (μm) (approximate)	Particle size distribution
Low-pressure water	Irregular	150–600	Wide
High-pressure water	Irregular	20–100	Medium
Low-pressure gas	Spherical	50–250	Medium
Medium- to high-pressure gas	Teardrop, spherical	25–100	Medium
Gas soluble	Spherical	20–150	Medium
VH technique	Spherical	50	Medium
Liquified gas	Spherical	20–100	Medium
Gas–water	Irregular spherical	20–150	Medium
EIGA	Spherical	30–100	Medium
Capillary wave	Spherical	20–100	Narrow
Vibrating electrode	Spherical	75–500	Narrow
Roller	Flaky	200	Narrow
Rotating electrode	Spherical	150–250	Narrow
Electron beam rotating disk	Spherical	30–100	Narrow
Centrifugal shot casting	Spherical, flaky	150–1000	Medium
Plasma microatomization	Spherical	5–50	Narrow

of powder droplets produced by centrifugal or rotating atomization processes. These processes are expected to grow very rapidly. Table 3.1 provides a brief comparison between some of these newly developed processes and some of the classic commercial processes such as gas and water atomization. The table also shows the typical average powder particle size, particle shape, and a qualitative particle size distribution obtained from some of the atomization techniques.

Microstructure control through various atomization processes is also an area where more attention will be focused in the future. Gas atomization forms the frozen droplets from molten materials by convective cooling. When the droplets are formed in vacuum, the cooling rate is much slower and the mode of cooling is through radiation. The dendritic arm spacing provides a rough estimate of the homogeneity and also the rate of cooling of an atomized powder. Generally, the smaller the powder particle size, the finer the secondary dendritic arm spacing, and the better is the homogeneity. Still higher cooling rates can produce equiaxed microstructures that have better homogeneity than their dendritic counterpart. With extremely fast cooling rates, it is possible to produce amorphous powders. It is expected that the future trend will be towards refined microstructures, which are also favored by smaller particle sizes.

Alternate powder atomization techniques with large potential, such as the capillary wave atomization and the standing wave technique, should receive more attention in the future. The processes are very attractive as they do not require a large cooling chamber and consume very little gas.

Lastly, the rapid developments in the area of spray forming will provide a tremendous boost for the melt atomization technique. The fact that shapes can be directly produced in one step without going through the second high-temperature consolidation step provides a tremendous motivation for continued development in this area. The application of this process in the fabrication of metal matrix composites is expected to witness significant growth in the near future.

With the turn of the century, atomization will still remain one of the most economical technique for producing large quantities of powders. The interest in fine atomized powders will be propelled by the demand exerted by powder injection molding, but there will be a considerable erosion of its field of application. Chemical powder processing will specially impact the high-temperature high-performance alloys and the reactive metals and alloys that are now produced by atomization techniques. However, for the lower melting and nonreactive materials, atomization will remain the process of choice well into the next century. An excellent book providing a detailed discussion on melt atomization is that of Lawley [89].

References

1. R.M. German, *Powder Metallurgy Science*, Metal Powder Industries Federation, Princeton, NJ, p.80, 1984.
2. Y. Seki, S. Okamoto, H. Takigawa, and N. Kawai, *Metal Powder Report*, vol. 45, p.38, 1990.
3. A. Kimura, K. Nakabayashi, and T. Shimura, *Metal Powder Report*, vol. 45, p.106, 1990.
4. R.J. Grandzol, *Water Atomization of 4620 Steel and Other Metals*, Ph.D. Thesis, Drexel University, Philadelphia, PA, 1973.
5. R.J. Grandzol and J.A. Tallmadge, *International Journal of Powder Metallurgy and Powder Technology*, vol. 11, p.103, 1975.
6. C.S. Wright, A.R. Nutton, and D.L. Jones, *World Conference on Powder Metallurgy*, The Institute of Metals, London, UK, and The Institute of Metals, Brookfield, VT, vol. 3, p.20, 1990.
7. N. Dautzenberg, H.J. Dorweiler, and R.-H. Lindner, *World Conference on Powder Metallurgy*, The Institute of Metals, London, UK, and The Institute of Metals, Brookfield, VT, vol. 1, p.227, 1990.
8. M. Umino et al., U.S. Patent 4,437,891, February 24, 1981.

9. M. Umino et al., U.K. Patent 2094834, February 23, 1982.

10. K.U. Kainer and B.L. Mordike, *Modern Developments in Powder Metallurgy*, compiled by P.U. Gummeson and D.A. Gustafson, Metal Powder Industries Federation, Princeton, NJ, vol. 20, p.323, 1988.

11. B. Schiborr, K.U. Kainer, and B.L. Mordike, *World Conference on Powder Metallurgy*, The Institute of Metals, London, UK, and The Institute of Metals, Brookfield, VT, vol. 2, p.103, 1990.

12. P.R. Bridenbaugh, F.R. Billman, W.S. Cebulak, and G.J. Hildeman, *Advanced High Temperature Alloys: Processing and Properties*, Proc. N.J. Grant Symposium, ed. S.M. Allen, R.M. Pelloux, and R. Widmer, ASM, Metals Park, OH, p.53, 1986.

13. C.-S. Lin, G.-J. Jih, C.-S. Chang, and S.-Y. Chiou, *Advances in Powder Metallurgy*, compiled by T.G. Gasbarre and W.F. Jandeska, Metal Powder Industries Federation, Princeton, NJ, vol. 2, p.57, 1989.

14. A. Lawley and R.D. Doherty, *Advanced High Temperature Alloys: Processing and Properties*, Proc. N.J. Grant Symposium, ed. S.M. Allen, R.M. Pelloux, and R. Widmer, ASM, Metals Park, OH, p.65, 1986.

15. D. Bradley, *Journal of Physics D: Applied Physics*, vol. 6, p.1724, 1973.

16. D. Bradley, *Journal of Physics D: Applied Physics*, vol. 6, p.2267, 1973.

17. H. Lubanska, *Journal of Metals*, vol. 22, p.45, 1970.

18. J.S. Coombs, G.R. Dunstan, and R.I.L. Howells, *World Conference on Powder Metallurgy*, The Institute of Metals, London, UK, and The Institute of Metals, Brookfield, VT, vol. 2, p.35, 1990.

19. M.K. Veistinen, E.J. Lavernia, J.C. Baram, and N.J. Grant, *International Journal of Powder Metallurgy*, vol. 25, p.89, 1989.

20. M.J. Couper and R.F. Singer, *Proc. 5th Int. Conf. Rapidly Quenched Metals*, ed. S. Steeb and H. Warlimont, p.1737, 1984.

21. T.W. Clyne, R.A. Ricks, and P.G. Goodhew, *International Journal of Rapid Solidification*, vol. 1, p.59, 1984.

22. L. Arnberg, *Scandinavian Journal of Metallurgy*, vol. 13, p.391, 1984.

23. O.P. Pandey and S.N. Ojha, *Powder Metallurgy International*, vol. 5, p.291, 1991.

24. S.N. Singh and S.N. Ojha, to be published in *Metals, Materials, and Processes*.

25. C.E. Seaton, H. Henien, and M. Glatz, *Powder Metallurgy*, vol. 30, p.37, 1987.

26. R. Wille and H. Fernholtz, *Journal of Fluid Mechanics*, vol. 23, p.801, 1985.

27. A. Unal, *Metallurgical Transactions*, vol. 20B, p.61, 1989.

28. G.B. Kenny and C.P. Ashdown, U.S. Patent 4,626,278.

29. C.P. Ashdown, J.G. Bewley, and G.B. Kenney, *Modern Developments in Powder Metallurgy*, compiled by P.U. Gummerson and D.A. Gustafson, Metal Powder Industries Federation, Princeton, NJ, vol. 20, p.169, 1988.

30. C. Aslund and T. Tingskog, *Modern Developments in Powder Metallurgy*, compiled by P.U. Gummerson and D.A. Gustafson, Metal Powder Industries Federation, Princeton, NJ, vol. 20, p.181, 1988.

31. C. Aslund, *Progress in Powder Metallurgy*, compiled by C.L. Freeby and H. Hjort, Metal Powder Industries Federation, Princeton, NJ, vol. 43, p.611, 1987.

32. J.M. Vetter, G. Gross, and H.W. Bergmann, *Metal Powder Report*, vol. 45, p.100, 1990.

33. D. Stock, P.R. Sahm, and P.N. Hansen, *World Conference on Powder Metallurgy*, The Institute of Metals, London, UK, and The Institute of Metals, Brookfield, VT, vol. 1, p.158, 1990.

34. *Metal Powder Report*, vol. 45, p.529, 1990.
35. M. Hohmann, S. Jonsson, and E. Lierke, *Modern Developments in Powder Metallurgy*, compiled by P.U. Gummerson and D.A. Gustafson, Metal Powder Industries Federation, Princeton, NJ, vol. 20, p.159, 1988.
36. G. Matei et al., *Modern Developments in Powder Metallurgy*, ed. H.H. Hausner and P.W. Taubenblat, Metal Powder Industries Federation, Princeton, NJ, vol. 9, p.153, 1977.
37. Y. Kito, T. Sakuta, T. Mizuno, and T. Morita, *International Journal of Powder Metallurgy*, vol. 25, p.13, 1989.
38. M. Hohmann and M. Ertl, *Advances in Powder Metallurgy*, compiled by E.R. Andreotti and P. J. McGeehan, Metal Powder Industries Federation, Princeton, NJ, vol. 1, p.37, 1990.
39. S. Jonsson, M. Ertl, and M. Hohmann, *Advances in Powder Metallurgy*, compiled by T.G. Gasbarre and W.F. Jandeska, Metal Powder Industries Federation, Princeton, NJ, vol. 2, p.69, 1989.
40. *Powder Metallurgy International*, vol. 22, p.66, 1990.
41. A.R.E. Stringer and A.D. Roche, *Modern Developments in Powder Metallurgy*, eds. H.H. Hausner and P.W. Taubenblat, Metal Powder Industries Federation, Princeton, NJ, vol. 9, p.127, 1977.
42. P.K. Dolmalavage and N.J. Grant, *Metal Powder Report*, vol. 38, p.555, 1983.
43. A.J. Aller and A. Losada, *Metal Powder Report*, vol. 46, p.45, 1991.
44. J. Perel, J.F. Mahoney, B.E. Kalensher, and R. Mehrabian, *25th Segamore Army Materials Research Conference*, ed. J.T. Burke, R. Mehrabian, and V. Weiss, Plenum Press, New York, p.79, 1981.
45. J. Perel, J.F. Mahoney, P. Dewez, B.E. Kalensher, and R. Mehrabian, *2nd Int. Conf. Rapid Solidification Process, Principles and Technologies*, ed. R. Mehrabian, B.H. Kear, and M. Cohen, Claitor's Publishing Division, Baton Rouge, LA, p.287, 1980.
46. J. Perel, J.F. Mahoney, S. Taylor, Z. Shanfield, and C. Levi, *Rapidly Solidified Amorphous and Crystalline Alloys*, ed. B.H. Kear, B.C. Giessen, and M. Cohen, Materials Research Society Symposium Proceedings, Elsevier Science Publishing Co., New York, p.131, 1982.
47. B. Champagne and R. Angers, *International Journal of Powder Metallurgy and Powder Technology*, vol. 16, p.359, 1980.
48. B. Champagne and R. Angers, *Powder Metallurgy International*, vol. 16, p.125, 1984.
49. B. Champagne and R. Angers, *Modern Developments in Powder Metallurgy*, ed. H.H. Hausner, H.W. Antes, and G.D. Smith, *Metal Powder Industries Federation*, Princeton, NJ, vol. 12, p.83, 1981.
50. K. Halada, H. Suga, and Y. Muramatsu, *World Conference on Powder Metallurgy*, The Institute of Metals, London, UK, and The Institute of Metals, Brookfield, VT, vol. 1, p.193, 1990.
51. R.M. German, *Powder Metallurgy Science*, Metal Powder Industries Federation, Princeton, NJ, p.84, 1984.
52. H. Schmitt, *Powder Metallurgy International*, vol. 11, p.7, 1979.
53. A. Horata, T. Kato, and K. Kusaka, *Progress in Powder Metallurgy*, ed. H.I. Sanderow, W.L. Giebelhausen, and K.K. Kulkarni, Metal Powder Industries Federation, Princeton, NJ, vol. 41, p.479, 1985.

54. A.J. Aller and A. Losada, *Metal Powder Report*, vol. 45, p.51, 1990.
55. A.R. Cox and E.C. Van Reuth, *Metals Technology*, vol. 7, p.238, 1980.
56. J.J. Ramon, D. Shechtman, and F.S. Dirnfeld, *Metal Powder Report*, vol. 46, p.36, 1991.
57. J.J. Ramon, D. Shechtman, S.F. Dirnfeld, and G. Staniek, *World Conference on Powder Metallurgy*, The Institute of Metals, London, UK, and The Institute of Metals, Brookfield, VT, vol. 2, p.16, 1990.
58. A. Henly, *Metal Powder Report*, vol. 46, p.40, 1991.
59. K.B. Patel and D.J. Lloyd, *Progress in Powder Metallurgy*, compiled by C.L. Freeby and H. Hjort, Metal Powder Industries Federation, Princeton, NJ, vol. 43, p.633, 1987.
60. J.L. Speier and T.M. Gentle, U.S. Patent 4,347,199, 1982.
61. J.L. Speier and D.T. Liles, U.S. Patent 4,419,060, 1983.
62. W.A. Johnson, N.E. Kopatz, and E.B. Yoder, *Progress in Powder Metallurgy*, compiled by C.L. Freeby and H. Hjort, Metal Powder Industries Federation, Princeton, NJ, vol. 43, p.139, 1987.
63. D.L. Houck, *Metals Handbook*, 9th ed., ASM Metals Park, OH, vol. 7, p.73, 1978.
64. *Plasma Jet Technology*, NASA Report No. SP-5033, October 1965.
65. D. Apelian, M. Paliwal, R.W. Smith, and W.F. Schilling, *International Metals Review*, vol. 28, p.271, 1983.
66. R.F. Cheney, U.S. Patent 4,502,885.
67. R.F. Cheney and R.H. Pierce, U.S. Patent 4,592,781.
68. R.F. Cheney and R.H. Pierce, U.S. Patent 4,613,371.
69. R.F. Cheney and R.H. Pierce, U.S. Patent 4,687,510.
70. W.A. Johnson, N.E. Kopatz, and E.B. Yoder, *Progress in Powder Metallurgy*, compiled by E.A. Carlson and G. Gaines, Metal Powder Industries Federation, Princeton, NJ, vol. 42, p.775, 1986.
71. R.F. Cheney and E.R. Seydel, Metal Powder Report, vol. 45, p.43, 1990.
72. R.F. Cheney, E.R. Seydel, and D. Kapoor, *Advances in Powder Metallurgy*, compiled by E.R. Andreotti and P.J. McGeehan, Metal Powder Industries Federation, Princeton, NJ, vol. 1, p.81, 1990.
73. R.F. Cheney, R.L. Daga, R.M. German, A. Bose, and J.W. Burlingame, *Modern Developments in Powder Metallurgy*, compiled by P.U. Gummeson and D.A. Gustafson, Metal Powder Industries Federation, Princeton, NJ, vol. 19, p.155, 1988.
74. M. Paliwal and R.J. Holland, *Advances in Powder Metallurgy*, compiled by T.G. Gasbarre and W.F. Jandeska, Metal Powder Industries Federation, Princeton, NJ, vol. 3, p.35, 1989.
75. H.E. Edgerton and J.R. Killion, *Moments of Vision, The Stroboscopic Revolution in Photography*, MIT Press, Cambridge, MA, 1979.
76. D.A. Gerdeman and N.L. Hecht, *Arc Plasma Technology in Materials Science*, Springer-Verlag, New York, 1972.
77. M. Paliwal and R.J. Holland, U.S. Patent 5,124,091, June 33, 1992.
78. A.R.E. Singer, *Metals and Materials*, vol. 4, p.246, 1970.
79. A.R.E. Singer, *Journal of the Institute of Metals*, vol. 100, p.185, 1972.
80. Osprey Metals Ltd., U.K. Patent 147239.
81. G. Gillen, P. Mathur, D. Apelian, and A. Lawley, *Progress in Powder Metallurgy*, compiled by E.A. Carlson and G. Gaines, Metal Powder Industries Federation, Princeton, NJ, vol. 42, p.753, 1986.

82. E.J. Lavernia, G. Rai, and N.J. Grant, *International Journal of Powder Metallurgy*, vol. 22, p.9, 1986.
83. R.H. Bricknell, *Metallurgical Transactions*, vol. 17A, p.583, 1986.
84. A.L. Morgan, W.A. Palko, C.J. Madden, and P. Kelley, *Advances in Powder Metallurgy*, Compiler E.R. Andreotti and P.J. McGeehan, Metal Powder Industries Federation, Princeton, NJ, vol. 1, p.553, 1990.
85. P. Kelley and A. Morgan, *Advances in Powder Metallurgy*, Compiler L.F. Pease III and R.J. Sansoucy, Metal Powder Industries Federation, Princeton, NJ, vol. 5, p.277, 1991.
86. H.-D. Steffens and R. Kaczmarek, *Powder Metallurgy International*, vol. 23, no.2, p.105, 1991.
87. G. Jiang, H. Henein, and M.W. Siegel, *Advances in Powder Metallurgy*, compiled by T.G. Gasbarre and W.F. Jandeska, Metal Powder Industries Federation, Princeton, NJ, vol. 2, p.43, 1989.
88. *Advanced Materials and Processes*, vol. 139, p.42, 1991.
89. A. Lawley, *Atomization: The Production of Metal Powders*, Metal Powder Industries Federation, Princeton, NJ, 1992.

Chapter 4

Mechanical Alloying

Introduction

Mechanical alloying (MA), a dry high-energy ball milling process, was first used for developing a nickel-based thoria dispersion-strengthened material for space applications. The process was later commercialized by International Nickel Company (INCO) [1]. The process has been used to produce composite powders with very fine controlled microstructures, intermetallic compounds, and amorphous metallic powders.

The commercial development effort in mechanical alloying was initiated by the desire to oxide dispersion-strengthen nickel base superalloys for gas turbines. One of the first commercial products developed was a Y_2O_3 dispersion-strengthened nickel base alloy, MA754, having a composition of $20Cr-0.3Al-0.5Ti-0.05C-0.6Y_2O_3$, and balance Ni (wt.%). The applications quickly spread to oxide dispersion-strengthened iron- and cobalt–based materials. Various light metals such as aluminum, titanium, and magnesium have also been subjected to mechanical alloying. In the process of mechanical alloying, surfactants are often used for alloying soft, ductile materials such as aluminum. Mechanical alloying is finding increasing use in the aerospace arena with the new alloys that are based on light metals.

Presently this process is being used to fabricate numerous exotic materials and alloys including amorphous materials, intermetallic compounds, heat-resistant copper alloys, and hierarchical composites. The list of new materials being fabricated by MA is continually increasing and the process is suitable for producing powders in large quantities.

Process and Equipment

The basic process of mechanical alloying depends on the repeated welding, fracturing, and rewelding of a mixture of powder particles in a high-energy ball mill. In brief, the process consists of microforging, cold welding, and fracturing of particles to gradually produce fine randomly oriented laminates of the materials. The fine laminate formation is aided by the kneading and impact action of the entrapped powders between colliding balls. One school of researchers emphasizes that the milling balls and the milling container are first coated with the powders, and the laminated structure refinement, homogenization, and even solid-state amorphization occurs in the coated layer. The coated layers often spall off and are further refined by the high-energy milling action. In these high-energy mills, the competing processes of welding and fracturing are somewhat balanced and the average particle size of the composite powder remains relatively coarse. This is in contrast to the low-energy milling process, which is meant to refine the ultimate particle size of the material by simple comminution.

The process of mechanical alloying can be used for a variety of materials such as ductile metals, ductile alloys, and ductile intermetallic compounds, brittle metals, brittle alloys, intermetallic compounds which are generally brittle, and the commonly used brittle ceramics. During mechanical alloying the starting material, which is in a powder form, is entrapped between two colliding balls. Figure 4.1 shows a schematic diagram of the changes that a material or a combination of materials will undergo after a few collision events. Ductile materials will tend to become flattened into plates, while brittle materials will fracture into smaller pieces. After few impacts, a ductile–ductile material combination will tend to produce a laminated composite type of powder; the ductile–brittle material combination will produce powders in which the brittle phase is embedded in the ductile matrix; and in a brittle–brittle material combination, smaller powder particles of both materials will be formed. These are discussed in more detail later in the chapter.

To provide readers with a better feel for the process of mechanical alloying, a sequence of microstructures showing the evolution of mechanical alloying has been depicted in Figures 4.2a through 4.2d. The materials chosen represent a common type of mechanical alloying combination, namely a ductile and a brittle material. In this case, the ductile material is a gas-atomized powder of an advanced Ni_3Al-based intermetallic compound and the brittle material is a titanium diboride powder. Attritor milling of 20 vol.% of TiB_2 powder with 80 vol.% of the intermetallic powder is carried out over an extended period of time. The process of milling is interrupted at various intervals to allow the removal of small portions of the milled material. For microstructural evaluation, the powder removed from the attritor mill is mixed with epoxy while the epoxy is still in a liquid form. The epoxy is then allowed to harden, and the flat end is polished to show the cross-sectional view of the powder mixture. Figure 4.2a shows the powder mixture before it has been mechanically alloyed. The

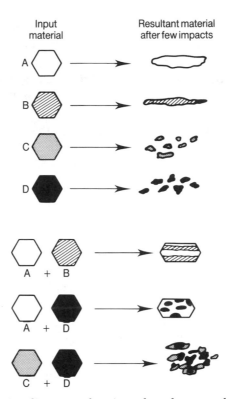

Figure 4.1: Schematic diagram showing the changes that a material or a combination of materials will undergo after a few collision events. A and B are ductile metals, alloys, or intermetallic compounds; C and D are brittle metals, alloys, ceramics, or intermetallic compounds. The left-hand side denotes the material put into an attritor mill; the right-hand side denotes the resultant material after a few impacts.

gas-atomized Ni$_3$Al powder is coarse and spherical in nature, while the TiB$_2$ powder is fine and irregular in shape. Figure 4.2b shows a similar cross section of the powders after they have been mechanically alloyed in an attritor mill for 108×10^2 s (3 hours). Note that most of the particles have become highly elongated in nature. There is, however, one large particle that could be the flat side of a platelike formation that would occur due to the flattening of a spherical particle. This large particle has been partially fractured due to the large plastic deformation it has undergone. There is clear evidence of fine second phase particles of TiB$_2$ embedded in the matrix of the ductile intermetallic compound. Figure 4.2c shows the microstructure after the material has been milled for 216×10^2 s (6 hours). The structure is far more

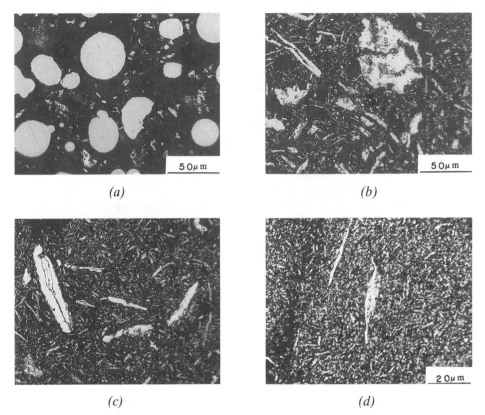

(a)   *(b)*

(c)   *(d)*

Figure 4.2: *Microstructural evolution during mechanical alloying of an advanced nickel aluminide intermetallic compound with 20 vol.% titanium diboride: (a) as-mixed powder; (b) milled for 3 hours; (c) milled for 6 hours; (d) milled for 15 hours.*

refined compared to that milled for 108×10^2 s. A few large elongated particles of the ductile intermetallic compound are still evident, with one of the larger particles showing a large crack through its center. This plastic deformation and subsequent cracking of the particles result in the size refinement of the material during mechanical alloying. In this picture, it is generally not possible to resolve the two different phases, except in a couple of the large particles that have a few TiB_2 particles embedded in them. Figure 4.2d shows the structure of the mechanically alloyed material after 540×10^2 s (15 hours). The magnification of this picture is higher than that in the previous three micrographs. In this picture, it is not possible to resolve the two phases. The powders are fine, though a few platelets of coarse particles are still present. Mechanical alloying is said to be complete once the two phases cannot be optically resolved. By that definition, mechanical alloying has been completed for this system between 216×10^2 s and 540×10^2 s (6 and 15 hours).

Mechanically alloyed powders can be produced by any high-energy mill that can yield repeated welding, fracture, and rewelding by imparting to the powders rapid compressive impact forces. Various types of milling equipment have been discussed in the ASM Handbook by Kuhn et al. [2]. Most of the discussion in this chapter will focus on two of the most popular types of mills, which are an attritor-type ball mill and a Spex shaker mill.

For large-scale powder production, the attritor mill is the most suitable equipment. Figure 4.3 shows a schematic of a high-energy attritor type ball mill such as a Szegvari grinding mill. The equipment essentially consists of a water-cooled chamber and a rotating impeller. The rotating impeller is a central vertical shaft that is concentric to the attritor chamber, with impeller rods radiating outward from the shaft. Another form of impeller arm that is also in use is L-shaped in nature. In this type of impeller, the arm instead of radiating outward from the shaft, first radiates out and then bends downward at a right angle. Attritor mills usually have a capacity between 0.04 and 0.4 m^3 (1 to 100 gallons), and the central shaft can rotate at speeds up to 250 rpm (4.2 Hz). The gas seal at the top ensures that the desired milling atmosphere can be maintained during mechanical alloying. A large number of impacting balls are used to impart the compressive impact loadings to the powder particles.

Unlike the attritor mill, which is used for large-scale powder production, the Spex shaker mills are used for producing small batches of powders (less than 10 g). In this mill a small grinding container with the total charge is agitated in three mutually perpendicular directions at approximately 20 Hz [3].

It should be pointed out that conventional ball mills can also be used for mechanical alloying as long as high-energy conditions are implemented. High rotational speeds cannot be used since this would result in the powder and balls being pinned to the walls by centrifugal force. The optimum condition would require the ball and powder charge to be released for free fall after attaining the maximum height possible in the milling chamber. The diameter of the chamber needs to be large (usually greater than 1 m) to allow sufficient impact conditions to occur. The ratio of the powder to the milling balls needs to be small. The vibratory ball mill, a variation of the conventional ball mill, has also been used to form mechanically alloyed powders. This alternate form of mechanical alloying equipment uses a stainless-steel container and hardened steel balls. In addition to the tumbling motion, a small gyratory motion is provided by a small unbalanced weight. This action provides additional milling energy and mechanical alloying can be achieved in this equipment. Though

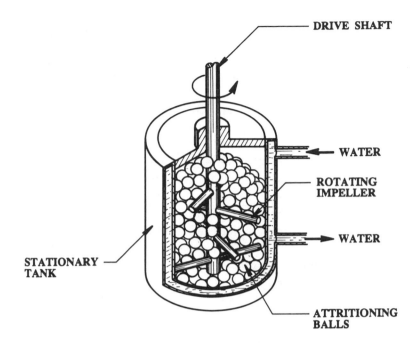

Figure 4.3: *Schematic of a high-energy attritor-type ball mill such as a Szegvari grinding mill.*

other equipment also may be used to produce mechanically alloyed powders, these will not be discussed in this chapter.

Contamination

The process of mechanical alloying relies on mechanical energy supplied by impacting balls to attain the desired results. The powders are subjected to intimate physical contact with the impacting balls as well as the container chamber walls. For milling softer powders, organic surfactants are generally used along with the powders. These surfactants are also entrapped within the powder mass. The atmosphere used during the process of mechanical alloying can also lead to material contamination. Thus, it is important to discuss some of the sources and effects of contamination in mechanically alloyed powders.

Usually the impacting balls are made of bearing steels. The ball charge usually consists of 52100-bearing-steel balls. The balls and the grinding chamber are one major source of contamination during mechanical alloying. In an effort to reduce contamination, sometimes the milling balls are made to a composition that closely matches the alloy being processed. However, owing to the myriad of exotic materials and compositions being fabricated from mechanically alloyed powders, it is almost impossible to have the milling balls made of the same composition. Thus, some amount of iron contamination has to be tolerated. It has been found that iron contaminations as high as 5 at.% can be reached when milling for long periods such as 360×10^2 to 108×10^3 s (10 to 30 hours) [4].

One conceivable way of economically decreasing the ball-related contamination problem is to coat the balls with a material that is extensively present in the material to be mechanically alloyed. For example, if the final mechanically alloyed powder is an oxide dispersion-strengthened nickel-based superalloy powder, the balls could be coated with a thin layer of nickel by allowing the balls to fall freely through a chamber in which decomposition of nickel carbonyl occurs. Coating of the balls will prevent the contamination problem to a great extent without having to rely on producing different ball materials every time a new material is being mechanically alloyed. A slightly more difficult solution, but not an insurmountable one, would consist of also coating the milling chamber with the desired material.

Though mechanical alloying is termed a dry milling process, surfactants are often used to prevent excessive cold welding. The surfactants aid in bringing about a balance between cold-welding and fracture, especially in soft, ductile materials like aluminum or tin. A typical organic compound used as surfactant is a wax with the trade-name Nopcowax-22DSP, $C_2H_2(C_{18}H_{36}ON)_2$. It has been determined that after some initial mixing time of less than 36×10^2 s (1 hour), the surfactant cannot be traced by the differential scanning calorimeter used to monitor the melting point of the wax [5]. It is believed that the surfactant is either incorporated in the alloy or is removed by the formation of gaseous species. This could provide another source of contamination for mechanically alloyed powders, though the addition of surfactants, in some instances, cannot be eliminated.

The milling atmosphere can also be another source of powder contamination, especially if air is used. Milling of reactive materials in air results in a great deal of oxygen pickup by the powder. It was observed that the oxide concentration in aluminum powder increased threefold after only 108×10^2 s (3 hours) of mechanical alloying [6]. The solution to this problem is to have an inert atmosphere during the milling process. However, the inert atmosphere should also have sufficient purity, depending on the material that is being mechanically alloyed. As a case in point, investigations carried out by Jang and Koch on mechanically alloyed nickel aluminide-based intermetallic compounds revealed that the use of impure argon could result in a twofold increase in the oxide concentration after 216×10^2 s (6 hours) of milling [7].

This brief discussion only points out some of the common sources of contamination in mechanically alloyed powders. It should be realized that different systems will have varying levels of tolerance for different contaminants. Thus, it would be extremely important to tailor the process such that contaminations that result in a concomitant drop in the properties of the material can be avoided or minimized.

Mechanism of Strengthening and Alloying

The mechanism of formation of mechanically alloyed powder and its effects on alloying and strengthening are quite unique. Events that occur during microstructural evolution of various systems such as ductile–ductile, ductile–brittle, and brittle–brittle materials are being extensively studied and will be appropriately covered in this chapter. The formation of intermetallic compounds and new amorphous materials by mechanical alloying also warrants detailed analysis. This section will deal with the various mechanisms of evolution of the microstructures and phases.

The physics of mechanical alloying has been covered by only a few research groups who have been involved in both modeling and some experimental verification of the models [8–11]. This is one aspect of mechanical alloying that has literally been neglected until recently. Thus, very little material is actually available on this important subject relating to the physics of this process. Hopefully, with better appreciation of the role of modeling of this process, interest in this area will increase. Presently the modeling mostly deals with at least one ductile material. This needs to be gradually extended to cover all material combinations.

1. Modeling of Mechanical Alloying

Modeling of mechanical alloying is an extremely difficult and complex problem that has received very little attention in the past. However, the importance of the physics of mechanical alloying is a vital link in the attempts of the materials community to achieve some fundamental understanding of and means of providing some predictive capabilities of this process. It would be outside the scope of this book to cover all the aspects of this area; thus, readers are referred to the references given earlier for a detailed discussion on the physics of mechanical alloying.

The process of mechanical alloying invariably involves the collision between minute powder particles and comparatively much larger tools (which could include the side of the mill and impacting ball or impacting balls). The end result of this collision process is powder fragmentation and coalescence. In the attritor mill described earlier, the majority of the impaction events will involve

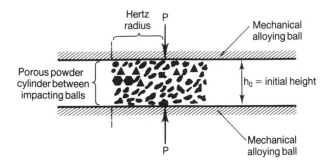

Figure 4.4: *Schematic of the conceptual "miniforging" event that occurs due to powder aggregate entrapment between two colliding balls. (Redrawn after Ref. 8.)*

powder particles that are trapped between two balls. When the constituent powders are ductile in nature, they will be plastically deformed and work-hardened to a very large extent and often cause rupturing of the material, exposing fresh, clean surfaces. Cold welding between two different materials can occur and at the end the flakelike powder will consist of laminates of the two materials. The laminate thickness is refined with continued milling, though an average steady-state powder particle size may be reached in a short period of time. This occurs due to a balance between the fragmentation and coalescence events.

As indicated before, the powder particles and their aggregates are much finer than the impacting balls between which they are trapped and the Hertz radius (as shown in Figure 4.4). Thus, the surface of the colliding balls can be considered to have an infinite curvature with respect to the powder aggregates trapped between them. The process can be visualized as akin to upset forging of powders between two parallel plates. Figure 4.4 shows a schematic of the conceptual "miniforging" event that occurs due to powder aggregate entrap-ment between two colliding balls [8]. The flat faces represent the surfaces of the two impacting balls (considered flat since the Hertz radius is much less than that of the impacting balls) and the volume of powder entrapped is considered to be a cylindrical slug with an initial height of h_0, and r_h the Hertz radius. This form of modeling allows the determination of various parameters such as strain, strain rate, and temperature of the powder subjected to the "miniforg-ing" event. The modeling of these parameters has been discussed in detail by Maurice and Courtney [8].

The results of the modeling show that the impact strains and temperature increases are moderate and the temperature increase is generally not sufficient to produce chemical alloying over a considerable part of the miniforging distance. The impact strain, which depends on the product of the precollision relative velocity of the impacting media, v, with the half-duration of the impact during the Hertzian collision, τ, divided by h_0. When $v\tau/2$ approaches h_0, the strain per collision increases rapidly, and could thereby result in higher temperatures. However, it should be realized that the process is stochastic in nature, and there will be a number of impacts in the regime where $v\tau$ is approximately equal to $2h_0$.

The temperature increase per impact and the impact strains vary from material to material and are dependent on the charge ratio (mass of grinding balls/mass of powder). Temperature rises of a macroscopic nature have been measured by thermocouples during mechanical alloying and found to be in the range 333 to 393 K [2,12], although it has been postulated that due to the intense localized plastic deformation, localized melting of the constituents could occur [13]. Computer-modeled calculations of the kinetics of milling showed that the approximate temperature rise was around 300 K [9,10]. The analysis was carried out by videotaping the vial and ball motion, which was later slowed down by a stroboscope. The analog motion is then converted into digital signals by a motion analysis computer translation system, allowing continuous monitoring of the displacement and velocities of the ball in the vial. The model predicts that most of the impacts will result in an energy dissipation in the range of 10^{-3} to 10^{-2} J. This model predicts that (depending on the thermal properties) within 18×10^2 s (30 minutes) of milling, each particle will have undergone at least one collision with enough energy to raise the bulk temperature by 100 to 350 K. However, localized melting is not predicted from their calculations. This form of temperature rise could substantially aid the process of diffusion and could be responsible for the alloying and compound formation during mechanical alloying. In the case of powder mixtures that form intermetallic compounds with a high associated exothermic reaction, local hot spots could result in the local initiation of the reaction and result in faster compound formation.

A model has also been developed for the milling times required to obtain the desired microstructural refinement. When the interphase spacing between two materials is reduced to a certain critical level, the alloying is deemed to be complete. On the assumption that an accumulated true strain of 3 to 5 is necessary for the completion of alloying, the time required for completion can be calculated. The time required to mechanically alloy is less in a Spex mill when compared to the attritor mill, owing to the fact that higher precollision impact velocities are generated in a Spex mill, and this results in a greater strain per impact. Thus, fewer impacts are required to complete the milling process. In this case also the charge ratio has an important influence.

Assuming a Hertzian impact model, impact stresses, powder strain, powder strain rates, and powder temperature increase during plastic deformation caused by entrapped powders can be roughly estimated. The milling times required to produce the desired level of microstructural refinement during MA have been estimated and found to be close to observed milling times. The authors expect that semiquantitative predictions of the effects of different parameters on mechanical alloying can be made from the models and calculations carried out by them are extensively discussed in their paper [8]. Most of the discussions cited above have been based on ductile materials. The physics of mechanically alloying ductile–brittle and brittle–brittle materials also needs to be developed on the basis of analogous modeling approaches to provide some degree of predictability.

Analysis of mass and energy transfer during the process of mechanical alloying has also been carried out by Lazarev et al. [11]. They considered the ball and powder system inside the attritor mill as a multicomponent, incompressible, non-Newtonian fluid. Their analysis indicated the formation of two large asymmetrical eddies, with the alloying intensity growing in zones with large velocity gradients. The analysis also predicts that at high attrition speeds the spread of a turbulent zone that reduces the time of contact of individual balls could ultimately stop the alloying. Thus, there exists an optimum attrition speed for facilitating the process of mechanical alloying.

The preceding discussion briefly summarizes the attempts that have been made to model the mechanical alloying process. Modeling does provide a degree of predictability to this extremely complex problem. However, more work needs to be done in this area. The subsequent discussions on different types of material combinations will be generally qualitative in nature and are intended solely to give readers a better idea about the variety of materials that can be processed by mechanical alloying.

2. MA with at Least One Ductile Phase

Depending on the starting material, a single collision of two balls with a powder particle trapped between them modifies the powder particle in different ways. Initially, mechanical alloying always included one or more ductile materials that would flatten out and deform heavily by plastic deformation. The ductile materials are microforged into plates and work-harden tremendously, while the more brittle particles are reduced in size by fragmentation. The classic picture showing the representative constituents of starting powders and their deformation characteristics after one collision is shown in Figure 4.5 [6].

When two or more ductile particles are trapped between two impacting balls, they are brought into intimate contact and form cold welds. This process

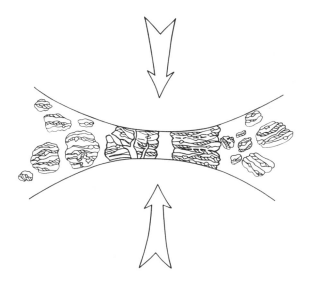

Figure 4.5: *The classic picture showing the representative constituents of starting powders and their deformation characteristics after one collision. (Reprinted with permission from Ref. 6.)*

continues to build up layered composite powder particles with continually decreasing lamellar spacings. The ductile work-hardened plates initialy form large lamellae. If brittle materials, like oxide dispersoids, are also introduced along with the ductile metals and alloys, the brittle material (dispersoid) is usually trapped and strung along the particle–particle weld that forms when two or more platelike surfaces are welded together. The individual particles are still easily distinguishable at this juncture and the chemical composition shows significant variation from particle to particle.

   With increasing milling times, the lamellae become finer, and more convoluted. The already work-hardened particles fracture and reweld. The dispersoids are further refined and begin to be distributed more homogeneously within the particles. At this point, the lamellar spacing is reduced and dissolution of solute elements forming pockets of solid solution usually occurs over a majority of the particles. The solid solution formation is aided by structural defects that provide sites for low-activation-energy diffusion events, heat produced by the kinetic energy of the impacting balls, and very small diffusion distances. It is at this juncture that metastable phases, precipitates, or complex oxides could also form within the powder particles. If dispersoids formed an initial part of the material, then the distance between dispersoids strung out along the ductile particle welds will increase, while the distance between the welds themselves (the interlamellar spacing) decreases.

During the final stage of mechanical alloying of ductile materials, the lamellae become extremely fine and convoluted, with the composition of each individual particle tending toward the overall starting composition. The individual constituents forming the composite are no longer optically identifiable. If dispersoids were initially present, they are now uniformly dispersed within the individual particles. In practice, the distance between the dispersoid particles along the welds becomes nearly equal to the weld spacing, which has been drastically reduced with the progression of mechanical alloying. The hardness of the mechanically alloyed material, which increases in nearly a linear fashion during the initial part of the process, also reaches saturation toward the end.

The above discussion covers the development of the mechanically alloyed material where the starting constituents consist of a combination of ductile–ductile or ductile–brittle materials. In these material systems, it has been shown that it is possible to obtain a true atomic level of alloying by MA. According to Benjamin, each time a powder particle is trapped between two impacting balls, it is plastically deformed to approximately a half or a third of its original thickness. Froes [14] has pointed out that when starting with a particle size of 100 μm, only six to eight mechanical alloying "events" are required to obtain lamella thickness in the range of 0.2 μm, and only eleven to thirteen "events" will reduce the spacing to 0.5 nm, which is in the atomic diameter range. Investigation on mechanical alloying of nickel and chromium powders indicated that, on completion of mechanical alloying, the magnetic properties of the alloy were similar to those of a cast alloy of the same composition, proving that true atomic-level alloying had occurred during the process [15].

The alloying behavior of brittle–ductile materials, however, is influenced by the mutual solid solubilities of the components. For example, dispersion of boron and silicon (brittle) in iron (ductile), presented different mechanical alloying behaviors. It was not possible to alloy the boron by mechanical alloying and the boron particles were dispersed in the metal matrix. Boron, which has negligible solid solubility in iron, could only be alloyed (for a Fe–Zr–B system) after annealing below the crystallization temperature of amorphous Fe–Zr alloy. However, the other brittle material, silicon, which has significant solid solubility in iron, formed an alloy after mechanical alloying. Also, the oxides in the oxide dispersion-strengthened materials did not go into solution or form an alloy with the base material and remained as discrete oxide particles. Thus, it follows that for a ductile–brittle system the structural refinement will continue via fragmentation of the brittle material and decrease in the interlamellar spacing; but whether alloying will occur or not depends on the solid solubility of the constituents.

There was some question about the role of thermal activation in alloy formation during mechanical alloying of materials constituting a ductile phase, since it was argued that a significant amount of temperature spiking could occur due to the impacts. However, it has been observed that alloying can be achieved in both the ductile–ductile and ductile–brittle systems at low temperatures. For example, alloying was achieved in a Ni–Ti system mechanically alloyed at a temperature of 233 K (−40°C) [16], and in a Nb–Ge system at 258 K (−15°C) [12]. Thus, thermal activation is perhaps not an essential condition for alloying with a ductile material.

This discussion briefly covers the general area of mechanical alloying of materials where at least one of the constituents is a ductile material. This type of system (with at least one ductile phase) constitutes the largest proportion of the material combinations that have been subjected to mechanical alloying. Some of the topics covered in following sections of this chapter, such as compound formation, phase redistribution process, hierarchical composite structure formation, and amorphous phase formation, are special cases where often one or more ductile phases are used. However, these specific cases are so important and their microstructural evolution is so different and interesting, that they are discussed separately in this chapter, in spite of the fact that one or more constituents could be a ductile material. Before entering into the discussion of these special cases, it will be appropriate to discuss the case where all the constituents used for mechanical alloying are brittle in nature.

3. MA with All Brittle Phases

The mechanism of microstructural evolution in a brittle–brittle material system is not very well understood. Logic suggests that milling will only reduce the particle sizes to a certain level while maintaining the discrete identity of individual constituents. However, the results of work on some nominally brittle–brittle material systems like Mn–Bi (intermetallic) and Si–Ge (solid solution) indicate the contrary.

Alloy formation was observed in a mechanically alloyed Si–28at.%Ge composition [17]. Lattice parameter measurements initially showed two discrete materials (Si and Ge). With increasing milling time, the two discrete lattice parameters began to converge until after around 180×10^2 s (5 hours) of milling only a single lattice parameter could be measured. This new lattice parameter corresponded to an alloy of the same material produced by conventional metallurgical processes including melting. Thus, even for brittle–brittle systems, alloy formation by mechanical alloying is possible.

The microstructural evolution does not follow the usual lamellae formation route as in the case of mechanical alloying with a ductile material. A granular morphology is observed, with the harder particles apparently embedded in the relatively softer one. In case of materials with similar hardness, both the phases can be "tacked" on to one another. Thus, this system is different in a number of ways from that with at least one ductile phase.

An interesting observation for the brittle–brittle material system is that alloying can be suppressed if mechanical alloying is carried out in liquid nitrogen. An example of this is the mechanical alloying of an Si–Ge combination in liquid nitrogen. Thus, for brittle–brittle materials, thermal activation is probably necessary for producing alloys. This is contrary to MA systems having at least one ductile system as discussed earlier. However, a large number of brittle–brittle systems have not been mechanically alloyed, and it is plausible that some combination of brittle–brittle phases may form an alloy by MA at low temperatures. Until this is demonstrated, it will be prudent to assume that thermal activation is a necessary part in alloy formation of mechanically alloyed materials having no ductile phases.

It is postulated that some form of cold-welding and diffusion process is also possible in this system during mechanical alloying, probably due to a combination of local temperature increases, surface deformations, and hydrostatic stresses in the powder [9]. It should also be realized that in brittle–brittle systems the comminution of the material often reaches a limit beyond which further refining stops due to the inability of the particles to undergo further fracture events. The particles often tend to aggregate as conditions become favorable for cold-welding or other forms of material transfer processes. More investigation is needed on brittle–brittle systems to gain a better understanding of the process by which mechanical alloying occurs in these materials.

4. Homogenization and Compound Formation

Mechanical alloying has been used to form a homogeneous mixture of mutually insoluble or sparsely soluble materials. Oxide dispersion-strengthened materials, Fe–Cu, Cu–Pb, Cu–W, Cu–Ru, Cu–Cr, Al–In, and Mg–Fe provide a few examples of such systems. Mechanical alloying was carried out over prolonged period of time to determine the lower size limit to which these immiscible systems can be milled. X-ray diffraction, TEM and SEM were employed to study the progress of milling. In systems like Ge–Sn, Ge–Al, and Ge–Pb the lattice parameters of the two materials remained distinct and constant over the entire milling times investigated. After 1152×10^2 s (32 hours) of milling, the average germanium interparticle distance was found to be around 0.02 μm (20 nm) and did not change with further milling time. Thus, this could be considered the limit of refinement possible for mechanical alloying in this system. This refinement limit is also dependent on the mechanical alloying temperature [18].

Intermetallic compound formation has also been achieved by mechanical alloying. Line compounds as well as compounds with some degree of compositional variation have been produced by mechanical alloying. For example, the intermetallic line compound formation of Mn–Bi has been reported by Davis and coworkers [9]. Compound formation was carried out in a Spex mill operated at approximately 1200 rpm. Elemental powders with 7.9-mm stainless-steel balls were loaded into a shock-resistant tool steel vial using a ball to powder ratio of 5 : 1. Using X-ray diffraction and electron microprobe analysis it was confirmed that the intermetallic line compound MnBi was formed after 288×10^2 s (8 hours) of milling.

Mechanical alloying was also used to produce β-brass in the Cu–Zn system [19]. Two compositions, Cu–50Zn and Cu–47.5Zn (wt.%) were prepared by mixing elemental powders of copper and zinc. Alloying was carried out in a Spex mixer/mill. The B2 structure formation was evident in this system only after 18×10^2 s (0.5 hour) of milling.

Another interesting use of mechanical alloying in the fabrication of intermetallic compounds is the production of initial powders by MA of elemental powders. The powders are mechanically alloyed to a point where an intimate mixture of the elements forming the compound is attained. In some cases it is also possible to obtain some amount of compound formation along with the intimate mixture of the elements forming the intermetallic compound. These MA powders can then be thermomechanically treated to produce the desired intermetallic compound [20]. Intermetallic compounds such as Ti_3Al, TiAl and Al_3Ti have been produced by MA stoichiometric compositions under an inert atmosphere with admixed process control agents. The as-milled powder, which was an intimate mixture of the elements, could be heat treated at 813 K (540°C) to produce the intermetallic compound Ti_3Al and at 873 K (600°C) to produce the compound TiAl. The formation of the intermetallic compounds has been confirmed by X-ray diffraction. It should be pointed out that it is extremely difficult to produce the compound Al_3Ti by the conventional melting processes. However, this compound can be fabricated by mechanically alloying a mixture of titanium and aluminum in the desired ratio followed by adequate heat treatment.

Mechanical alloying of some intermetallic compounds such as Ni_3Al, which has some compositional range away from the stoichiometric composition, showed evidence of a disordered alloy formation along with amorphization.

The Ni_3Al is first formed as a highly defective metastable crystalline compound, which is destabilized by subsequent milling. The ordered $L1_2$ structure of Ni_3Al is first transformed to a disordered f.c.c. solid solution which, on further milling, is converted to a partially amorphous material.

An interesting investigation on the mechanically alloyed Nb_5Si_3 compound has been described by Kajuch and Vedula [21]. This compound has a low density and a high melting point, which makes it attractive as a high-temperature structural material. Producing this material by conventional means is extremely difficult. Continuous milling of elemental niobium and silicon powders resulted in compound formation in approximately 126×10^2 s (3.5 hours). However, with interrupted cooling during the mechanical alloying process, compound formation could be achieved in 45×10^2 to 72×10^2 s (75–120 minutes). The interrupted cooling is attained by mechanically alloying the elements for 18×10^2 s (30 minutes), followed by cooling the material, then remilling the material for another 18×10^2 s (30 minutes) followed by cooling and remilling. In this interrupted mechanical alloying process, X-ray diffraction patterns indicate that the peaks of niobium and silicon have disappeared and peaks of the compound Nb_5Si_3 have appeared. It is observed that exothermic reactions occur during MA, causing local "hot spots" to develop. The reaction is expected to be self-sustaining and occurs in a matter of seconds. During the cooling, the powder is embrittled and remilling results in the brittle hard particles, causing local impact points to rapidly heat up and initiate the exotherm. This phenomenon has also been described by Atzmon, who suggested that lower operating temperatures could promote compound formation due to the exothermic compound formation mechanism [22].

5. Amorphization by Mechanical Alloying

Amorphous metals and alloys have been fabricated by rapid solidification where a molten stream of metal is subjected to rapid cooling by various means or by vapor deposition on cold substrates. Recently amorphization has been achieved by solid-state processing such as mechanical alloying. The first conclusive study of amorphization by MA was by Koch and coworkers in the Ni–Nb system [23]. Elemental powders of nickel and niobium (60Ni–40Nb at.%) were mechanically alloyed for various times. With continued milling, the individual peaks of the elements continued to broaden. After 360×10^2 s (10 hours) of mechanical alloying, the material showed an X-ray diffraction pattern that was comparable to an amorphous $Ni_{60}Nb_{40}$ alloy of the same composition produced by rapid solidification.

Schwarz and coworkers postulated that the excess point and lattice defects that are created during the severe plastic deformation process associated with

MA, and the rapid instantaneous rise in temperature due to the collision of the balls, result in rapid interdiffusion and formation of the amorphous alloys. They have described the process as a solid-state amorphization reaction, or SSAR [16]. The essential requirements for the binary system that will form an amorphous alloy by SSAR have been outlined by Schwarz and coworkers [24–26]. According to these authors, the two metal constituents of the binary phase diagram must have vastly different diffusivities in each other; the two constituents should also have vastly different diffusivities in the amorphous alloy that is formed; and the two must have a large negative heat of mixing in the amorphous state. Generally the formation of crystalline intermetallic structure from two totally different metals is favored when the atomic motions of both the metals are involved. They selected gold and lanthanum alloys since gold is known to diffuse very fast in crystalline lanthanum and they have a large negative heat of mixing.

A detailed description of the SSAR process with the help of the schematic phase diagram shown in Figure 4.6 has been given by Schwarz and coworkers in their paper on amorphization of Al–Hf [26]. Figure 4.6a shows the schematic binary phase diagram of a system that exhibits a negative heat of mixing in the liquid stage. The primary crystalline solutions are α and β, and γ is the crystalline intermetallic compound. Figure 4.6b shows the free-energy variation with composition at a reaction temperature T_r for the phases α,β,γ, and the amorphous phase λ (assumed to be the free energy of the molten alloy undercooled to the reaction temperature). The free energy for the mixture of elements A and B is denoted by the broken line that joins the free energies of the pure elements. Under equilibrium conditions, the reaction products would have consisted the equilibrium phases: α, β, and γ. But the reaction temperature chosen is too low for the nucleation and the growth of the crystalline compound γ. To determine the product next in line, it is necessary to draw the tangents between the phases α,β, and the amorphous phase λ. This is shown in Figure 4.6b. The diagram predicts that five reaction products will formed and the single-phase amorphous alloy, λ, will be produced for a compositional range lying between the compositions x_2 and x_3 shown in Figure 4.6b. These free-energy diagrams have been able to predict the reaction products with reasonable accuracy [16].

Often the mechanism of the amorphization reaction for compositions that exhibit intermetallic compound formation follows the following steps:

1. Milling of mixed elemental powders result in the particle size refinement to a great extent.
2. Continued milling results in homogenization and formation of the crystalline intermetallic compound.

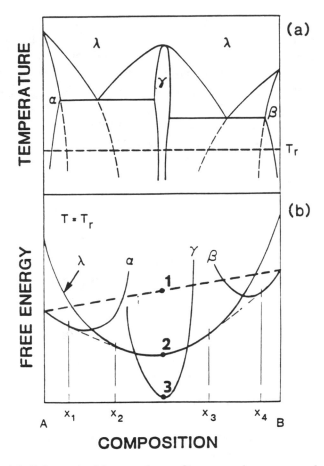

Figure 4.6: (a) Schematic binary phase diagram of a system that exhibits a negative heat of mixing in the liquid stage. The diagram shows two primary crystalline solutions, α and β, and one intermetallic compound, γ. (b) The variation of free-energy with composition at a reaction temperature T_r for the phases α, β, γ, and the amorphous phase λ. (Reprinted with permission from Ref. 26.)

3. Further milling results in the transformation of the intermetallic compound to an amorphous structure.

A case in point where these steps of solid-state amorphization are clearly exemplified is shown in the work of Kim and Koch [4]. Elemental niobium and tin powder blended to a composition of 75 at.% niobium and 25 at.% tin, was milled for various lengths of time. The progress of milling was monitored by X-ray diffraction, the result of which is shown in Figure 4.7. The as-blended

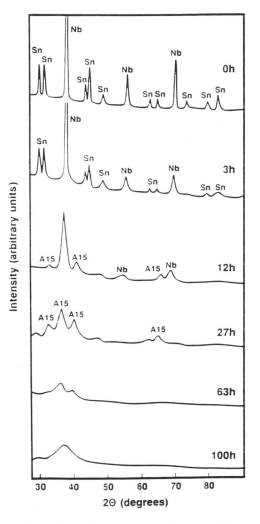

Figure 4.7: Progress of alloy formation and amorphization by mechanical alloying of 75 at.% niobium and 25 at.% tin powder. The progress of milling was monitored by X-ray diffraction. (Reproduced, with permission, from the Annual Review of Materials Science *vol. 19, © 1989 by Annual Reviews Inc., Ref. 18.)*

elemental powders showed clear peaks of niobium and tin that were retained even after 108×10^2 s (3 hours) of milling, which only refined the microstructure. After 432×10^2 s (12 hours) of milling, the crystalline A15 intermetallic structure along with some niobium peaks were observed in the X-ray diffraction pattern. Further milling resulted in the disappearance of the remaining niobium peak and finally the pattern was that of an amorphous phase. It is also possible that a range of intermetallic compounds might be formed during the early stages of mechanical alloying, even though the starting elemental composition is that of one intermetallic phase. This is demonstrated from the work of Kenik et al. [27] on 75 at.% niobium and 25 at.% germanium. During the initial stages of milling, a number of intermetallic compounds were observed by transmission electron microscopy (TEM). With continued milling times the material was seen to homogenize and result in the formation of the crystalline A15 structure within an amorphous matrix.

Milling of the long-range ordered intermetallic compound of Ni$_3$Al provided an interesting observation [28]. It was found that the long-range order parameter, S, decreased to zero after 180×10^2 s (5 hours) of milling, but the amorphous phase is only obtained after 180×10^3 s (50 hours) of milling time. This leads to the conclusion that, in this case, a rapid loss of the long-range ordering is obtained before amorphization can occur. Thus, milling of the ordered intermetallic compound first results in a disordered solid solution long before any amorphization can occur.

Another intriguing observation on the milling of equiatomic titanium and aluminum elemental powders was recently described by Watanabe et al. [29]. Titanium and aluminum powders were vibratory ball milled for various times using different ball sizes. Their X-ray diffraction results are depicted in Figure 4.8. A rather curious phenomenon can be observed when the vibratory milling was carried out with 12.7-mm-diameter balls. After 2160×10^2 s (60 hours) of milling with 12.7-mm-diameter balls, there was evidence of amorphous phase formation in the X-ray results. However, further milling for 4320×10^2 s (120 hours) resulted in the reappearance of the intermetallic compounds, which again disappeared after 7200×10^2 s (200 hours) of milling, resulting in the reappearance of the amorphous phase. It can also be observed from Figure 4.8 that using 4.76-mm balls resulted in amorphous powder formation, while the 19.1-mm balls produced crystalline intermetallic compounds. It is postulated that the relatively larger contribution of the frictional energy in case of smaller-diameter balls results in the formation of amorphous powders, while the dominance of impact energy results in the formation of crystalline intermetallic compounds when using larger balls. The phase change instability with the 12.7-mm diameter balls is, however, quite surprising. Thus, there remain a number of unanswered questions that need to be investigated, and these should provide a tremendous challenge to materials scientists involved in this area of research.

A number of other amorphous alloy powders formed from various combinations such as iron–zirconium, nickel–titanium, nickel–niobium, titanium–aluminum, manganese–titanium, manganese–zirconium, and aluminum–

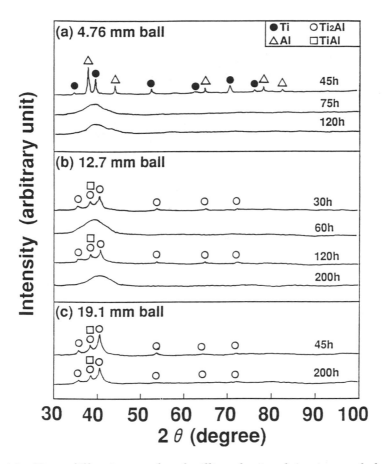

Figure 4.8: *X-ray diffraction results of milling elemental titanium and aluminum powders in a vibratory ball mill using (a) 4.76-mm balls, (b) 12.7-mm balls, and (c) 19.1-mm balls. (Reprinted with permission from Ref. 29.)*

hafnium have been reported by various investigators [18,26,30–34]. Although it will not be possible to discuss all the amorphous powder production methods and their consolidation techniques for each individual system mentioned above, it is worth while to discuss one individual system (Al–Hf) to bring out some of the generic principles of processing of such material systems. Synthesis of equiatomic aluminum–hafnium elemental powders by mechanical alloying resulted in the formation of an amorphous alloy powder [26]. Hexane was used as the dispersant to prevent excessive cold welding and agglomeration, thus decreasing the time to form the single-phase amorphous alloy. This amorphous alloy, when heated to 1003 K (800°C), converts to a crystalline orthorhombic AlHf. The mechanically alloyed powder was hot-consolidated by heating rapidly to a temperature close to the crystallization temperature. The consolidated material retained its amorphous nature and exhibited a very high hardness of 1025 DPH. The material exhibits a high crystallization temperature that is approximately 48.4% of the congruent melting temperature of crystalline AlHf. This material exhibits one of the highest devitrification temperatures known to date.

An interesting process of re-amorphization of a crystallized metallic glass ribbon by mechanical alloying has been described by Calka et al. [35]. They have pointed out that mechanical alloying has not been successful in producing amorphization in multicomponent systems. However, multicomponent systems can be produced in the amorphous form by rapid solidification processes. Melt extraction or melt spinning provides a way of producing amorphous ribbons of multicomponent materials. But to process bulk materials it is best to have the starting material in the powder form. Conventional powder-producing techniques such as atomization do not provide cooling rates that are rapid enough to produce amorphous materials. Thus, the comminution of the amorphous ribbons provides some means of producing the powders from the ribbons. However, in a large number of cases the amorphous ribbons are quite ductile in nature, which makes the process of comminution to powder extremely difficult. To embrittle the material the ribbons should be heated to a temperature above the threshold of brittleness temperature, which is usually lower than the crystallization temperature. However, according to the authors, the brittleness obtained in a $Co_{70.3}Fe_{4.7}Si_{15}B_{10}$ metallic glass ribbon held below the crystallization temperature was not sufficient to make the ribbons brittle for easy comminution. Thus, it had to be heated above the crystallization temperature. This in turn destroys the amorphous nature and results in grain growth. By mechanically alloying such a heat-treated ribbon it is possible to regain the amorphous nature in a powdered form. However, it was found that crystal sizes below $0.01\,\mu m$ (10 nm) can be transformed into the amorphous material. Thus, controlling of heat treatment conditions to produce grain sizes in the correct range is an important step in the success of this process. For the $Co_{70.3}Fe_{4.7}Si_{15}B_{10}$ alloy produced in ribbons by a single roller quenching technique, a heat treatment at 873 K (600°C) produces a fully crystallized structure with average grain size in the range 0.008 to $0.016\,\mu m$ (8 to 16 nm). The process of mechanical alloying was able to produce 80% of amorphous phase in the comminuted powders. The remaining crystalline material is agglomerated into 5- to 10-μm sized particles.

This type of re-amorphization process provides an avenue for utilizing the best of both rapid solidification and mechanical alloying processes. The finding that only crystal grains less than $0.01\,\mu m$ (10 nm) could be transformed to the amorphous state is an extremely interesting one and also warrants further investigation.

6. Nanocrystalline Materials by Mechanical Alloying

The process of mechanical alloying (by any high-energy milling) is capable of producing nanocrystalline metals, resulting in grain sizes that are around the 0.01 μm (10 nm) range. A systematic investigation on the processing of nanocrystalline metals by high-energy deformation processes has been reported by Fecht et al. [36]. It was found that the soft f.c.c. materials were too soft for effective storage of energy that was necessary for producing these nanocrystalline materials. However, b.c.c. materials such as iron, chromium, niobium, and tungsten, as well as some of the h.c.p. metals such as zirconium, hafnium, cobalt, and ruthenium, were amenable to such processing.

To investigate the above theory, elemental powders with particle sizes finer than 125 μm were Spex-milled in steel containers using argon as the protective atmosphere. The milling was carried out for 864×10^2 s (24 hours). Around 2 at.% of iron was picked up as impurity. X-ray investigations of the milled powders showed a considerable amount of peak broadening, which was due to the crystal size refinement and the atomic-level strains. The average crystal size has been measured by separating out the two effects. It was determined that after 864×10^2 s (24 hours) of milling, the average grain size of the b.c.c. metals was around 0.009 μm (9 nm) and that of the h.c.p. metals was around 0.013 μm (13 nm). There were, however, areas that had crystal sizes of 0.01 to 0.03 μm (100 to 300 nm). Mechanical alloying for times longer than 1080×10^2 s (30 hours) produced a relatively uniform crystal size. The average microcrystalline sizes obtained for various metals are shown in Figure 4.9.

The stored enthalpy of the material after mechanical alloying for 864×10^2 s (24 hours) varies from material to material with no correlation with the melting point or the grain size of the material. The stored enthalpy can be measured by differential scanning calorimetry and is also shown in Figure 4.9 along with the grain sizes obtained for the various materials. Usually, severe cold working results in a maximum stored enthalpy of 1 to 2 kJ/mole, which is normally a very small fraction of the heat of fusion. But with milling, the stored enthalpy can be substantially higher. However, no correlation was found between the stored enthalpy and the heat of fusion. The specific heats of the materials are seen to increase after 864×10^2 s (24 hours) of ball milling. The percentage change in the specific heat shows a linear variation with the stored enthalpy divided by the enthalpy of fusion for the investigated metals. It is postulated

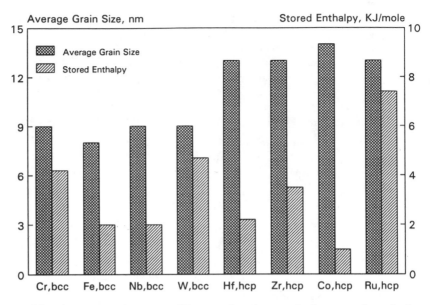

Figure 4.9: Average microcrystalline grain size and the stored enthalpy for various b.c.c. and h.c.p. metals mechanically alloyed for 24 hours.

that the changes in the stored enthalpy and specific heat occur due to the high density of grain boundaries and lattice defects in the crystals.

The microstructure obtained by the high-energy milling process has the characteristics of nanocrystalline metals produced by gas-condensation methods. The intense cold working goes to decrease the thermodynamic stability of the material by increasing imperfections such as dislocations and interfaces. The small crystals are separated by grain boundaries that do not exhibit any long- or short-range order. The grain boundary energies in these nanocrystalline materials are extremely high, and are not obtained by conventional processing. Thus, the dominant deformation mechanism at high strain rates consists of the formation of shear bands with dense networks of dislocations. The localized shear instability is believed to be caused by material inhomogeneities and nonuniform heating in certain zones. During milling, the atomic level of strain increases due to the dislocation density increase, which after a certain level causes the crystal to disintegrate into subgrains with low-angle grain boundaries. With continued milling, the size of the subgrains in existing bands is further reduced and the relative orientation of the subgrains becomes random. Once a complete nanocrystalline structure is attained, further reduction in grain size becomes almost impossible. The final structure consists of randomly oriented nanosized grains almost akin to gas-condensed nanocrystalline metals.

A number of recent reports have also shown that nanocrystalline materials can be produced by mechanical alloying [37–40]. Suryanarayana and Froes [39] have reported the synthesis of a titanium–magnesium alloy by mechanical alloying. The binary alloy system under equilibrium conditions exhibits very small solid solubility of magnesium in titanium. Through mechanical alloying, approximately 6 at.% of magnesium was taken into solid solution in titanium, producing a metastable f.c.c. structure. A large number of nanocrystalline grains were observed in the hot-pressed microstructure. The average grain size range was 0.01 to 0.015 μm (10 to 15 nm) and the grains were randomly oriented. The microstructure also contained some large grains, around 0.4 μm (400 nm) in size. It was interesting that the composition of the large grains revealed only pure titanium while the nanocrystalline grains exhibited a composition that showed between 2.5 and 3 wt.% magnesium, with the rest being titanium. The extended solubility that has been obtained in this sample is due to the high reactivity and high diffusivity of the nanocrystalline grains, where the open volumes of grain boundaries helped in allowing the 3 wt.% solubility of magnesium in titanium.

The recent flurry of activity in processing of nanocrystalline materials provides an exciting direction for mechanical alloying. The production of these exotic materials in any significant volume is perhaps best accomplished by mechanical alloying. Thus, it is expected that mechanical alloying will play an important role in the future development of nanocrystalline materials.

7. Hierarchical Composite Structure

The concept of hierarchical composite structure has been discussed by Benjamin [15,41] and a schematic diagram of the basic principle is shown in Figure 4.10. The concept relies on the addition of different powders at different times during the mechanical alloying process. This form of inter-rupted addition results in unique microstructures that would not normally be available by any other form of processing. In the case described in Figure 4.10, the starting powders are tungsten and zirconium oxide. Mechanical alloying of the two powders for some time results in the fine dispersion of zirconia in tungsten. This can be termed a single-level composite. The introduction of nickel powder by interrupting the mechanical alloying process after a disper-sion of the zirconia in tungsten has been attained, followed by remilling, results in tiny particles of zirconia-dispersed tungsten embedded in nickel. This unique form of hierarchical microstructure represents a two-level composite as shown in Figure 4.10. It is envisaged that numerous unique composite powders can be produced by this form of mechanical alloying.

This form of alloying can also be applied to ease the process of mechanical

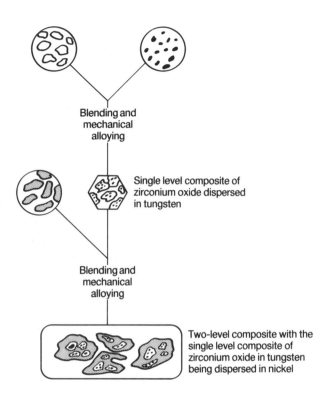

Figure 4.10: *Schematic diagram of the hierarchical composite structure as discussed by Benjamin [41].*

alloying. For instance, in the dispersion-strengthening of NiAl intermetallic compound with oxides, the brittle intermetallic compound creates composite degradation. This can probably be eliminated by using the present concept. In this process, the oxide can be mechanically alloyed separately with both aluminum and nickel, and then subsequently the oxide-dispersed nickel and aluminum powders can be mixed together in the desired ratio (to form the intermetallic compound NiAl) and mechanically alloyed [21].

8. *Phase Redistribution Processing*

To obtain the properties desirable in modern materials it is often necessary to use a combination of two or more advanced processing techniques. As the trend increases toward the use of nonequilibrium structures, coupling of the desirable attributes of various processes is becoming increasingly important. As a case in point, both mechanical alloying and rapid solidification techniques have been used to develop immiscible alloy systems. Both powder production techniques have their attributes and their shortfalls. For example, it would be extremely difficult to produce rapidly solidified W–Cu material with a uniform distribution of the two phases because of the very high melting temperature of tungsten compared to that of copper. However, processing difficulties will also prevent the production of Cu–Pb systems by the process of mechanical alloying since the extremely soft nature of lead makes it difficult to process by MA. These two systems (Cu–W and Cu–Pb) bring out the inherent processing difficulties that are unique to the two processing techniques. Thus, if three materials such as copper, tungsten, and lead are required to be homogeneously distributed to synthesize an advanced material, one would need to resort to a sequential processing that will incorporate the utility of both rapid solidification and mechanical alloying techniques.

The concept discussed above has led to the development of the phase redistribution process (PRP). This process has been used to produce a Cu–23Pb–3Sn (wt.%) alloy powder by the sequential use of rapid solidification and mechanical alloying [42,43]. The molten alloy of the above composition is first rapidly solidified by melt spinning. Prior work on gas atomization of this material had led to large inhomogeneous distribution of lead in the matrix; using the melt-spinning process, the lead particle sizes were considerably reduced. However, even this was considered too coarse a size distribution. Thus, the ribbon produced by melt spinning was chopped and then Spex-milled in dry argon. After 12×10^2 s (20 minutes) of milling, the powders formed were equiaxed in nature and did not show any resemblance to the starting feedstock material, which was the chopped ribbon. The major advantage gained by the secondary mechanical alloying treatment after rapid solidification was the homogeneous redistribution of the lead in the alloy matrix. This was confirmed by X-ray mapping of lead, which showed the material was uniformly distributed. The alloy was later vacuum hot-pressed at 583 K for 36×10^2 s (1 hour) using a pressure of 83 MPa. The fine lead distribution was still retained after the vacuum hot-pressing operation.

There are some differences between the phase redistribution process and conventional mechanical alloying. In conventional MA, usually at least two dissimilar materials are used, while the PRP process uses chopped ribbons of a single material. In PRP, the homogeneous redistribution of the second phase is easily attained within a very short period of time. This provides the advantage of lower contamination pickup, which usually occurs during prolonged mechanical alloying. Thus, this process utilizes both rapid solidification and mechanical alloying, and could yield new and exciting advanced materials.

9. *Displacement Reactions by Mechanical Alloying*

Simple solid-state displacement oxidation–reduction type reactions by mechanical alloying have recently been reported by Schaffer and McCormick [44]. Mechanical alloying was successful in increasing the chemical kinetics of the oxidation–reduction reaction in which a solid metal oxide is reduced to its metallic form while the other pure metal is oxidized. The repeated comminution and cold-welding provides intimate contact between the reacting species. The high defect density acts to enhance the diffusion rates through the product phases, and the extremely small nanocrystalline particle sizes reduce the diffusion distances significantly, thereby pushing the reaction to a critical condition in which ignition occurs and the reaction proceeds as a self-propagating combustion front. A minimum adiabatic temperature of 1300 K (1027°C) is required to initiate the combustion in these systems. Thus, the reaction enthalpy is a critical factor in solid-state displacement reaction by mechanical alloying.

In the investigation by Schaffer and McCormick [44], the reduction of cupric oxide to pure copper was accomplished by mechanically alloying it with other metals such as aluminum, calcium, magnesium, titanium, manganese and iron. Cupric oxide and other reducing powders (with approximately 10% of reducing powder in excess of that required to totally reduce the CuO) were mechanically alloyed in a Spex 8000 mill that had a special vial with rounded corners. The radius of curvature of the corners was equal to the curvature of the three 10-mm diameter, 8-g steel balls. This special vial was necessary because the flat-ended vial trapped some of the powders in the corners, which isolated those powders from the reaction. The powders with the balls were loaded and sealed in an argon atmosphere. Milling time was varied from a few seconds to 1080×10^2 s (30 hours). Various analytical tools such as SEM, TEM, EDAX, and X-ray diffraction were used to analyze the course of the reaction. The different elements and their reactions causing a reduction of the copper oxide to its elemental form are as follows:

Reaction 1 $3CuO + 2Al = 3Cu + Al_2O_3$
Reaction 2 $CuO + Ca\quad = Cu + CaO$
Reaction 3 $CuO + Mg\quad = Cu + MgO$
Reaction 4 $2CuO + Ti\quad = 2Cu + TiO_2$
Reaction 5 $CuO + Mn\quad = Cu + MnO$
Reaction 6 $4CuO + 3Fe = 4Cu + Fe_3O_4$
Reaction 7 $CuO + Ni\quad = Cu + NiO$

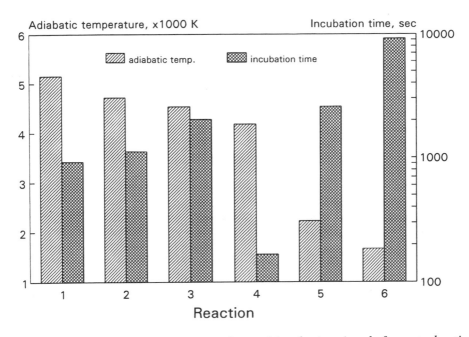

Figure 4.11: Adiabatic temperature rise and incubation time before combustion for the different reactions between CuO and various metals.

All the CuO/metal combinations except for CuO/Ni showed some incubation time before the combustion began. The CuO/Ni combination did not show any combustion. The adiabatic temperature rise for the reaction between CuO and various metals (except nickel) is illustrated graphically in Figure 4.11. The incubation time before combustion occurred has also been plotted in the same figure. Only when the adiabatic temperature rise was below 1288 K (1015°C) as in the case of CuO/Ni did combustion not occur. Except for titanium, where the incubation time before combustion occurred was only 167 s, incubation times generally decreased with increasing adiabatic temperature. Right after combustion, the powders formed a mixture of relatively coarse spherical particles of 1 to 5 mm in diameter and fine particles about 0.5 to 10 μm in diameter. The spherical nature of the copper particles suggested that they had been melted and re-solidified from the molten state.

When a process control agent was added, the combustion did not occur, but the reaction proceeded slowly. In the reaction between CuO and Ti with toluene as the process control agent, combustion was initiated after approximately 72×10^2 s (2 hours) and completed within 1080×10^2 s (30 hours) of milling.

The factors that affect the kinetics of the displacement reactions include the charge ratio, collision frequencies, adiabatic temperature, fracture characteristics, addition of process control agents that could contaminate the reaction, and defect thermodynamics. This form of solid-state displacement reaction could open up new areas of applications for mechanical alloying, such as novel, inexpensive oxide dispersion-strengthened materials.

The above discussion covers the generic aspects dealing with the mechanisms of microstructure, compound, alloy, reaction, and phase formation during mechanical alloying. The application of MA to specific material systems will also be briefly discussed in the next section.

Some Specific Systems Produced by Mechanical Alloying

It is worth while to discuss a few general systems that have been produced by the process of mechanical alloying. Some of the MA materials discussed are already in commercial use, while some exhibit great potential for commercial applications.

1. Nickel-based Materials

Nickel-based alloys were the first material that was used for dispersion strengthening by MA. For producing the desired material, Al, Ti, B, and Zr (the reactive materials) in the form of crushed nickel-based master alloys were added to elemental powders of fine Ni, Cr, W, Mo, and Ta. For dispersion strengthening, usually yttria agglomerates of approximately 1 μm in size were used as the starting material. For thoria-dispersed (TD) nickel, pure nickel was used with ThO_2 powder. Some of the mechanically alloyed nickel-based materials that have been developed are shown in Table 4.1 [3]. The powders are mechanically alloyed to a point where the individual constituents can no longer be optically resolved. An approximately 36 kg powder charge is mechanically alloyed in a Szegvari attritor mill. The MA powder is canned, hot-extruded, and hot-rolled to produce the desired material. An additional heat treatment step is incorporated to form large, elongated grains with an aspect ratio of at least 10. The main structural feature that differentiates the mechanically alloyed oxide dispersion-strengthened (ODS) nickel-based superalloy from its conventional ingot counterpart is the presence of a fine (usually 0.035 μm) dispersion of Y_2O_3, large elongated grain sizes, and a recrystallization texture.

Table 4.1. Nominal compositions of some nickel- and iron-based alloys produced by mechanical alloying. Some of the compositions (in wt.%) have trace amounts of boron and carbon that have not been included in the table

Alloy	Fe	Ni	Al	Cr	Ti	Mo	W	Ta	Zr	Y_2O_3/ThO$_2$
Inconel MA 6000	–	bal.[a]	4.5	15	2.5	2	4	2	0.15	1.10/–
Inconel MA 754	–	bal.	0.3	20	0.5	–	–	–	–	0.60/–
Inconel Ma 758	–	bal.	0.3	30	0.5	–	–	–	–	0.60/–
Inconel Ma 760	–	bal.	6	20	–	2	3.5	–	–	0.95/–
NAVAIR Alloy 51	–	bal.	8.5	9.5	–	3.4	6.6	–	0.15	1.10/–
TD nickel		bal.	–	–	–	–	–	–	–	–/0.5
Incoloy MA 956	bal.	–	4.5	20	0.5	–	–	–	–	0.50/–
Incoloy MA 957	bal.	–	–	14	1.0	0.3	–	–	–	0.25/–
ODM 331	bal.	–	3	13	0.6	1.5	–	–	–	0.50/–
ODM 361	bal.	–	6	13	0.6	1.5	–	–	–	0.50/–
ODM 031	bal.	–	3	20	0.6	1.5	–	–	–	0.50/–
ODM 061	bal.	–	6	20	0.6	1.5	–	–	–	0.50/–

[a]bal. = balance.

The typical ODS nickel-based alloy like Inconel alloy MA 6000 obtains its characteristics from various constituents. The refractory additions of tungsten and molybdenum strengthen the alloy by solid-solution strengthening. The gamma prime hardening for intermediate temperature resistance is provided by intermetallic compounds. At higher temperatures, where the precipitate will start to lose its effectiveness due to growth and dissolution, the dispersion of Y_2O_3 increasingly begins to play an important role by obstructing dislocation motion. Aluminum and chromium provide the oxidation resistance. The sulfidation resistance is provided by chromium, tantalum, and titanium collectively [3].

Perhaps the most successful applications of mechanically alloyed materials are the Inconel alloys MA 754, MA 758, and MA 760. The nominal compositions of these alloys are outlined in Table 4.1. The Inconel alloy MA 754 has been used for turbine vanes and bands in jet engines. The material is processed by unidirectional hot rolling of the mechanically alloyed powder.

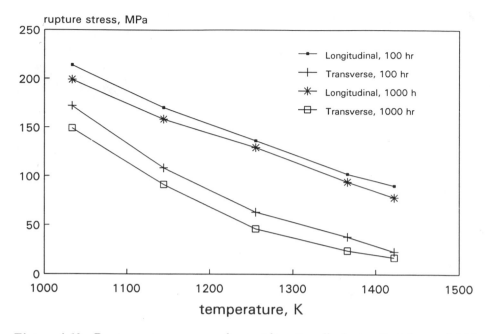

Figure 4.12: Rupture stress curves for unidirectionally hot rolled Inconel MA 754 tested in the longitudinal and transverse directions for 100 and 1000 hours.

Since the rolling is carried out in only one direction, the material develops nonisotropic properties which vary from the longitudinal to the transverse direction. The stress rupture properties of Inconel MA 754 when tested in the longitudinal and transverse directions are shown in Figure 4.12. The nonisotropic properties of the hot-rolled material are quite clear from this figure. With the improvements in net-shaping processes, the various alloys mentioned above have been used in a variety of applications [45]. For example, Inconel Alloy MA 758, which has a high chromium content, is being used in the glass melting industry and as furnace fixtures. The material has excellent resistance to molten glass and has high oxidation and elevated-temperature creep resistance. Inconel MA 760, which has a greater extended-time creep strength than advanced single-crystal alloys and oxidation and corrosion resistance equal to the best cast alloys, has been used as the blade and vane material in industrial gas turbines. This alloy is a modification of the MA 6000, with higher chromium and aluminum contents for superior corrosion and oxidation resistance.

Mechanical alloying of nickel-based intermetallic compounds has been investigated by various research groups [20,46–50]. Different dispersoids such as alumina, yttria, and thoria have been used by the various research groups. A prealloyed Ni_3Al-based intermetallic compound having the composition Ni–23.5Al–0.5Hf–0.2B (at.%), was mechanically alloyed with different volume fractions of alumina, thoria, or yttria additions [50] in a Spex mill (Model 8000). Initial experiments indicated that between 25% and 50% of the oxide dispersoids were lost due to cold welding to the vial and the milling media. Stearic acid, 1 wt.%, was added as the process control agent. The materials were mechanically alloyed for 216×10^2 s (6 hours) followed by vacuum canning and hot isostatic pressing at 1573 K (1300°C) for 108×10^2 s (3 hours) at a pressure of 103.4 MPa. One control sample of the $Ni_3Al(Hf,B)$ powder was hot isostatically pressed without any milling.

The results indicated that the material not subjected to MA had a grain size of approximately 35 μm, whereas the mechanically alloyed materials had a grain size around 1 to 2 μm. Electron microscopy indicated that the oxide dispersion was generally uniform (Figure 4.13a) throughout the matrix, though in some areas the oxides were present in a network form (Figure 4.13b), which probably represented the oxides present at the prior particle boundaries. The average dispersoid diameter was around 0.08 μm (80 nm) with a variation of $\pm$0.02 μm (20 nm). However, the oxide dispersoid content had been considerably increased during the milling process, and even species of oxides such as hafnia, which had not been intentionally added, were found in the microstructure. The dispersoids approximately doubled the room temperature yield strength of the alloys, mainly by grain size refinement. The ductility of the alloys was poor, with approximately 3% ductility being observed in samples that were processed in argon atmosphere. Oxygen contamination was one of the major problems encountered during mechanical alloying of this ductile intermetallic compound.

Similar investigations were also carried out by other research groups. The intermetallic compound $Ni_3Al(Hf,B)$ was mechanically alloyed with alumina for 360×10^2 s (10 hours) and was consolidated by hot pressing. The hot hardness of the oxide-dispersed material was actually lower than that of the Ni_3Al intermetallic compound without the dispersoids at temperatures above 1273 K (1000°C). In general, the oxide additions provided a tremendous increase in the yield strength and rupture strength at ambient temperatures with a severe degradation in the ductility. Oxygen contamination has been identified as one of the main sources of embrittlement. Often, the mechanically alloyed material had more oxides than the amount of oxide intentionally added for dispersion strengthening. The oxides formed in situ are often coarse in nature.

Investigations carried out by Vedula and coworkers [49] on mechanical alloying of NiAl with Y_2O_3 revealed some interesting aspects of oxide

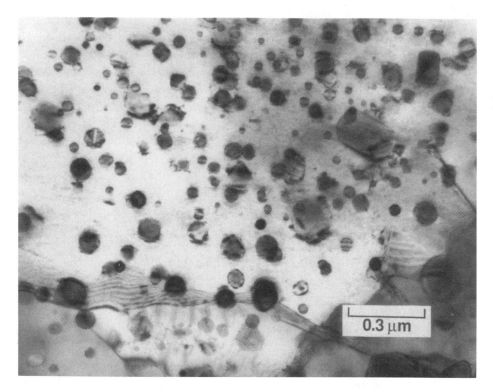

Figure 4.13: (a) TEM bright-field image showing the oxide dispersion in the matrix and the grain boundaries of Ni₃Al(Hf,B) mechanically alloyed with 2.5 vol.% alumina. (Reprinted with permission from Ref. 50.)

dispersion in brittle intermetallic matrix composites. The mechanical alloying was carried out in Szegvari attritors. The final composition was a Ni–50at.% Al with 2 vol.% Y_2O_3 dispersion. Different blends were used to attain the same final composition. The first blend was prepared by milling a Ni–45at.% Al prealloyed powder with elemental aluminum and Y_2O_3 powders to attain the final composition of Ni–50at.% Al with 2 vol.% Y_2O_3 dispersion. The second blend increased the volume fraction of the ductile phase by milling the prealloyed Ni–45at.% Al and Y_2O_3 powders with elemental nickel and aluminum powders. The total weight fraction of the prealloyed aluminide powder was around 30 wt.%. In the third blend, the amount of the ductile phase was increased still further by using only 20 wt.% of the prealloyed aluminum phase. The last blend was prepared by milling elemental nickel and aluminum powders with Y_2O_3 particles. Around 1 to 2 wt.% of methanol, which served as the process control agent, was added to the powder blend prior to milling.

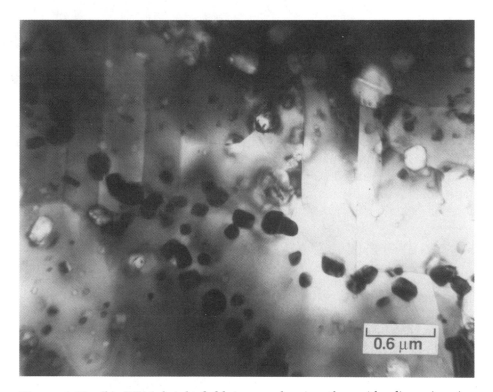

Figure 4.13: *(b) TEM bright-field image showing the oxide dispersion in a networklike structure in the matrix of Ni$_3$Al(Hf,B) mechanically alloyed with 2.5 vol.% alumina. (Reprinted with permission from Ref. 50.)*

The results indicated that severe composite degradation occurred in all the blends when mechanical alloying was carried out for an extended period of time. Composite degradation was attributed to the brittle NiAl intermetallic compound. Even in the last blend, which did not have any prealloyed nickel aluminide powder, extending the milling time beyond 144×10^2 s (4 hours) resulted in severe composite degradation. In this case, the elemental powders did form the NiAl intermetallic compound at the expense of the elemental powders. However, once a certain volume of the brittle NiAl intermetallic compound is formed, composite degradation sets in. Hence, optimization of the amount of ductile phase addition and stopping of the milling process when an optimum level of the brittle compound is formed, could provide the only means of processing oxide dispersion-strengthened brittle intermetallic matrix compounds. However, premature cessation of the milling process would probably result in a nonuniform dispersion of the oxides and also the retention of coarse oxide particles in the microstructure. The alternative suggestion is to mill the Y_2O_3 particles separately with both elemental nickel and elemental aluminum, which are soft and ductile, and obtain a fine and uniform dispersion of the Y_2O_3 particles in both the blends. The two blends can then be mechanically alloyed for a short period of time to produce the desired oxide dispersion-strengthened NiAl matrix composite. This process has good potential for processing ODS materials of brittle intermetallic compound matrices.

Bulk amorphous $Ni_{60}Nb_{40}$ compound is another material that has been successfully processed by mechanical alloying of elemental powder blends. The process of mechanical alloying provides the opportunity of producing large quantities of the desired powders in an inexpensive manner. The bulk forms made from these amorphous materials could find a number of new applications in various areas.

2. Iron-based Materials

The development of iron-based oxide dispersion-strengthened alloys followed on the heels of the nickel-based ODS alloys. It has been known that ferritic materials, which are suitable as irradiation damage-resistant materials, are not strong enough at the operating temperatures (873 to 973 K; 600–700°C) of the fast neutron breeder reactors. An iron-based alloy, Incoloy MA 957 [14Cr, 1Ti, 0.3Mo, 0.25 Y_2O_3, rest iron (wt.%)], produced by INCO, was specifically developed as a fuel cladding material for fast neutron breeder reactors because of its lower swelling characteristics compared to the conventional AISI 316 stainless steel when subjected to irradiation. Dispersion strengthening provides the required strength at the high temperatures and the material can also maintain its strength over long periods of time. This alloy also exhibits a flat rupture stress curve.

The other commercially important mechanically alloyed iron-based material, Incoloy MA 956, is also produced by INCO. This alloy has a higher chromium content than MA 957. This material can be easily cold and hot worked. One interesting aspect of this material is the fact that it may be cold rolled to produce 90% reduction in thickness without intermediate annealing. Due to the ductile to brittle transition temperature (DBTT) of this ODS material at 348 K (75°C) the material needs to be slightly warmed before cold working. Once the rolling process has been initiated, the internal friction of the material will maintain the temperature above the DBTT. Cross cold rolling is necessary to maintain some isotropic properties of the material. The material can be formed into bars, rods, thin wires, and sheets. Using the process of bare extrusion, near net-shape parts can also be obtained from this material. Turbine blade vanes and tubes have been processed successfully. For the bare extrusion process, the canned mechanically alloyed powder is first upset to full density in the extrusion press. The can material is removed by machining, and the fully dense bare alloy billet is then re-extruded to form tubes or other contours [45].

Magnetic material based on the FeNdB system has also been produced by mechanically alloying elemental powders taken in the desired proportions [51]. Elemental powders of iron (5 to $40\,\mu$m), neodymium (smaller than $10^3\,\mu$m), and amorphous boron (less than $1\,\mu$m) were mixed to yield a final composition of $Nd_{15}Fe_{77}B_8$. The powders are sealed under argon in a steel container and milled in a planetary ball mill. During the microstructural evolution, first iron and neodymium form a layered structure with the relatively undeformed boron embedded in it. With continued milling for times greater than 72×10^2 s (2 hours) further refinement of the layers occurs. Milling is continued for 1080×10^2 s (30 hours), after which the iron and neodymium can no longer be resolved by optical microscopy. The X-ray diffraction shows broadened iron peaks and smeared out neodymium peaks. The milled powders were heated to different temperatures for different times, with the best properties being obtained after reacting the mechanically alloyed powder at 973 K (700°C) for 18×10^2 s (30 minutes). The boron particles are completely dissolved in the FeNd. X-ray analysis clearly identified the $Fe_2Nd_{14}B$ phase. Magnetically isotropic NdFeB magnets could, thus, be obtained by reacting the mechanically alloyed powder in the solid state. Typical properties include $B_r = 0.8$ T and $BH_{max} = 12.8$ MGOe. Due to the extremely short reaction time, the magnetically isotropic particles have a very fine microstructure that is comparable to rapidly quenched material.

The processing of amorphous materials by mechanical alloying has added a new dimension to this processing technique. An iron–tantalum amorphous powder has been successfully produced by mechanical alloying of elemental powders [52]. The iron content was varied from 0.2 to 0.8 atom fraction. Mechanical alloying was carried out by sealing the desired powder mixture in a steel vessel with steel balls in an argon atmosphere. The powder-to-ball ratio was 1 : 15. The material was mechanically alloyed in a high-energy ball mill for times varying from 180×10^2 s to 900×10^2 s (5 to 25 hours). With increasing milling time, the degree of amorphization increased. The stability of the amorphous phases was measured by differential thermal analysis (DTA). For a composition of $Fe_{30}Ta_{70}$, the crystallization temperature was determined to be around 1307 K (1034°C), which is considered to be very high. Thus, this material can be consolidated and used in the form of a bulk amorphous material that could have hitherto unknown properties.

A multiphase microstructure consisting of hard and fine dispersions in a tough matrix is an ideal material that undergoes sliding wear. This sort of material is expected to find uses in the automotive industry, especially with modern combustion engines. Mechanical alloying has been used to produce a sintered steel structure that has a fine dispersed oxide phase intentionally introduced into the microstructure by the process of mechanical alloying [53]. From the economic standpoint, the use of alumina as the oxide powder seemed quite appropriate. In this investigation, elemental iron carbonyl powder was mixed with alumina and iron phosphide powders, and the mixture was mechanically alloyed in an alcohol medium. The powders were then dried in vacuum and annealed in hydrogen at 873 K (600°C). The purpose of the hydrogen anneal is to soften the hard mechanically alloyed powder and increase its compressibility, and also to reduce the oxygen content of the powders. Iron phosphide in the form of Fe_3P is added to allow the formation of a liquid phase during sintering. The annealed MA powders are mixed with a pressing aid and 0.6% carbon. The mixture is pressed at 550 MPa, and sintered in hydrogen at temperatures varying from 1543 to 1613 K (1270 to 1340°C). The mechanically alloyed material showed better densification when contrasted with a material of similar composition that has not been mechanically alloyed. This is attributed to the better distribution of the Fe_3P particles and a greater powder activation. The strength and elongation were also increased substantially. When the wear properties of an alloy having the composition of Fe–0.6C–0.6P (wt.%), with 10 vol.% alumina was compared with a similar alloy with 10 vol.% NbC dispersion, it was observed that the wear of the two materials was nearly the same when the hard material particle size was the same. Thus, the mechanically dispersed alumina in the steel matrix can substitute the NbC hard phase. A post-sintering HIP treatment substantially reduced the amount of wear. Mechanical alloying provides an excellent iron-based oxide dispersion-strengthened material with good sliding wear resistance, which could have application in the cam-tappets of modern automobiles.

Dour Metal of Belgium has developed a range of iron-based oxide dispersion-strengthened alloys that could have uses in energy conversion systems such as gas turbines or heat exchangers [54]. The concept of fine dispersion of stable oxides provides the material with high-temperature stability that is essential for such applications. A series of iron-based alloys with chromium, aluminum, molybdenum, titanium, and Y_2O_3 additions, were prepared by mechanical alloying of powdered materials with master alloys. Five alloys, ODM 331, ODM 361, ODM 031, ODM 061, and ODM 751, with essentially varying aluminum and chromium contents (with 1.5Mo, 0.6Ti, and $0.5Y_2O_3$, wt.%) were processed by mechanical alloying of the powders in a dry high-energy ball mill. The mechanically alloyed materials are densified into large billets. Two successive heat treatments were used to obtain a large grain structure and to create a protective oxide surface layer.

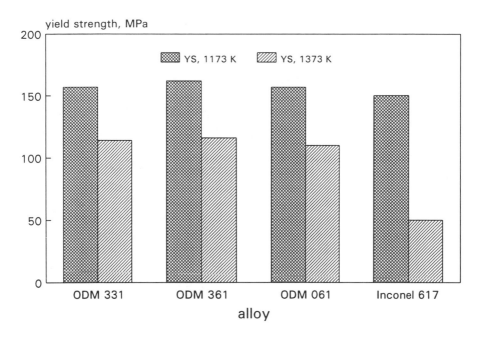

Figure 4.14: *The yield strength of different ODM alloys and Inconel 617 at 900°C and 1100°C.*

The ODS materials attain their excellent high temperature properties due to the fine 0.02 to 0.05 μm (20 to 50 nm) dispersion of the stable oxides in a tough ferrite matrix and due to the adherent protective oxide layer. The fine-grained ODM alloys have excellent hot forming ability, and in the fine-grained condition the strength of the ODS alloys surpasses that of the conventional FeCrAl alloys without the oxide dispersion. The ODM alloys compare favorably in terms of their high-temperature strength properties when compared to Inconel 617. In fact, at 1373 K (1100°C), which is about the upper limit of the superalloy, the strength of the ODS alloys is much higher with poor elongations. The yield strengths of different ODM alloys and Inconel 617 at 1173 and 1373 K (900 and 1100°C) are shown graphically in Figure 4.14. The elongation of the Inconel material is, however, larger than that of the ODM materials. The creep properties of the ODM alloys were compared to those of a number of high-temperature alloys like MA 956, F10, Inconel 617, and Haynes 230, and the creep properties of the different alloys are shown in Figure 4.15. It can be observed that the creep resistance of ODM 751 surpasses that of the superalloys tested at 1253 K (980°C). Compared to MA 956, the ODM alloy shows superior creep properties with increased temperature for the

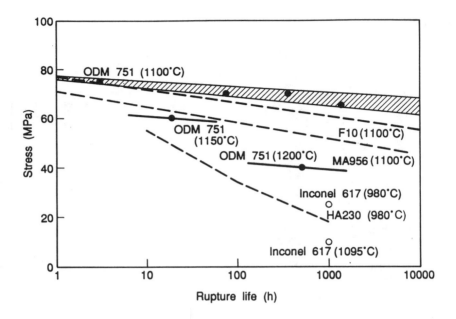

Figure 4.15: *The creep properties of some high-temperature alloys along with the ODM alloys. (Redrawn after Ref. 54.)*

same rupture life by 50 K, or increased rupture time under similar conditions by approximately two orders of magnitude. The oxidation resistance of the new ODM alloys showed that it retained the excellent oxidation resistance imparted to it by its FeCrAl matrix. Thus, this class of iron-based mechanically alloyed material opens up new horizons in high-temperature oxidation-resistant alloys for use in energy conversion applications. The nominal compositions of some of the oxide dispersion-strengthened iron-based mechanically alloyed materials are shown in Table 4.1.

Iron aluminide-based intermetallic compounds also have attractive moderate- to high-temperature properties, including oxidation and sulfidation resistance at elevated temperatures. Attempts to oxide dispersion-strengthen this material have been made by various research groups [55]. The investigation carried out by Seybolt consisted of ball milling 12 vol.% of alumina powder in a Fe–40Al (at.%) alloy. The material, with a grain size between 0.1 and 0.4 μm, exhibited a fourfold increase in stress rupture strength over the binary alloy processed by conventional methods. Strothers and Vedula mechanically alloyed a prealloyed Fe–40Al–0.1Zr (at.%) powder with 1 vol.% yttria and 0.2 at.% boron. The milling was accomplished in an attritor mill. The mechanically alloyed powders, which had layers of Y_2O_3 in the prealloyed iron aluminide particles, were hot extruded at 1250 and 1450 K (977 and 1177°C). The extruded grain sizes were around 3 to 5 μm with uniformly distributed Y_2O_3 particles. The room temperature ductility of the ODS alloy was 10% and it also had superior strength at both room and elevated temperatures. Coarse-grained microstructures with the fine dispersoids could provide excellent materials for use in high-temperature applications.

3. Light Metals

The mechanical alloying of light metals has been covered extensively by a number of authors. Since light metals like aluminum have a tendency to form excessive cold welds, it is imperative that organic process control agents be used during mechanical alloying. It is often necessary to carry out mechanical alloying under an inert atmosphere since the majority of light metals like aluminum, titanium, and magnesium are quite reactive.

Aluminum-based alloys are probably the leading materials in mechanically alloyed light materials. Mechanical alloying usually results in the formation of an extremely fine dispersion of alumina and carbide. The alumina that is available in the form of the thin oxide skin present in aluminum alloy powders is usually broken and uniformly dispersed in the matrix. Also, some amount of oxygen pickup occurs due to the milling atmosphere. The carbides can come from the reaction of aluminum with the carbon available from the organic process control agents added to prevent excessive cold welding.

Three new aluminum alloys with small amounts of magnesium and other elemental additions, developed by INCO, are now commercially available. The Incomap® AL-9052 (INCO Mechanically Alloyed Products), an alloy with 4Mg and 0.8 to 1.1C (wt.%), is a high-strength corrosion-resistant alloy with possible marine applications. The combination of excellent strength, corrosion resistance, and stress corrosion cracking resistance of this MA material makes it a preferred material over the popular high-strength 6061-T6 aluminum alloy. The mechanically alloyed material is approximately 45% stronger, has a 10% higher elastic modulus, 40% higher toughness, and slightly improved corrosion resistance. Incomap® AL-905XL, with 4Mg, 1.3Li and 1.2C (wt.%), is a low-density alloy that looks promising as a forged aerospace material. The Incomap® alloy AL-905XL, due to its low density, high strength, and stiffness, is designed to replace the alloy 7075-T73 in aerospace forgings [56]. The mechanically alloyed material is 8% lighter, 13% stiffer, and exhibits a hundredfold improvement in corrosion resistance, while the strength, ductility, and toughness of the two are comparable. In addition, the MA material can be used in the as-worked condition and is nonaging in nature [57]. The third alloy, Incomap® AL-9021 which has 1.5Mg, 4Cu and 0.8 to 1.1C (wt.%), is being considered as a highly rigid matrix for composite materials, such as SiC particulate-reinforced metal matrix composite.

Another group of investigators working on mechanical alloying of the aluminum–lithium–magnesium system used vacuum-dehydrogenized powders of $AlLiH_4$, and elemental magnesium and aluminum powders [58,59]. The difficulties in mechanical alloying of the desired constituents relate to the difference in the properties of the three in their pure elemental form. The aluminum powder is soft and ductile, while the magnesium is comparatively brittle. Lithium is very soft and has to be kept under protective conditions (such as under paraffin oil). The material also has a high rate of sublimation at moderately high temperatures. Thus, a master alloy mixed with elemental powders was used as the initial starting material. The master alloy powder was produced by grinding a mixture of 44.4 wt.% of coarse magnesium with 55.6 wt.% of fine AlLi powder. To this, deoxidized aluminum powder was added at the rate of 10.1 g per gram of the master alloy powder, and milled. The powders were vibratory ball-milled in a vacuum atmosphere followed by vacuum hot pressing, solution treatment in a salt bath, and aging in an oil bath.

With the magnesium content (2.2 wt.%) remaining the same, the variation in the lithium content from 1 to 3 wt.%, resulted in increased hardness, yield and fracture stress, and decreased ductility and energy applied until fracture. The formation of the coherent Al_3Li precipitates as a result of aging is responsible for the improvement in the properties of this material with increasing lithium contents. The alloy density decreased from 2.65 g/cm^3 for the Al–1Li–2.2Mg (wt.%) alloy to 2.58 g/cm^3 for Al–3Li–2.2Mg (wt.%). With the lithium content being constant at 2 wt.%, increasing the magnesium content from 2.2 to 4.4 wt.% produced some improvements in yield and fracture strength, and practically no effect on the ductility and elastic modulus. The improvements were, however, not as pronounced as that obtained by increasing the lithium content. With magnesium, the precipitate after aging is transformed to Al_2MgLi, which is not coherent with the matrix and does not contribute to the increase in hardness.

Another interesting dispersion-strengthened aluminum alloy, known as Dispal, was developed by Sintermetallwerk Krebsoge of Germany. This process utilizes reaction milling without the use of any process control agents that are normally used in mechanical alloying of aluminum alloys. The material is already in use in piston preforms and high-temperature electrical conductors and interferometers.

A mechanically alloyed Al–6Ti (wt.%) material reported by Mirchandani and coworkers [60] has shown an excellent combination of elevated-temperature strength, creep resistance, thermal stability, stiffness, and corrosion resistance when compared to the conventional series of 2xxx and 7xxx aluminum aerospace alloys produced by ingot metallurgy. In the development work, aluminum and titanium powders with a process control agent were mechanically alloyed to produce a fine dispersion of titanium within the aluminum along with alumina and Al_4C_3 dispersoids. The mechanically alloyed powder was consolidated by high-temperature vacuum degassing to remove impurities and react the fine dispersion of titanium to form an Al_3Ti dispersion in aluminum, and extrusion. The final structure consists of mixed ultrafine grains of aluminum and Al_3Ti (0.15 to 0.25 μm) with 0.03 to 0.04 μm (30 to 40 nm) dispersoids of Al_2O_3 and Al_4C_3 mainly present along the grain boundaries. This new mechanically alloyed material is expected to find applications in the skin, wheels and structure of advanced aircraft. Production scaleup for commercial use of this material is in progress.

Similar mechanically alloyed work on Al–Ti systems with Nb additions have been discussed by another research group [61]. This investigation also reports on the addition of yttria and alumina particles as dispersoids. Some of the compositions used are as follows: Al, 8.6 Ti, 1.9C, $0.8O_2$ (wt.%); Al, 6 Ti, 1.4C, $0.9O_2$, $0.9Y_2O_3$ (wt.%); Al, 6 Ti, 1.4C, $0.8O_2$, $5.0Y_2O_3$ (wt.%); Al, 5 Ti, 5Nb, 1.4C, $0.8O_2$ (wt.%). Depending on the volume fraction of the total dispersoids, the experimental alloys had good creep properties in the temperature range 673 to 803 K (400 to 530°C).

Mechanical alloying of titanium-based alloys has also been investigated. Sundaresan and Froes have reported the mechanical alloying of titanium with magnesium, a Ti–Ni–Cu alloy, Ti–1Al–8V–5Fe–1Er (wt.%) alloy, and Ti–24V–10Cr–5Er (wt.%) alloy [62]. The Ti–Mg powders showed considerable variation from run to run. The normally immiscible titanium and magnesium have been alloyed together. The resulting solid solution alloy has a density only 10% greater than that of aluminum [14]. It has been possible to retain up to 3.1 wt.% (6.6 at.%) magnesium in solid solution. However, fully dense material has not been achieved due to the high vapor pressure of magnesium. In the Ti–18Ni–15Cu system, formation of an amorphous phase has been detected. Mechanical alloying of atomized Ti–1Al–8V–5Fe–1Er (wt.%) powders led to the formation of 0.03 to 0.05 μm (30 to 50 nm) dispersion in the consolidated alloy. A Ti–24V–10Cr–5Er (wt.%) alloy in the as-atomized state was segregated with large chunks of erbium present in the microstructure. This material after 1728×10^2 s (48 hours) of mechanical alloying consisted of very fine and coarse regions. The fine-grain region had dispersions with an average size of 0.01 μm (10 nm). Further optimization is required for consolidation of the material to full density.

Mechanically alloyed magnesium-based materials have been subjected to some of the most unusual applications. An interesting use of mechanical alloying was in the processing of supercorroding magnesium-based alloys. The

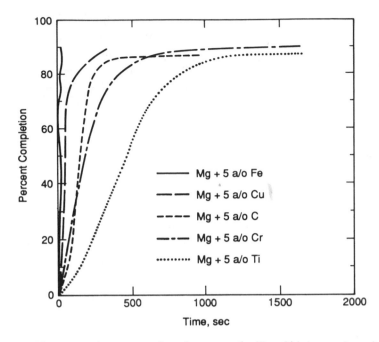

Figure 4.16: *The corrosion rates of various metals (5 at.%) in conjunction with magnesium. (Redrawn after Ref. 63.)*

supercorroding materials produce short-circuited galvanic cells that react rapidly when placed in an electrolyte such as sea water. The rapid corrosion (reaction) produces heat and evolves hydrogen gas [63], which can be used as a heat source in deep-sea diving, a gas generator for buoyancy control, in fuel cells and fuel in hydrogen-based engines. The corrosion rates of various metals (5 at.%) in conjunction with magnesium (mechanically alloyed) are shown in Figure 4.16.

To maximize the desired characteristics of the supercorroding materials, a number of conditions need to be satisfied. Thus, it is desirable to maximize the cathode and anode surface area that is exposed, provide a short electrolyte path between the two, have a strong bond between the cathode and the anode, and minimize resistance to the flow of external currents through the corroding pair. Mechanical alloying of magnesium with more noble metals such as iron, copper, carbon, titanium, etc., accounts for all of the above-mentioned desirable properties. Since various combinations of magnesium with other noble metals corrode at different rates, they can be used in different applications. Rapid corrosion rates are necessary when generating hydrogen gas or underwater heat. For those applications it would be best to consider iron or copper mechanically alloyed with magnesium. However, for applications

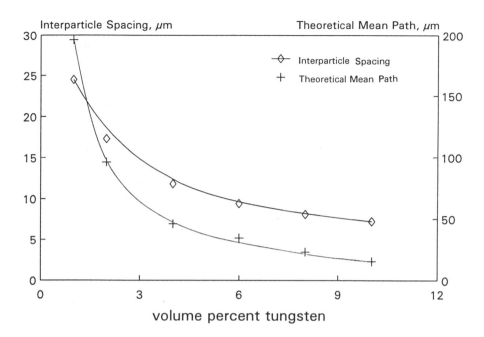

Figure 4.17: *The variation of the theoretical mean path and the interparticle spacing with different volume fractions of tungsten. (Data from Ref. 67.)*

like release of deep-sea equipment from the ocean bed, it would be beneficial to use a slowly corroding link. This type of link can be produced using mechanically alloyed magnesium with titanium.

4. Copper-based Materials

Elevated-temperature applications of copper-based materials are quite frequent, and include microwave and X-ray parts, electrical switches, electrical contact materials, welding electrodes, etc. Wrought copper-based alloys based on chromium and zirconium rely on the precipitates formed by thermomechanical treatment to provide the elevated-temperature properties. Oxide dispersion strengthening of copper-based materials has also proved to be extremely useful for elevated-temperature applications. The introduction of oxide dispersion in a copper-based alloy matrix through mechanical alloying was a natural evolution in the search for improved copper-based materials suitable for elevated-temperature use. Copper with 2 vol.% alumina and copper with 4 vol.% zirconia have been produced by MA [64]. Milling of the copper powder with the oxides was carried out in a Szegvari Model 1-S attritor mill with 4.8-mm-diameter steel balls. Milling for 252×10^2 s (7 hours) produced a uniform distribution of the fine oxide dispersion in the copper matrix. Around 0.05 μm (50 nm) alumina and around 0.03 μm (30 nm) zirconia particle size dispersion was attained by MA. The materials were hot-swaged to form bars. The grain sizes were found to vary from 0.2 to 0.5 μm, with the oxide particles uniformly dispersed in the matrix as well as in the grain boundaries. Usually the dispersion consisted of single particles, but a few agglomerated particles were also observed. The agglomerated particles served as pinning agent and were usually found at grain boundaries or grain triple junction points. The 0.02 to 0.05 μm (20 to 50 nm) particles stabilized the fine grain size against grain growth even when the material was repeatedly heated to 973 K (700°C).

Investigations have also been carried out on mechanical alloying of Cu–Cr and Cu–Mo systems. Copper with molybdenum (Cu–10Mo), after 72×10^2 s (2 hours) of milling, produced large composite powder particles that are coarser than the original starting powder particles. The composite particles showed a fine lamellar structure that continued to be refined with additional milling time. Milling for 648×10^2 s (18 hours) produced powders in which the molybdenum was uniformly distributed as submicrometer-sized globules in the structure.

An interesting high-current-capacity copper–ruthenium electrical contact material has been described by Green et al. [65]. Mechanical alloying of insoluble copper and ruthenium powders produced a uniform dispersion of the hard and refractory ruthenium particles (1–2 μm) in the copper matrix. The final consolidated material, which is obtained by cold pressing and warm rolling of the mechanically alloyed powder, is given an etching treatment to remove the surface layer of copper. The resultant structure consists of protruding hard, conductive ruthenium particles that do not wear or erode quickly, embedded in a highly conductive copper matrix. This produces a unique high-current-capacity electrical contact material.

Mechanically alloyed materials are also being considered for use in automotive applications like bearings [43] and in electrical contact materials [66,67]. Copper with 1 to 10 vol.% of tungsten was produced by mechanical alloying the materials in an attritor mill. The theoretical mean path and the interparticle spacing are calculated from the two formulas given below [68]:

$$\text{Theoretical mean path} = \frac{2d(1-f)}{3f}$$

$$\text{Interparticle spacing} = \left(\frac{2d^2(1-f)}{3f}\right)^{1/2}$$

where d is the mean tungsten particle size (μm) and f is the volume fraction of tungsten. The variation of the theoretical mean path and the interparticle spacing with different volume fractions of tungsten is shown in Figure 4.17.

The mechanically alloyed powders are compacted in a floating die at 300 MPa and then sintered at 1273 K (1000°C) in a dry hydrogen atmosphere for 144×10^2 s (4 hours). The resultant sintered density of the composites is less than 80% of theoretical. This sintered material is then cold extruded and the resultant sintered density is increased to 97.5% of theoretical. Subsequent annealing of the material results in a rapid decrease in the hardness due to recrystallization. The high hardness, good electrical conductivity, and high abrasion resistance of the mechanically alloyed Cu–W composites makes them ideal candidates for low-temperature high-wear electrical applications.

A number of other materials that have not been discussed above have also been processed by mechanical alloying. It would be impossible to cover the details of all these material systems; this section was meant to provide readers with a brief glimpse of the common areas where mechanically alloyed materials have been used successfully.

A large number of intermetallic compounds have been produced by mechanical alloying. Intermetallic compounds such as Nb_3Ge, Nb_5Ge_2, $NbGe_2$, Nb_3Sn, Nb_6Sn_5, Ni_3Al, Ni_2Al_3, Al_3Ni, Ti_3Al, $TiAl$, Al_3Ti, Nb_3Al, and some silicides have been produced by mechanical alloying of elemental powder constituents. Continued milling in a number of cases produces an amorphous material. Intermetallic compound formation can be the direct result of mechanical alloying. In many instances, however, mechanical alloying can produce an intimate mixture of the two elemental constituents, which on further heating can react to form the intermetallic phase. Detailed discussion of this class of materials has been covered by a number of authors and has also been covered in the earlier part of this chapter. Mechanical alloying of cobalt-based materials has not been included in this section due to its similarity with the applications of the nickel-based composites. Similarly, investigations on mechanical alloying of refractory metals such as tungsten, niobium, etc., have not been discussed as their commercial use is still extremely limited. For more material on this area, the readers are referred to the paper on MA of reactive and refractory metals by Suryanarayana et al. [69]. This section does not in any way cover all the materials that have been produced by mechanical alloying, but generalizes the discussions into broad categories of materials that have been commercially exploited or are on the verge of commercial exploitation. It is expected that other classes of mechanically alloyed materials will gradually come into the limelight and will become commercial products in the near future.

The Future

A wide spectrum of possibilities can be realized by utilizing mechanical alloying for producing powders that are difficult or almost impossible to produce by other conventional means. The capability of producing an extremely fine uniform dispersion of one material in another is certainly not attainable by conventional ingot metallurgy processes.

The process of mechanical alloying, which is a nonequilibrium process, has nearly all the attributes of rapid solidification methods. Mechanical alloying can produce materials with far superior homogeneity and extremely fine grain sizes; extend the solid solubility of one material in another; form intermetallic compounds; produce amorphous or glassy materials; exhibit a continuous and wide homogeneity range of the amorphous phase centered near the equimolar composition compared to the narrow ranges of composition usually divided into areas centered near the deep eutectics in the binary phase diagram; and produce nanocrystalline structures and mixtures of crystalline and amorphous phases.

This wide range of possibilities exposes new horizons in materials design. A large number of unique applications based on mechanically alloyed materials have been successfully developed. Supercorroding materials used to produce hydrogen gas and controlled-release links for marine applications, copper–ruthenium electrical contact materials that can withstand high currents and have very high erosion resistance, and the new amorphous Hf–Al alloy that has an extremely high crystallization temperature are only a few of the unique areas where mechanically alloyed materials are being utilized. The recent investigation on displacement reactions by mechanical alloying also creates new options for producing oxide dispersion-strengthened powders.

Mechanical alloying is not only being used for producing unique materials but is also encroaching into areas that were the domain of conventional metallurgical processes. Some examples where mechanically alloyed materials are replacing the conventional cast and wrought alloys are the oxide dispersion-strengthened superalloys and the new Al–6Ti aerospace material. In the future, it is expected that mechanical alloying will increasingly be used to produce unique materials with a large variety of property requirements that are not attainable by conventional processes. The volume of mechanically alloyed materials that will be used in more conventional applications like gas turbines, bearings, and aerospace structures will also continue to grow at the expense of other conventional processing techniques. The relative ease of producing mechanically alloyed materials in large quantities compared to the other nonconventional material production techniques such as rapid solidification, plasma spraying, etc., will be the key driving force in propelling this process into the next century. The other competing process that has similarly diverse capability as mechanical alloying is the chemical precursor route for producing alloy powders. It is visualized that these two processes will compete with each other in the area of powder precursors for producing advanced materials for the next century.

References

1. J.S. Benjamin, *Metallurgical Transactions*, vol. 1, p.2943, 1970.
2. W.E. Kuhn, I.L. Friedman, W. Summers, and A. Szegvari, *Metals Handbook*, vol. 7, *Powder Metallurgy*, ASM, Metals Park, OH, p. 56, 1985.
3. P.S. Gilman and J.S. Benjamin, *Annual Reviews in Material Science*, vol. 13, p.279, 1983.
4. M.S. Kim and C.C. Koch, *Journal of Applied Physics*, vol. 62, p.3450, 1987.
5. M.L. Ovecoglu and W.D. Nix, *International Journal of Powder Metallurgy*, vol. 22, p.17, 1986.
6. P.S. Gilman and W.D. Nix, *Metallurgical Transactions*, vol. 12A, p.813, 1981.
7. J.S.C. Jang and C.C. Koch, *Scripta Metallurgia*, vol. 22, p.677, 1988.
8. D.R. Maurice and T.H. Courtney, *Metallurgical Transactions*, vol. 21A, p.289, 1990.
9. R.M. Davis, B.T. McDermott, and C.C. Koch, *Metallurgical Transactions*, vol. 19A, p.2867, 1988.
10. B.T. McDermott, MS Thesis, North Carolina State University, Raleigh, NC, 1988.
11. A.S. Lazarev, A.A. Kolesnikov, and V.A. Korol, *Soviet Powder Metallurgy and Metal Ceramics*, vol. 25, p.613, 1986.
12. C.C. Koch and M.S. Kim, *Journal de Physique*, vol. 46, p.C8–573, 1985.
13. A.E. Ermakov, E.E. Yurchikov, and V.A. Barinov, *The Physics of Metals and Metallography*, vol. 52, p.50, 1981.
14. F.H. Froes, *Journal of Metals*, vol. 41, p.25, 1989.
15. J.S. Benjamin, *Scientific American*, vol. 234, p.40, 1976.
16. R.B. Schwarz, R.R. Petrich, and C.K. Saw, *Journal of Non-Crystalline Solids*, vol. 76, p.281, 1985.
17. R.M. Davis and C.C. Koch, *Scripta Metallurgia*, vol. 21, p.305, 1987.
18. C.C. Koch, *Annual Reviews in Material Science*, vol. 19, p.121, 1989.
19. B.T. McDermott and C.C. Koch, *Scripta Metallurgia*, vol. 20, p.669, 1986.
20. R.C. Benn, P.K. Mirchandani, and A.S. Watwe, *Modern Developments in Powder Metallurgy*, compiled by P.U. Gummeson and D.A. Gustafson, Metal Powder Industries Federation, Princeton, NJ, vol. 21, p.479, 1988.
21. J. Kajuch and K. Vedula, *Advances in Powder Metallurgy*, compiled by E.R. Andreotti and P.J. McGeehan, Metal Powder Industries Federation, Princeton, NJ, vol. 2, p.187, 1990.
22. M. Atzmon, *Solid State Powder Processing*, The Minerals, Metals, and Materials Society, Warrendale, PA, 1990.
23. C.C. Koch, O.B. Calvin, C.G. McKamey, and J.O. Scarbrough, *Applied Physics Letters*, vol. 43, p.1017, 1983.
24. R.B. Schwarz and W.L. Johnson, *Physical Review Letters*, vol. 51, p.415, 1983.
25. R.B. Schwarz, *Materials Research Bulletin*, vol. 11, p.55, 1986.

26. R.B. Schwarz, J.W. Hannigan, H. Shienberg, and T. Tiainen, *Modern Developments in Powder Metallurgy*, compiled by P.U. Gummeson and D.A. Gustafson, Metal Powder Industries Federation, Princeton, NJ, vol. 21, p.415, 1988.
27. E.A. Kenik, R.J. Bayuzick, M.S. Kim, and C.C. Koch, *Scripta Metallurgia*, vol. 21, p.1137, 1987.
28. J.S.C. Jang and C.C. Koch, Unpublished research results (cross-referenced from reference 18).
29. R. Watanabe, H. Hashimoto, and Y.-H. Park, *Advances in Powder Metallurgy*, compiled by L.F. Pease III and R.J. Sansoucy, Metal Powder Industries Federation, Princeton, NJ, vol. 6, p.119, 1991.
30. E. Hellstern and L. Schultz, *Philosophical Magazine B*, vol. 56, p.443, 1987.
31. P.Y. Lee, J. Jang, and C.C. Koch, *Journal of Less-Common Metals*, vol. 140, p.73, 1988.
32. E. Hellstern and L. Schultz, *Journal of Applied Physics*, vol. 63, p.1408, 1988.
33. F. Petzoldt, *Journal of Less-Common Metals*, vol. 140, p.85, 1988.
34. L. Schultz, *Materials Science and Engineering*, vol. 97, p.15, 1988.
35. A. Calka, A.P. Pogany, R.A. Shanks, and H. Engelman, *Materials Science and Engineering*, vol. A128, p.107, 1990.
36. H.J. Fecht, E. Hellstern, Z. Fu, and W.L. Johnson, *Metallurgical Transactions*, vol. 21A, p.2333, 1990.
37. W. Shlump and H. Grewe, *New Materials by Mechanical Alloying Techniques*, ed. E. Arzt and L. Schultz, Deutsche Gesellschaft für Metallkunde, Oberursel, Germany, p.307, 1989.
38. P.H. Shingu, B. Huang, J. Kuyama, K.N. Ishihara, and S. Nasu, *New Materials by Mechanical Alloying Techniques*, ed. E. Arzt and L. Schultz, Deutsche Gesellschaft für Metallkunde, Oberursel, Germany, p.319, 1989.
39. C. Suryanarayana and F.H. Froes, *Journal of Materials Research*, vol. 5, p.1880, 1990.
40. H.J. Fecht, G. Han, Z. Fu, and W.L. Johnson, *Journal of Applied Physics*, vol. 67, p.1744, 1990.
41. J.S. Benjaminn, *Modern Developments in Powder Metallurgy*, compiled by P.U. Gummeson and D.A. Gustafson, Metal Powder Industries Federation, Princeton, NJ, vol. 21, p.397, 1988.
42. A.N. Patel and S. Diamond, *Material Science and Engineering Proceedings*, RQ6, Montreal, vol.98, p.329, 1987.
43. S. Diamond and A.N. Patel, *Modern Developments in Powder Metallurgy*, compiled by P.U. Gummeson and D.A. Gustafson, Metal Powder Industries Federation, Princeton, NJ, vol. 21, p.445, 1988.
44. G.B. Schaffer and P.G. McCormick, *Metallurgical Transactions*, vol. 21A, p.2789, 1990.
45. J.J. Fischer and R.M. Haeberle, *Modern Developments in Powder Metallurgy*, compiled by P.U. Gummeson and D.A. Gustafson, Metal Powder Industries Federation, Princeton, NJ, vol. 21, p.461, 1988.
46. R.J. Lauf and C.A. Wells, *ECUT Quarterly Report*, Oak Ridge National Laboratory, July 1 to September 30, 1985.
47. C.C. Koch, *High Temperature Ordered Intermetallic Alloys II*, ed. N.S. Stoloff, C.C. Koch, C.T. Liu, and O. Izumi, Materials Research Society Symposium, Pittsburgh, PA, vol. 81, p.369, 1987.

48. A.U. Seybolt, *Transactions of ASM*, vol. 59, p.860, 1966.
49. K. Vedula, G.M. Michal, and A.M. Figueredo, *Modern Developments in Powder Metallurgy*, compiled by P.U. Gummeson and D.A. Gustafson, Metal Powder Industries Federation, Princeton, NJ, vol. 20, p.491, 1988.
50. J.S.C. Wang, S.G. Donnelly, P. Godavarti, and C.C. Koch, *International Journal of Powder Metallurgy*, vol. 24, p.315, 1988.
51. L. Schultz, J. Wecker, and E. Hellstern, *Journal of Applied Physics*, vol. 61, p.3583, 1987.
52. C. Veltl, B. Scholz, and H.-D. Kunze, *Modern Developments in Powder Metallurgy*, compiled by P.U. Gummeson and D.A. Gustafson, Metal Powder Industries Federation, Princeton, NJ, vol. 20, p.707, 1988.
53. E. Kohler, C. Gutsfeld, and F. Thummler, *Powder Metallurgy International*, vol. 22, No.3, p.11, 1990.
54. B. Kazimierzak, M. Prigon, and D. Coutsouradis, *Metal Powder Report*, vol. 45, p.699, 1990.
55. S.D. Strothers and K. Vedula, *Progress in Powder Metallurgy*, compiled by C.L. Freeby and H. Hjort, vol. 43, p.597, 1987.
56. R.D. Schelleng, J.H. Weber, and G.A.J. Hack, *Metal Powder Report*, vol. 45, No.10, p.709, 1990.
57. D.O. Gothard, *Modern Developments in Powder Metallurgy*, compiled by P.U. Gummerson and D.A. Gustafson, Metal Powder Industries Federation, Princeton, NJ, vol. 21, p.511, 1988.
58. A. Layyous, S.N. Nadiv, and I.J. Lin, *Powder Metallurgy International*, vol. 19, p.11, 1987.
59. A.A. Layyous, S. Nadiv, and I.J. Lin, *Modern Developments in Powder Metallurgy*, compiled by P.U. Gummerson and D.A. Gustafson, Metal Powder Industries Federation, Princeton, NJ, vol. 20, p.691, 1988.
60. P.K. Mirchandani, D.O. Gothard, and A.I. Kemppinen, *Advances in Powder Metallurgy*, compiled by T.G. Gasbarre and W.F. Jandeska, Jr., Metal Powder Industries Federation, Princeton, NJ, vol. 3, p.161, 1989.
61. J.A. Hawk, J.K. Briggs, and H.G.F. Wilsdorf, *Advances in Powder Metallurgy*, compiled by T.G. Gasbarre and W.F. Jandeska, Jr., Metal Powder Industries Federation, Princeton, NJ, vol. 3, p.285, 1989.
62. R. Sundaresan and F.H. Froes, *Modern Developments in Powder Metallurgy*, compiled by P.U. Gummeson and D.A. Gustafson, Metal Powder Industries Federation, Princeton, NJ, vol. 21, p.429, 1988.
63. S.S. Sergev, S.A. Black, and J.F. Jenkins, U.S. Patent 4,264,362, 1981.
64. J.G. Schroth and V. Franetovic, *Journal of Metals*, vol. 41, p.37, 1989.
65. M.L. Green, E. Coleman, F.E. Bader, and E.S. Sproles, *Materials Science and Engineering*, vol. 62, p.231, 1984.
66. J. Kaczmar, *Powder Metallurgy*, vol. 32, p.171, 1989.
67. B.L. Mordike, J. Kaczmar, M. Kielbinski, and K.U. Kainer, *Powder Metallurgy International*, vol. 23, no.2, p.91, 1991.
68. B.I. Edelson and W.M. Baldwin, *Transactions of ASM*, vol. 55, p.230, 1962.
69. C. Suryanarayana, R. Sundaresan, and F.H. Froes, *Advances in Powder Metallurgy*, compiled by T.G. Gasbarre, and W.F. Jandeska, Jr., Metal Powder Industries Federation, Princeton, NJ, vol. 3, p.175, 1989.

Chapter 5

Intermetallic Compounds

Introduction

Intermetallic compounds have emerged as the next generation of oxidation-resistant materials for high-temperature applications. With ceramics not living up to expectations as possible replacements for superalloys, attention has focused more and more on intermetallic compounds [1,2]. Some of the intermetallic compounds, especially those based on aluminum and silicon, have the attractive combinations of low density, high strength, good corrosion and oxidation resistance. They usually do not contain strategic elements and are relatively inexpensive. Some of the intermetallic compounds exhibit improved strength with increasing temperature. These attributes, coupled with the high melting point of a number of intermetallics, make these materials an ideal choice for high-temperature applications.

In spite of the attractive properties, intermetallic compounds have had limited applications due to their extreme brittleness. The recent surge of interest in the intermetallics was precipitated after Aoki and Izumi [3] in the late 1970s observed that the addition of trace amounts of boron to Ni_3Al increased the room-temperature ductility from 0 to about 50%.

Powder metallurgy is emerging as one approach ideal for the fabrication of complex-shaped, high-performance intermetallic compounds [4,5]. Indeed, all of the obvious powder metallurgy approaches are under exploration, including hot isostatic compaction, hot pressing, powder extrusion, reactive sintering, and powder injection molding.

This chapter will briefly describe some of the unique characteristics of these materials, processing techniques, their potential applications, and the scope for future research on this evolving material system. The majority of the chapter will focus around the most widely studied group of intermetallics, namely the aluminides. Other emerging intermetallic compounds such as silicides, beryllides, shape memory alloys like TiNi, and a few exotic high-temperature intermetallic compounds will also be briefly discussed.

In order to conceive the unique characteristics of these compounds, one must have an understanding of the differences between the intermetallics and the other metallic materials such as pure metals and alloys, normally having disordered solid solutions. The long-range order exhibited by these intermetallic compounds sets them apart from other alloys, and is also primarily responsible for some of the unique properties of intermetallic compounds.

What Are Intermetallic Compounds?

Intermetallic compounds are ordered alloys, normally consisting of two elements. They are characterized by the long-range order that they exhibit, which is caused by the stronger bonding between the unlike atoms when compared to the like atoms in the system.

The best way to get a clear understanding of the long-range ordering phenomenon is to consider the phase diagram of the classic Ni–Al system, which is shown in Figure 5.1 [6,7]. At the extreme right-hand side of the phase diagram is pure nickel, which has a face centered cubic (f.c.c.) structure. With the addition of small amounts of aluminum, the aluminum atoms initially substitute the f.c.c. nickel atoms in a random manner. This is known as disordered solid solution, and it forms the basis for the majority of the alloy systems presently in use. As one continues to add aluminum atoms to the system, after a certain point the randomly disordered suubstitution of the aluminum in the nickel lattice changes and a new ordered structure develops. In case of the nickel–aluminum system, the maximum solubility of aluminum in disordered nickel solid solution is around 20 at.%. Further addition of aluminum causes the formation of a long-range-ordered compound of the $L1_2$ type where the aluminum atoms take up positions at the corners of the cube and the nickel atoms are situated at the cube faces. The $L1_2$ type of structure is shown in Figure 5.2a, and is representative of an ordered f.c.c. structure. For the perfectly ordered $L1_2$ structure in the nickel–aluminum system, the atomic ratio is 25% aluminum and 75% nickel. With aluminum contents close to 0.5 atom fraction, a second form of ordered nickel–aluminum intermetallic compound is formed. This form of ordered structure is classified as the B2 type and is shown in Figure 5.2b. Here the aluminum atoms position themselves at the center of the cube and the nickel atoms at the cube corners. The ideal atomic ratio for this compound is 1 : 1 and it is often described as an ordered body centered cubic (b.c.c.) structure. With still further aluminum additions, other intermetallic compounds with different structures, such the Ni_2Al_3 and

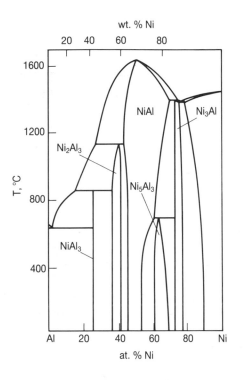

Figure 5.1: Binary nickel–aluminum phase diagram. (After Refs. 6 and 7 and with permission from Ref. 37.)

NiAl$_3$, are formed. In the binary Ni–Al system, the compounds Ni$_3$Al and NiAl have been studied extensively, and their crystal structures are reproduced in Figures 5.2a and 5.2b.

It is not unusual to find intermetallic compounds that exhibit small or significant deviations from the ideal stoichiometric ratio. Examples of both are seen in the nickel aluminum phase diagram shown in Figure 5.1. The L1$_2$ ordered compound Ni$_3$Al has a phase field that extends over a few atomic percent around the ideal 75Ni : 25Al atomic ratio. The B2 ordered NiAl compound, however, shows a significant extension in the phase field around the ideal 50Ni : 50Al atomic ratio. In contrast, the compound NiAl$_3$ is almost a line compound. The deviations in the stoichiometry are usually accommodated by three possible arrangements (for excess of nickel atoms): some of the nickel atoms sit on aluminum sites; some of the aluminum sites remain vacant; and some of the nickel atoms take up interstitial positions. However, combinations of these arrangements are also possible and often do exist. It is expected that the intermetallic compounds that have extended phase fields are those that

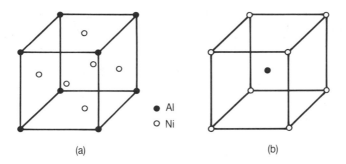

Figure 5.2: Crystal structures of two intermetallic compounds based on nickel and aluminum: (a) The $L1_2$ compound Ni_3Al; (b) The B2 compound NiAl.

could have potential commercial applications. The "line" compounds, which have practically no solubility for their constituent elements beyond the fixed atomic ratio, would be extremely difficult to fabricate due to their fixed chemistry.

In some systems, the long-range-ordered intermetallic compound is actually present below a certain temperature region, above which the material is a disordered solid solution. This occurs if the ordering energy is low, as in case of the $L1_2$ intermetallic compound Ni_3Fe. The ordered compounds in these systems are usually obtained by holding for very long times at temperatures slightly below the ordering temperature. Materials such as Ni_3Fe offer the possibility of studying the effect of ordering on the properties since the same stoichiometry can exist in both the ordered and the disordered state. However, when the ordering energy is very high, the ordered structure could persist all the way to the melting point of the compound. Compounds such as Ni_3Al, NiAl, and CoAl are examples where the material remains ordered until its melting point is reached.

The majority of binary intermetallic compounds of elements A and B can be classified into different stoichiometric compositions such as A_3B, A_2B, A_2B_3, A_5B_3, A_7B_6, and AB. A number of such intermetallic compounds are present in the binary nickel–aluminum system, and they are shown in Figure 5.1. Within each stoichiometric group, intermetallic compounds with different crystal structures can occur [8]. The ordered structure of the intermetallic compounds is responsible for the majority of the unusual properties exhibited by this class of materials through the changes in the nucleation and motion of dislocations resulting from the constraints imposed due to the ordered structures.

The relationship between dislocations and ordering can be observed from the schematic diagram in Figure 5.3 [9]. It is apparent that there are two distinct types of antiphase domains possible, such that one type of domain may be entirely surrounded by the domain of the other type (see the dashed square at the middle of the picture). A unit dislocation moving in a superlattice structure will result in a disorder in the form of antiphase boundary (APB). The APB can terminate on the surface of the crystal or at a grain boundary (lower part of Figure 5.3). In order to have no net change in order behind the dislocations (thereby eliminating the additional APB energy), it becomes necessary for the dislocations to move in pairs. A schematic of a pair of dislocations of like sign connected by a strip of APB (termed a superlattice dislocation) in a simple cubic lattice is shown at the top of Figure 5.3 [9]. The two dislocations exert a mutual repulsive force upon each other that is balanced by the surface tension of the APB. The formation of superlattice dislocations (which consist of two ordinary dislocations), their motion, and their interactions with one another as well as with grain boundaries, precipitates, and grown-in APBs controls the mechanical properties of the intermetallic compounds. In this way, the long-range-ordered nature of the intermetallic compounds distinguishes them from other classes of metallic alloys, and is also responsible for some of the characteristic properties exhibited by these compounds.

Importance of Intermetallic Compounds

Intermetallic compounds offer the potential of meeting the challenge posed by the increased demands for advanced high-temperature structural components. This class of material could be the answer to the lightweight, high-stiffness, high-strength, good toughness, and high-temperature oxidation, corrosion, and creep resistance requirements of the aerospace industry. The unique properties of this class of materials might also make it an attractive candidate for other areas of application.

The Integrated High Performance Turbine Engine Technology Initiative (IHPTET), the turbine engine industry, and the recent initiative in the development of manned hypersonic vehicles represent the major driving forces for the development of intermetallic compounds in the United States. Extensive research and development efforts in the area of intermetallic compounds are also taking place in Japan, Europe, and some of the Asian countries. In the United States, the ambitious goal of the IHPTET initiative is to double turbine propulsion capability by the turn of the century [10]. The idea of hypersonic manned vehicles, which has recently been brought to the forefront by the National Aerospace Plane Program (NASP), is designed to attain velocities greater than Mach 5. The goal of the NASP program is to build hypersonic vehicles capable of taking off and achieving earth orbit with a

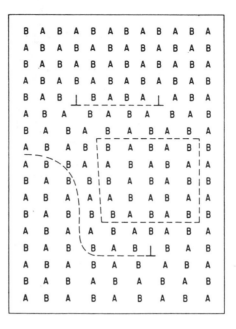

Figure 5.3: Schematic diagram of superlattice dislocation in a simple cubic lattice. (Reprinted from Ref. 9.)

single-stage air-breathing propulsion system. Intermetallic compounds such as titanium aluminides and their composites are expected to play a crucial role in meeting the challenges and the new requirements posed by these developments [11].

Demands for reliable high-temperature components are by no means confined to the aerospace industry. For land-based turbines, advanced fossil fuel conversion systems, nuclear plants, heating elements, and even in the automotive industry, the need for new reliable hot components is on the rise. Intermetallic compounds are increasingly being tried in a number of the applications mentioned.

A number of features of intermetallic compounds makes them almost a "natural" for high-temperature applications. The extensively studied, reasonably high-temperature (melting point around 1660 K; 1387°C) intermetallic compound Ni_3Al (the principal strengthening phase in high-temperature nickel-based superalloys) has already proved to be a very attractive material for the intermediate temperature range applications. The other encouraging aspect is the fact that there are a number of other intermetallic compounds that have much higher melting points than Ni_3Al, and whose properties have not yet been established. According to Fleischer et al. [12], there exist more than 300 binary intermetallic compounds with melting points greater than 1773 K (1500°C). High melting temperatures usually imply higher strength, elastic modulus, and creep properties. Also the strong bonding between the unlike neighboring atoms responsible for the ordering, results in higher elastic moduli and thus high strengths when compared to the individual constituents. For example, the intermetallic compound Ti_3Sn has a Young's modulus of 190 GPa compared to 115 GPa for titanium and 48 GPa for tin. Many intermetallic compounds, therefore, have the combination of high stiffness, low specific gravity, and high melting point. Also, due to their ordered nature, the intermetallic compounds have much lower self-diffusion coefficients compared to disordered alloys. This generally improves the room- and elevated-temperature creep properties and also the microstructural stability of intermetallic compounds.

A number of intermetallic compounds, including Ni_3Al and a number of other $L1_2$ compounds, some B2, $L1_0$, and $D0_3$ type of compounds, exhibit an anomalous strengthening behavior with increasing temperature. With increasing temperature, the strength of nearly all known alloys starts to decrease, whereas the strength of some intermetallic compounds actually increases up to a certain maximum temperature. For Ni_3Al-based compounds, the maximum strength is reached at approximately half the melting point before decreasing. Half the melting point is considered to be the maximum use temperature for the majority of high temperature single-phase materials. Thus, even the remote possibility that intermetallic compounds with higher melting points could also possess the anomalous strength behavior is a tremendous source of encouragement. The variation in the compressive flow stress with temperature for a number of intermetallic compounds that show an increase in strength with temperature is shown in Figure 5.4 [13]. Other intermetallic compounds tend to lose their strengths at higher temperatures at a much slower rate than the general metals and alloys. As an example, the compound TiAl has a strength of 450 MPa up to approximately 873 K (600°C) before this begins to drop with increasing temperature.

Nearly all high-temperature applications require environmental protection. The materials should not degrade in the operating environment. Thus, high-temperature oxidation is a major concern in high-temperature structural components. In certain applications such as advanced fossil fuel conversion systems, both high-temperature oxidation and sulfidation resistance are important. Intermetallic compounds such as aluminides and silicides have the ability to quickly form an adherent, protective alumina or silica layer that offers excellent protection against high-temperature oxidation. Some of the silicide-based intermetallic compounds have good resistance against both oxidation and sulfidation. Thus the two major classes of intermetallics that have been studied

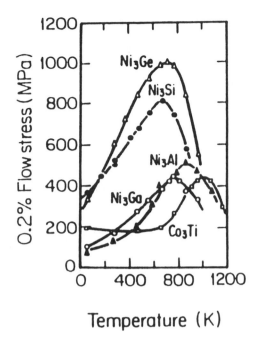

Figure 5.4: *Variation of compressive flow stress with temperature for some L1₂ compounds. (Reprinted with permission from Ref. 13.)*

so far offer good environmental protection. However, the environmental degradation of most intermetallic compounds has not been adequately studied and still remains a very attractive research area.

The effect of hydrogen embrittlement on the properties of intermetallic compounds is increasingly becoming a burning issue. It has been demonstrated in some alloy systems that ordering results in increased hydrogen embrittlement, though some of the results were inconclusive. The development of microstructures resistant to hydrogen and the problem of hydrogen embrittlement have in the past received very little attention and much work needs to be directed toward solving this problem.

Perhaps one of the most important property requirements for high-temperature materials is the ability to resist creep deformation. The mechanism of creep for the majority of the intermetallic alloys is dislocation creep that follows a power law [14]. This form of creep mechanism can be described mathematically by the Dorn equation:

$$\dot{\varepsilon} = A\left(\frac{DGb}{kT}\right)\left(\frac{\sigma}{G}\right)^n$$

where

$\dot{\varepsilon}$ is the secondary strain rate (s^{-1});
A is a dimensionless factor;
D is the effective diffusion coefficient (m^2/s);
G is the shear modulus (MPa);
b is the Burgers vector (μm);
k is the Boltzmann's constant (J/K);
T is the temperature (K);
σ is the applied stress (MPa);
n is the stress exponent (usually having a value of 3, 4, or 5).

Dislocation creep in disordered alloys occurs by dislocation glide and climb consecutively [14]. The rate-controlling step is the slower of the two and that determines the stress exponent value, which is 4 or 5 for dislocation climb and 3 for viscous dislocation glide. Dislocation climb results in a well-defined subgrain structure, whereas viscous glide leads to dislocation tangles without subgrain structure formation. Binary NiAl shows a well-defined substructure unlike FeAl.

Another creep mechanism that usually comes into play at lower stresses is diffusion creep [14], given by

$$\varepsilon_{\text{diff}} = A_{\text{diff}}\left(\frac{\Omega D}{kTd^2}\right)\sigma$$

where

$\varepsilon_{\text{diff}}$ is the diffusion creep;
A_{diff} is a dimensionless parameter;
Ω is the atomic volume;
D is the effective diffusion coefficient, considering both diffusion through grains and the grain boundaries;
d is the effective diffusion length;
σ is the stress.

The applied stress is the driving force for dislocation creep as well as diffusion creep. The two work in parallel, and the faster of the two controls the creep behavior. Thus the two constitutive equations can adequately describe the secondary creep of intermetallic compounds. Knowledge of the diffusion coefficients and the elastic constants of the intermetallic compounds is of great importance in predicting the creep performance of the alloys at elevated temperatures.

Extensive discussion of the creep behavior of various intermetallic compounds has been given by Jung et al. [14]. Their study describes a number of interesting findings that need to be taken into consideration when choosing intermetallic compounds for high-temperature applications. They found that an $(Fe_{0.2}Ni_{0.8})Al$ alloy that had the same B2 structure as the NiAl phase had the greatest creep resistance among the ternary alloys obtained by varying the Ni:Fe ratio. Interestingly, the composition $(Fe_{0.2}Ni_{0.8})Al$ was the one that showed the minimum estimated diffusion coefficient at nearly all temperatures. Thus, admixing a small amount of the softer FeAl to the harder NiAl phase can cause an increase in the creep resistance properties. It has been found that the creep resistance of CoAl is much higher than that of NiAl, and the reason for this behavior is under investigation. Also, the creep resistance of CoAl can be further improved by the addition of the softer NiAl phase.

Another often used method of increasing the creep resistance is to increase the atomic order. If ordering in a b.c.c. lattice is increased from two sublattices of the B2 structure to four sublattices, the resultant binary structure is the DO_3 structure represented by the A_3B type of compound and in the ternary case it converts to the $L2_1$ structure represented by the compound A_2BC. The DO_3 structure of Fe_3Al has higher strength at elevated temperatures compared to the B2 structure of the compound NiAl. Similarly, the compound obtained by substituting half of the aluminum with titanium in NiAl results in the $L2_1$ compound Ni_2AlTi, which has a creep resistance about twice that of the B2 compound NiAl. However, much larger increase in the creep resistance is possible by incorporating a second phase precipitate or adding dispersoids in the compound matrix. This has been achieved by a two-phase structure in which precipitates of NiAl in the Ni_2AlTi resulted in creep resistance comparable to that of the superalloy MAR-M-200. Thus, there are three possible means of improving the creep in intermetallic alloys:

1. Alloying while keeping the crystal structure the same
2. Alloying to change the crystal structure
3. Incorporating fine precipitates or a second phase.

These general principles can be applied to design better creep-resistant intermetallic compounds.

It is clear from the preceding that the intermetallic compounds offer a pool of potential and largely unexplored materials that might serve as the next generation of high-temperature materials. However, the extreme room-temperature brittleness of most of these compounds has prevented their extensive use. Composites based on intermetallic matrices are expected to be the answer to the general lack of toughness in these materials.

Another source of great uncertainty in the performance of these compounds as reliable engineering structural components is their processing. Given the virgin nature of the area itself, very little is known about the effects of processing on the structure and properties of most of the intermetallic compounds. This is an area that is expected to receive the greatest attention during the next few years, as this will be vital if mankind is to start utilizing these materials for high-temperature structural applications in the near future. Powder processing is expected to play a key role in the processing of these materials.

Although the $L1_2$ compound Ni_3Al has been studied extensively, it is unlikely that it will be the future replacement for superalloys due to its melting point and density being close to the superalloys currently in use. It is expected that major benefits could be gained from other very high-temperature intermetallic compounds whose ductility and room-temperature toughness are normally very poor. In that respect, it is likely that the intermetallic compound NiAl is more likely to find applications in true high-temperature oxidizing environments. The challenge for the future will be to develop intermetallic compounds or intermetallic matrix composites with adequate ductility and toughness at ambient temperatures, high strengths and creep resistance at elevated temperatures; and the resultant material should provide adequate protection against the working environment. Protection against hydrogen embrittlement is also becoming an increasingly important issue in the area of long-range-ordered intermetallic compounds. The search for materials that can expand the current envelope of existing high-temperature materials is expected to fuel the demand for and growth of intermetallic compounds. The following sections will discuss some of the extensively studied classes of intermetallic compounds.

Nickel Aluminides

The intermetallic compound Ni_3Al is definitely the most extensively studied material among all known intermetallic compounds. The interest in this compound led to renewed research efforts on intermetallics during the early 1980s. This compound has long been recognized as the principal high-temperature strengthening phase in nickel-based superalloys. The other extensively studied intermetallic compounds are the NiAl-based materials, which will also be discussed during the course of this chapter.

1. Ni_3Al

The nickel aluminides, Ni_3Al, based on nickel and aluminum, have the attractive combination of high melting point, low density (10% lower than nickel-based superalloys), high strength, and good corrosion and oxidation resistance, and they exhibit improved strength with temperature (up to a maximum temperature). Also, the simple form of this alloy does not contain any strategic elements. In spite of the attractive properties of Ni_3Al, the use of the compound was severely restricted due to the extreme brittleness of the bulk polycrystalline compound.

Interestingly, single crystals of Ni_3Al in all orientations are ductile at room temperature, but the inherent weakness of the polycrystalline grain boundaries usually results in an extremely brittle material with essentially no ductility at room temperature. Early research attributed the room temperature brittleness of the polycrystalline compound to the embrittling effect of impurity segregation (especially of sulfur) to the grain boundaries. However, later studies by Liu et al. [15] on high-purity polycrystalline Ni_3Al with clean grain boundaries showed that the compound still failed in a brittle manner.

The pioneering work by Aoki and Izumi [3] (reported in 1979) opened the floodgates of research on intermetallic compounds. They observed that the addition of trace amounts of boron to Ni_3Al significantly increased the room-temperature ductility. The microalloying with boron was found to be effective only in the substoichiometric aluminum composition range (around 24 at.%). Liu and Sikka [16] reported that microalloying with boron increased the ductility of polycrystalline Ni_3Al from 0 to around 50% by the addition of only 0.08 wt.% boron, and the fracture mode changed from totally intergranular to a transgranular nature.

It has been observed [15,17–20] that the boron segregates to the grain boundary regions and influences the fracture mode of the alloy, thereby imparting the desired ductility to this extremely attractive high-temperature compound. Unlike embrittling impurity segregation, boron segregation strengthens the grain boundary regions. It is interesting that boron is found to segregate to the grain boundaries and not to the free surfaces in Ni_3Al compounds. It has been generally concluded that boron increases the grain boundary cohesive strength and thereby produces the ductility in the compound [20,21]. It has also been suggested that the addition of boron results in the grain boundaries being enriched with nickel, which makes the grain boundaries disordered [22,23] and hence imparts some ductility to the material.

Boron addition also goes to increase the strength of the compound to some degree in addition to the improvement in the ductility. The beneficial effect of the boron addition is practically lost for compositions close to or greater than 25 at.% aluminum and the extent of boron segregation to the grain boundaries also decreases with increasing aluminum. Thus, boron is effective in increasing the ductility in substoichiometric aluminum-containing Ni_3Al intermetallic compounds (around 24 at.%). The effect of boron addition on the tensile elongation and strength of a Ni–24Al compound is shown in Figure 5.5 [24].

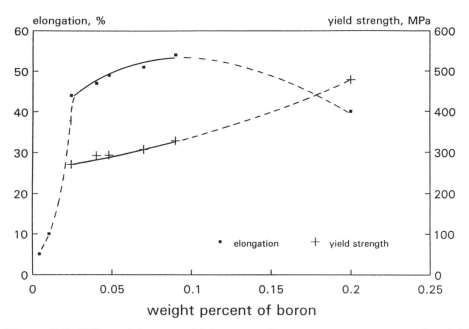

Figure 5.5: Effect of boron additions on the room-temperature tensile yield strength and room-temperature tensile elongation of a Ni–24Al (at.%) intermetallic compound. (After Ref. 24.)

Takasugi et al. [25] demonstrated that the addition of substitute solute elements such as iron and manganese could result in room-temperature ductility of Ni_3Al without the addition of any boron. Similar compositions were also tested by Dimiduk et al. [26] with varying heat treatment conditions that resulted in either a two-phase or single-phase alloy. They determined that the ductility did exist in the two-phase alloys but dropped to less than 0.1% in the single-phase alloy. It was concluded that the presence of small amounts of γ-phase is responsible for the ductility without the boron addition.

The elevated-temperature ductility of Ni_3Al is found to depend on the test environment. The binary compound exhibits severe loss in ductility at elevated temperatures when tested in air (versus in vacuum). The ductility essentially reached zero in the temperature range of 873 to 1073 K (600 to 800°C) for Ni_3Al compounds with or without boron, when testing was carried out in air [27,28]. Studies at Oak Ridge National Laboratory (ORNL) [16] have concluded that the embrittlement is caused by the dynamic oxidation effect. The problem was adequately solved by the addition of chromium to Ni_3Al. The chromium effectively stopped the dynamic oxygen embrittlement by rapidly forming a chromium oxide film, which decreases the influx of oxygen to the grain boundaries. The effects of test environment and chromium addition on the tensile elongation of Ni_3Al-based alloys are shown in Figures 5.6a and 5.6b [16], respectively.

To explain the anomalous strengthening behavior with temperature, Kear and Wilsdorf [29] proposed a cross slip mechanism. In the $L1_2$ ordered compounds, the superlattice dislocations are mobile on (111) planes. The antiphase boundary is more stable on the (100) planes. Thermal activation at high temperatures permits the dissociated partial dislocations to cross-slip onto the (100) planes. This produces sessile segments that resist further gliding on the primary slip plane by re-dissociating on other (111) planes.

Thornton et al. [30] carried out an interesting experiment by which the temperature dependence of the microstrain yield stress was determined. Microplastic yield stress for selected strains between 10^{-6} to 10^{-2} was determined for a range of temperatures up to nearly 1273 K (1000°C). The peak flow stress was found to increase with increasing magnitude of the corresponding microyield strain. This shows that dislocations become mobile at constant stress irrespective of the test temperature.

The anomalous behavior of increasing strength of Ni_3Al with temperature can be further enhanced by alloying additions. This compound can be strengthened by the classic solid solution hardening mechanisms because it can take into solution a substantial amount of alloying elements without disturbance of the long-range order. The alloying elements can be divided into three categories: elements that exclusively occupy the nickel sublattice sights like Co, Pt, etc.; elements that exclusively occupy the aluminum sublattice sights like Si, Ti, V, Hf, etc.; and elements that can occupy both the lattice sites like Cr, Fe, and Mn. Upon extensive review of the mechanical properties of alloyed Ni_3Al, Rawling and Staton-BeVan [31] found that effective hardening could be produced by elements that substitute the aluminum at its sites. The strengthening was most effective in the stoichiometric case or on the aluminum-rich side of the stoichiometric compound. It should be remembered that the boron ductilizing effect is applicable primarily to the nickel-rich side of the stoichiometric Ni_3Al compound.

Rapidly solidified boron-doped Ni_3Al also imparts considerable solid solution strengthening. Boron provides strengthening by an interstitial solid solution mechanism that provides large lattice dilations. Addition of zirconium and hafnium is also quite effective and these have been used as alloying elements in some advanced nickel aluminide alloys. Hafnium, due to its large atomic size misfit, was found to be the most potent for both ambient and elevated temperature strengths [16,32]. Hafnium addition is more effective at elevated temperatures. The addition of hafnium increased the temperature at which the peak in the yield strength occurred (from 873 to 1123 K; 600 to 850°C) and also increased the magnitude of the peak yield strength (from

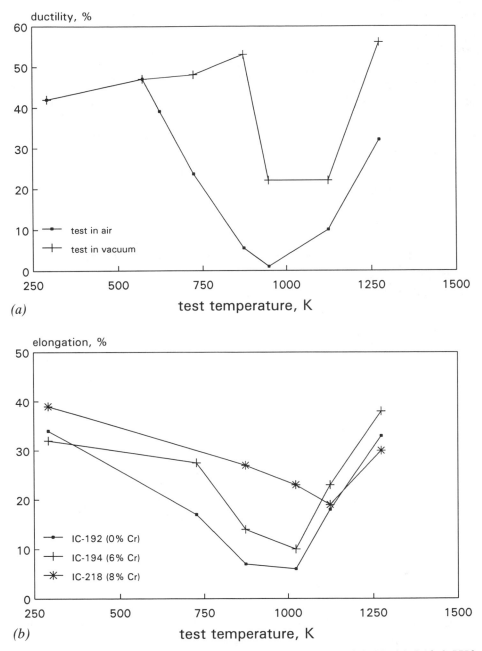

Figure 5.6: *Variation of ductility with test temperature. (a) Ni–21.5Al–0.5Hf–0.1B (at.%) compound with no chromium additions tested in air and in vacuum. (b) tests carried out in air on some Ni₃Al-based alloys with varying chromium contents. (After Ref. 16.)*

around 550 to 900 MPa) [28]. It is also found to decrease the creep rate significantly and provide creep resistance that is almost comparable to that of Waspaloy.

Fatigue crack propagation in a powder-processed advanced aluminide has been reported by Chang et al. [33]. The alloy had a nominal composition (wt.%) of 11.5Co–11.5Al–1.75Hf–1Mo–0.5Nb–0.05Zr–0.05B, rest Ni. The atomized powder of the material was sieved and the powder finer than 106 μm was canned, evacuated, sealed, hot isostatically pressed, given a 20% cold rolling reduction, and then annealed at 1273 K (1000°C) for 3600 s (1 hour). Fatigue crack propagation behavior of the alloy was studied at 673 K (400°C) in vacuum and in air. The crack growth rate in air exhibited a strong time dependence. For a fixed ΔK and temperature, the crack growth rate increases with the fatigue cycle period when testing was carried out in air. This was not the case when the testing was carried out in vacuum. It was concluded that the environmental embrittlement mechanism is responsible for the fast crack growth rates in air. It is interesting to note that this alloy also did not contain any chromium.

Based on the above discussions, new advanced nickel aluminides have been designed. ORNL has played a key role in this area and has developed new alloys that carry the nickel aluminides to the point where the material is on the verge of major commercial exploitation. The nominal composition range of some of the alloys that have been developed at ORNL with their designations are given below (atomic percent):

IC-50   23.0 Al, 0.5 Hf, 0.1 B, rest Ni
IC-218 16.7 Al, 0.5 Zr, 8.0 Cr, 0.1 B rest Ni
IC-221 16.0 Al, 1.0 Zr, 8.0 Cr, 0.05 B, rest Ni
IC-328 17.0 Al, 0.2 Zr, 8.0 Cr, 0.3 Ti, 0.1 B, rest Ni
IC-405 18.0 Al, 0.2 Zr, 8.0 Cr, 12.2 Fe, 0.1 B, rest Ni with 0.005 wt.% Ce

In the advanced aluminides, zirconium or hafnium improves the high-temperature mechanical properties by solid solution hardening; boron addition provides room-temperature ductility; and chromium reduces the elevated-temperature dynamic embrittlement in oxidizing environments. In some alloys small amounts of carbon and cerium are added to improve the hot ductility and to control the grain size.

Oxide dispersion strengthening of Ni_3Al with 1 wt.% alumina has been reported by Lauf and Walls [34]. The ODS alloy was produced by mechanically alloying the Ni_3Al with the alumina powder, which was subsequently vacuum hot-pressed at 1573 K (1300°C) for 900 s (15 minutes) using a pressure of around 28 MPa. The material was a fine-grained fully dense alloy that exhibited much higher hardness than the non-ODS alloy. At high temperatures, the increased hardness dropped off drastically and the material was actually softer than the non-ODS alloy at temperatures around 1273 K (1000°C).

The results of investigations on ODS alloys of Ni_3Al have also been reported by Jang and Koch [35]. Prealloyed powder of the composition Ni–23.5Al–0.5Hf–0.2B (at.%) was mechanically alloyed with 2.5 vol.% of alumina. The resultant powder was cold pressed, vacuum canned, and hot isostatically pressed at 1573 K (1300°C), 103.4 MPa, for 10 800 s (3 hours). The tensile yield strengths of the ODS samples were much larger and the grain sizes much smaller than the non-ODS samples. The major part of the increase due to ODS was found to be the grain size effect induced by the Hall–Petch effect. The other mechanism that contributed a smaller fraction of the strengthening was the Orowan mechanism in which nonshearable dispersoids acts as impediments to dislocation motion, forcing the dislocations to bow between the particles. The increase in the yield strength due to this effect can be represented by the Orowan–Ashby expression

$$\Delta\sigma = \frac{0.13Gb}{\lambda} \ln\left(\frac{r}{b}\right)$$

where

r = dispersoid radius (nm);
G = shear modulus (GPa);
b = Burgers vector (nm);
λ = dispersoid spacing (μm).

It was determined that simply by the Hall–Petch relationship, the yield strength of the 1- to 2-μm grain-sized specimens without the dispersoid addition was about twice that of the sample with a grain size of 35 μm without dispersoid. The dispersoids themselves accounted for only 13% of the strengthening by the Orowan mechanism.

A relatively novel process of reactive sintering of bulk Ni_3Al intermetallic compounds to near full density has recently been patented [36]. The process of reactive sintering as applicable to Ni_3Al has been described by Bose et al. [37] and is shown schematically in Figure 5.7. In contrast to the established methods of fabricating these compounds, reactive sintering requires very low sintering temperature (around half the melting point of the final compound) and short time (a few minutes). It involves the mixing and compaction of elemental powders (which are inexpensive and easily compressed) and heating them to a temperature at which the exothermic reaction occurs. The exotherm

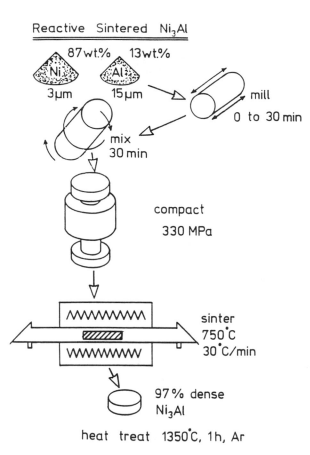

Reactive Sintered Ni₃Al

Figure 5.7: Schematic diagram of reactive sintering of Ni₃Al. (Reprinted with permission from Ref. 37.)

occurs at a temperature just above the first eutectic in the Ni–Al phase diagram (Figure 5.1), and this can be observed from the differential thermal analysis (DTA) scan of a Ni–25Al powder mixture reproduced in Figure 5.8 [37]. The low-melting aluminum melts, spreads, reacts, and causes very rapid densification and compound formation (often in a matter of seconds). However, processing variables like powder particle size, powder size ratio, sintering temperature, heating rate sintering atmosphere, and powder mixing have to be controlled carefully in order to obtain high as-sintered densities. The wrong powder size ratio can result in final samples with large pores, as shown schematically in Figure 5.9, and high porosities (i.e low densities), as shown in Figure 5.10 [37].

The recent results on pressureless reactive sintering of Ni₃Al suggest that

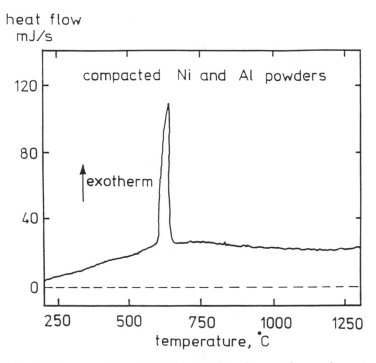

Figure 5.8: *DTA scan of a Ni–25Al (at.%) elemental powder mix during the heating cycle, showing the sudden exotherm that occurs when the elemental powders react to form the compound. (Reprinted with permission from Ref. 37.)*

material systems that have high exotherms and form liquid during the reaction could be reactively sintered to high densities. In such systems the compound formation and densification occur simultaneously, with the highly transient molten aluminum providing the capillary action for the rapid densification. The resultant sample is around 95% to 97% dense, with a trace of a second phase identified as Ni_5Al_3. The properties of reactively sintered compounds are often comparable to those reported in literature. The reactively sintered samples can be containerless hot isostatically pressed to produce fully dense materials since the retained porosity is not interconnected.

A variation of the pressureless reactive sintering process involves the simultaneous application of pressure as the reaction is in progress. This process has been termed reactive hot isostatic pressing or RHIP and has been used to produce Ni_3Al matrix composites [38]. The process offers the potential of inexpensive production of intermetallic alloys based on Ni_3Al. Materials processed by this route could be used in relatively noncritical hot components for the autmotive industry and hot furnace furniture.

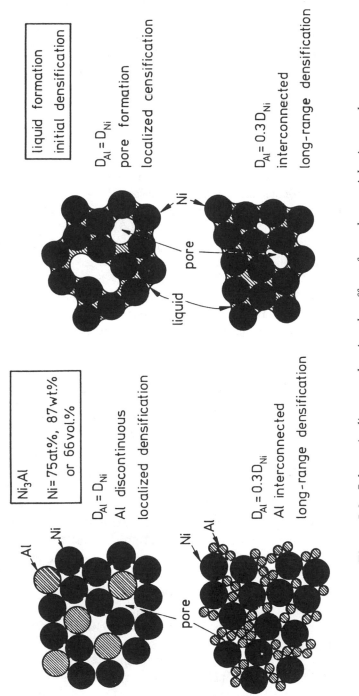

Figure 5.9: Schematic diagram showing the effect of powder particle size on the reactive sintering of Ni₃Al. (Reprinted with permission from Ref. 37.)

Variations of this reactive sintering process have been referred to in the literature as self-propagating high-temperature synthesis (SHS) or gasless combustion synthesis (GCS), and a variety of high-temperature compounds such as carbides, nitrides, borides, etc., have been produced. In most cases, the reaction product is a porous mass that is comminuted to form the desired powder, which is subsequently consolidated to full density.

2. NiAl

The B2 intermetallic compound NiAl has been used primarily as a high-temperature, oxidation-resistant coating material. This material is more attractive as a high-temperature material than the extensively studied Ni_3Al intermetallic compound due to its lower density of 5.9 g/cm^3 and higher melting temperature of 1911 K (1638°C). The compound has a wide composition range over which it remains a single phase ordered compound (see Figure 5.1, which shows the binary Ni–Al phase diagram), has a high modulus (189 GPa), and is extremely oxidation resistant. The compound is ordered up to its melting point. Only the extremely poor room-temperature ductility and toughness of polycrystalline NiAl intermetallic compounds have prevented their extensive use in high-temperature structural components.

The typical B2 crystal structure of NiAl is shown in Figure 5.2b. This compound deforms primarily by slip on the $\langle 100 \rangle (110)$ system. The single crystals of NiAl, like single crystals of Ni_3Al, exhibit considerable plasticity, but the polycrystalline materials have practically no ductility [39] except at very fine grain sizes [40]. Studies on the grain size and temperature dependence of ductility clearly indicate that reasonable ductility at room temperature can be obtained if the grain size is extremely fine. At 673 K (400°C) using strain rates of $10^{-4} s^{-1}$, the tensile ductility is observed to increase from 2% to 3% for grain sizes larger than 20 μm to over 40% for grain sizes around 10 μm. This is shown in Figure 5.11 [40]. There is, therefore, the existence of a critical grain size, d_c, above which the material becomes extremely brittle. Schulson [39] postulates that cracks in coarse-grained aggregates may be so large that they propagate as soon as they nucleate, and thus prevent any macroscopic ductility. This can be remedied by decreasing the grain size below a critical level at which the cracks formed will be stable. Propagation of these stable cracks requires some hardening through plastic flow, thereby rendering some ductility in tension. The smaller the grain size below the critical level, the greater will be the hardening required to propagate the cracks and, thus, greater will be the ductility. Rapid solidification by melt spinning has been tried with limited success [41]. However, for high-temperature applications, the extremely fine

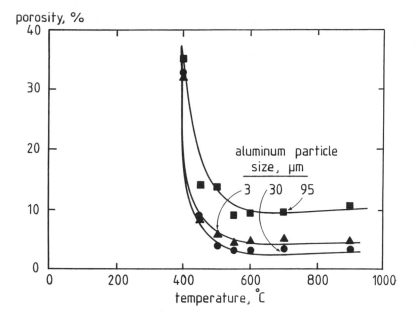

Figure 5.10: *Experimental results of reactive sintering Ni₃Al with different aluminum powder particle sizes and different reaction temperatures. (Reprinted with permission from Ref. 37.)*

grain sizes will tend to coarsen and the material is apt to lose the beneficial effects on ductility.

Powder processing has been applied to produce ternary alloys based on the NiAl composition [42,43]. These investigations resulted in the identification of a few additions that could strengthen the compound significantly. The ternary alloys were prepared by blending elemental additives to a prealloyed Ni–Al powder, followed by either hot extrusion or hot isostatic pressing of canned powders. The additives used were Ti, Ta, Nb, Mn, Mo, W, Re, Fe, and Hf. Some of the additives, such as Ti, Mn, and Fe, were completely soluble and did not show any major improvements in the properties over the binary alloys when tested in compression at 1300 K (1027°C) in air. Additives such as Ta, Nb, and Hf did not dissolve totally even after long homogenization times. These additives produced significant improvement in the hot compression tests at 1300 K (1027°C) in air. The mechanism of strengthening was postulated to be some form of interaction between dislocations (dislocation pinning) and the precipitates richer in the additives of Ta or Nb.

Argon gas atomization of NiAl–5Ti (at.%) and NiAl–5Nb (at.%) have recently been carried out [44]. Cold consolidation of the atomized powders was unsuccessful. Attritor-milled powders could, however, be cold compacted and sintered to 95% theoretical density at 1673 K (1400°C). Sintering at 1773 K (1500°C) resulted in only a slight increase in the density but a large increase in the grain size. The as-sintered material had closed porosity and is containerless hot isostatically pressed to a density of 98% theoretical. Alloying with 5 at.% Nb or Ti increased the hardness by 17% and 53%, respectively. The toughness was increased approximately 4-fold and the creep resistance of the material was increased significantly by both these additions. Titanium was the more effective strengthener for both room and elevated temperatures. The creep resistance results obtained from this study contradict the results of the previous study, where it was found that niobium addition was more effective than titanium in creep. The apparent reason could be the better homogeneity of the powders obtained in the gas-atomized powders.

The grain size effects on the hot strengths of these alloys are unusual because the fine-grained alloys near the stoichiometric composition exhibit much higher strengths than the coarse-grained samples of similar composition [45], whereas metals do not exhibit similar grain boundary strengthening effects at low strain rates and high temperatures. Similar results have also been observed in other ordered alloy systems [46]. The strengthening effect is also found to peak at the stoichiometric composition and drops off away from the 1:1 stoichiometry. The complex interrelationship between the strength, grain size, and alloy stoichiometry needs more attention for future alloy development efforts.

3. Applications

The nickel aluminides have recently received a great deal of attention and are already in commercial use. The list of applications keeps growing and not surprisingly the majority of uses are not aerospace related. A partial list of companies that are licensing the technology from ORNL are given below [47]:

Cummins   Diesel engine components
Metallamics   Powder metallurgy products
Armada   Resistance heating wire
Valley-Todeco   Aircraft fasteners
Armco   Mill products

The advanced Ni_3Al alloys often have additives to increase their density to a level almost comparable to that of the nickel-based superalloys. The nickel-based superalloys are so well developed and have such a strong technology base that, unless the density-compensated properties of a new class of material are phenomenally better, the acceptance in the aerospace industry of the

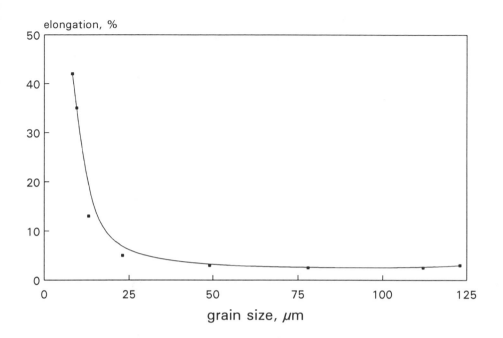

Figure 5.11: *Dependence of tensile elongation on the grain size of NiAl tested at 400°C and a strain rate of $10^{-4}s^{-1}$. (After Ref. 40.)*

replacement is extremely doubtful. Thus, it would seem unlikely that Ni_3Al-based alloys will see direct applications in the aerospace industry in the near future, except for noncritical components. However, there will be host of other applications where this material can find immediate use [45], including radiant heaters, dies and molds, pump impellers, hot furniture for furnaces, land-based steam turbines, oil and gas well components, and hot components in automotive applications.

The compound NiAl, on the other hand, with its extremely high melting point and low density, would be an ideal choice for engine applications. However, the problems of low toughness and ductility need to be solved, and long-range reliability studies need to be conducted before this material can even be considered for any structural applications. Oxidation-resistant high-temperature coatings are expected to be the major use of this compound.

Titanium Aluminides

The current and proposed advances in the aerospace arena are heavily dependent on the development of new lightweight high-temperature-resistant

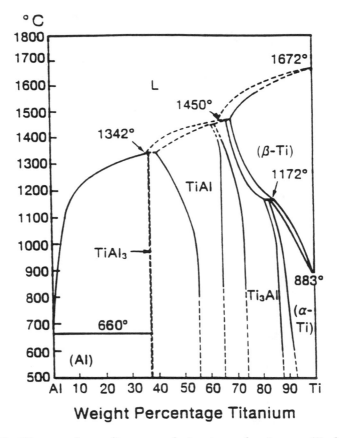

Figure 5.12: *Binary phase diagram of titanium–aluminum. (Redrawn after Ref. 48.)*

materials. The titanium aluminides are among the prime candidates in strong contention for being the materials of choice for these advanced structural aerospace components. The combination of low density, high stiffness, and good high-temperature strength, creep, and environmental resistance makes titanium aluminide an attractive candidate for replacing most conventional titanium alloys. The potential of titanium aluminide matrix composites is far greater than that of the pure monolithic titanium aluminides and is being investigated by a number of research laboratories.

The Ti–Al phase diagram is shown in Figure 5.12 [48]. The intermetallic compounds in the Ti–Al phase diagram that have been extensively studied are Ti_3Al ($\alpha2$ phase) and TiAl (γ phase). The other intermetallic compound, $TiAl_3$, which has a very narrow composition range compared to Ti_3Al, has

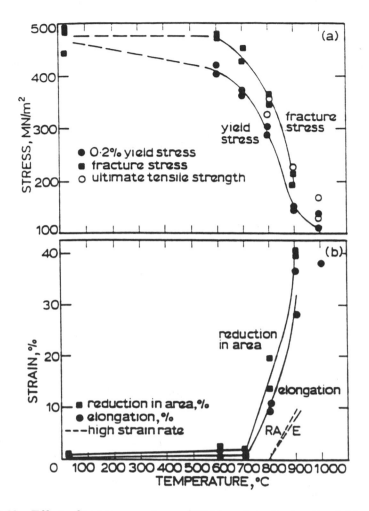

Figure 5.13: *Effect of test temperature of TiAl on (a)   the tensile yield stress and fracture stress; (b) reduction in area and elongation. (Reprinted with permission from Ref. 63.)*

been, for the most part, neglected in spite of its slight density advantage and better oxidation resistance.

1. Ti_3Al

The intensive development of the DO_{19} hexagonal close-packed compound Ti_3Al began in the early 1970s, though work on this alloy had been carried out even before the existence of the compound was confirmed. The material possesses good strength, high modulus and good creep behavior at elevated temperatures of 1073 K (800°C) [49,50]. However, the ordered structure is responsible for the low ductility of this compound. The low ductility is attributed to the lack of an adequate number of independent slip systems. Many static engine parts require high stiffness at elevated temperatures, and Ti_3Al, which possesses a higher modulus at around 1073 K (800°C) than pure titanium at room temperature, is an excellent candidate.

Various alloying approaches have been taken to modify the strength and ductility of the compound. Among the alloying additives, niobium addition in sufficient quantities to stabilize the β phase proved to be the most interesting and has been studied extensively [49–55]. The addition of niobium was seen to increase both strength and ductility of Ti_3Al by reducing the planarity of slip and by increasing the nonbasal slip activity [49]. The study also indicated that the addition of niobium slowed the domain growth kinetics. It has been determined that room-temperature ductility and toughness are enhanced by low aluminum content and a higher amount of β stabilizing additive, while the high-temperature properties are improved by higher aluminum contents.

Once it was realized that the $\alpha_2 + \beta$ alloy could generate sufficient low-temperature ductility [52], the Ti–24Al–11Nb (at.%) alloy became the standard composition around which a major part of further research was concentrated. This has resulted in a wealth of database and reliability studies of prime importance for materials to be used for aerospace applications.

The addition of molybdenum and vanadium has also been tried by various research groups with varying degrees of success. Jackson et al. [56] have described the rapidly solidified microstructures of a Ti–15Al–11Mo (wt.%) alloy. A super α_2 titanium aluminide with a composition of Ti–15Al–20Nb–3.2V–2Mo (wt.%) has also been reported. The molybdenum-containing alloy has a two-phase structure consisting of lenticular plates of α_2 and a B2 matrix whereas the super α_2 is also a dual-phase alloy with coarse plates or rods of α_2 in a β matrix. Later work on a Ti_3Al alloy with niobium and tungsten additions showed the precipitation of a fine b.c.c. β phase and considerable refinement of the α_2 plates that resulted in increased strength and ductility [51].

The microstructure of the alloys can be extensively varied by modification of processing and heat treatment. The final microstructure has a large bearing on the final properties of these alloys. The titanium aluminides are susceptible to hydrogen embrittlement, which has recently turned out to be one of the key impediments to the use of titanium aluminides for the NASP program. Hydride formation (which disappears on vacuum outgassing) is observed in both Ti–24Al–11Nb (at.%) alloy and stoichiometric Ti_3Al alloys when the material is charged with hydrogen at elevated temperatures. It is also reported that the two-phase TiAl-based intermetallic compounds are also susceptible to hydrogen embrittlement. These findings seriously jeopardize the chances of using these materials for the NASP program, which would mean reverting to the high-density superalloys, considered by many to be unacceptable. There is, however, a ray of hope in the recent finding that not all microstructures are susceptible to the hydrogen embrittlement effect [57].

Low oxygen content is considered to be an important factor in the improved room-temperature ductility of the titanium aluminide alloys with niobium additions [10]. A ductility maximum for the β solutioned materials with low oxygen content is observed to occur in the range 300 to 773 K (500°C), after which the ductility is seen to decrease. The decrease in the ductility at elevated temperatures is thought to result from a combination of environmental degradation and β decomposition. The environment has a strong effect on the elevated-temperature properties of the alloys [58]. When testing at elevated temperatures of 923 K (650°C) was carried out in air, surface cracks formed on an embrittled layer presumably by the inward diffusion of oxygen. Under tensile loading conditions, cracks developed in the embrittled layer normal to the loading direction and propagated right through the specimen. However, when the testing was carried out in vacuum, the specimens did not show the surface cracks and the fracture surface showed considerable ductile tearing.

The decrease in the yield strength with temperature is gradual up to 923 K (650°C) for the Ti–24Al–11Nb (at.%) alloy. Some of the new alloys like Ti–24Al–17Nb–1Mo (at.%) and Ti–26Al–17Nb–1Mo (at.%) show a gradual decrease in yield strength up to 1023 K (750°C), after which it decreases rapidly. Interestingly, the yield strength of the two new alloys described above was more than twice that of the Ti–24Al–11Nb (at.%) alloy at 923 K (650°C).

Another area of great interest is the dispersion of rare-earth oxides and sulfides in Ti$_3$Al-based intermetallic compounds. One of the means of solving the low ductility of the Ti$_3$Al compound was postulated to be decrease in the slip length, which could be achieved by refining grain size, by transformation (as in quenching to obtain martensite), or by the incorporation of fine, stable dispersoids that could interact with the dislocations [59]. The dispersoids should be uniform and fine, and should not coarsen at elevated use temperatures. By using rapid solidification techniques, rare-earth elements like erbium or cerium with minor impurities like oxygen and sulfur have been introduced to form metastable supersaturated alloys [59–62]. With proper heat treatment, the rare-earth oxides or sulfides can be precipitated. In the case of pure titanium, temperatures above the α to β transformation, 1146 K (873°C), cause rapid coarsening of the Er$_2$O$_3$ particles, probably due to increased diffusion in β

titanium. However, in the Ti$_3$Al compound, the transition of α_2 to α and α_2 occurs at 1373 K (1100°C). Thus, the precipitates are expected to be more stable in Ti$_3$Al. Rowe and coworkers [60–62] have studied the dispersions of Er$_2$O$_3$, Ce$_2$S$_3$ or Ce$_4$O$_4$S$_3$ in Ti$_3$Al or Ti$_3$Al–Nb matrices. The materials were consolidated by hot isostatic pressing, in some cases followed by extruding the melt-spun ribbons of the desired compositions at 1198 K (925°C). Extrusion caused much larger coarsening of the Er$_2$O$_3$ oxide particles compared to the Ce$_2$S$_3$ or the Ce$_4$O$_4$S$_3$ dispersoids. These results indicate that the rapid solidification approach needs to be explored in greater detail.

The ability to generate reproducible microstructures and improved properties that would warrant the expensive production route will determine whether this form of processing will become a commercial success. The RS technology for rapidly solidified titanium aluminide alloys is still in its infancy, but already numerous attractive possibilities are looming. It is expected that future improvements in the powder production techniques for economically producing rapidly solidified powders of titanium aluminides will result in phenomenal growth in the use of this material.

Powder processing has been used extensively in the area of fiber-reinforced titanium aluminide (Ti$_3$Al+Nb) matrix composites. One such process includes the powder processing of a titanium aluminide matrix composite that is reinforced with continuous silicon carbide fibers. The process in general consists of producing fiber mats by winding the fibers on a drum and then cutting the mats to the desired size. The powder is produced by the plasma rotating electrode process (PREP) that has been discussed in Chapter 3 on powder atomization. A powder cloth technique is used to form sheets of the powder, which can be cut to the desired size. In the powder cloth technique, the powder is mixed with an organic material that acts as a glue and allows the material to be rolled into thin sheets. The powder cloth and the fiber mats are laid up alternately in a hot press die and consolidated to full density. The matrix was reinforced with 40 vol.% of SiC fiber. The composite was fully dense and the tensile properties of the composite compared favorably with the predictions obtained by the rule of mixtures. The experimental values were, however, significantly lower than the predicted values at room temperature and at temperatures above 900 K. The reason for the deviation from the rule of mixtures is postulated to be the limited matrix ductility at room temperature and the reaction zone–matrix debonding at elevated temperatures. The composite properties far exceed those of a superalloy in a strength/density comparison. The rapid growth of the reaction zone at 1475 K (1202°C), however, is expected to limit the practical use temperature of the composite to below 1273 K (1000°C).

2. TiAl and TiAl₃

The other important and widely studied intermetallic compound in the Ti–Al system is the $L1_0$ compound TiAl. This compound has high oxidation resistance, specific strength, and modulus. There is still some confusion about the low-temperature phase boundary of the compound, with some research groups claiming that the equiatomic composition is a truly single-phase material whereas other reports suggest that the equiatomic composition results in a two-phase structure [48]. The Ti–Al phase diagram shown in Figure 5.12 shows a two-phase composition for the equiatomic TiAl.

The room-temperature ductility of the single-phase compound TiAl is poor. The dislocation mechanism of TiAl showed that many of the superdislocations are pinned by an unknown source that becomes ineffective at temperatures above 973 K (700°C), causing a rapid increase in ductility above that temperature. The effect of test temperature on the yield stress, fracture stress and elongation of TiAl is shown in Figure 5.13 [63]. It can be observed that the ductility increases sharply above 973 K (700°C), and both the yield and fracture stress start to decrease sharply above 873 K (600°C). In contrast to the single-phase TiAl, the two phase-intermetallic compound close to the stoichiometric TiAl composition is known to have some ductility.

Studies of alloying additions on the properties of the TiAl intermetallic compounds have been in progress for some time. It was found that a small amount of vanadium addition removed much of the pinning effect (believed to be responsible for the low ductility of the compound) even at ambient temperatures [64]. A single-phase alloy of the composition Ti–48Al–1V (at.%) provided some ductility at ambient temperatures. The mechanism by which some ductility is imparted to the brittle compound TiAl by the addition of vanadium is still an open question. Several other alloying additives such as Ta, Mn, Nb, etc., have been tried by various research groups with varying degrees of success [10, 12, 65].

The production of TiAl compound from elemental powders have been demonstrated by Schafrik [66]. The elemental powders mixed in the desired ratio could be canned and extruded followed by homogenization to yield a TiAl product. A modified form of reactive hot isostatic pressing of elemental powders to form the TiAl compound has also been successfully employed [67]. The process consists of prereacting the elemental powders in the form of loose cakes, which are subsequently milled, sieved, cold isostatically pressed and HIP treated. The compressive properties of the alloy produced by reactive processing are comparable to or better than the properties of monolithic TiAl compound reported in the literature. Most of the properties have been evaluated under compressive loading conditions, and more research is required for evaluation of tensile properties of this intermetallic alloy.

The creep deformation of a pure titanium aluminide (TiAl) and a TiAl + W intermetallic compound have been also reported by Martin et al. [68]. The activation energies for creep were higher than for the usual disordered materials. The compound with the tungsten addition had still higher activation energies, probably due to both solid solution and dispersoid effects.

The fatigue properties of both TiAl and TiAl+Nb-based intermetallic compounds were studied from room temperature to 1173 K (900°C) by Sastri and Lipsitt [69]. It was found that the ratio of the fatigue strength (at 10^6 cycles) to the ultimate tensile strength was larger than 0.5 for all temperatures. The fracture mode was observed to change from cleavage to intergranular when the test temperature was increased from ambient to above 873 K (600°C).

The oxidation behavior of TiAl compounds with and without alloying additions has been reported by Perkins et al. [70]. The TiAl shows good resistance to oxidation and alumina-forming kinetics up to 1223 K (950°C). Above 1223 K the material tends to form titanium oxide. In air, the alloy oxidizes at a more rapid rate. It is theorized that the high solubility and diffusivity of oxygen and low diffusivity of aluminum, coupled with the rapid growth of transient titanium-rich oxides, is the cause of the inability to form external alumina scales above 1223 K (950°C). The iron and nickel aluminides, in contrast, can produce the alumina film up to 1373 K (1100°C) in both air and oxygen. It has been suggested that the formation of external alumina scales can be increased by increasing the diffusivity of aluminum which is about two orders of magnitude higher in the b.c.c. β phase than the diffusivity in any other phase. Thus, the retention of β as the major phase, and the decrease of oxygen solubility and diffusivity in the β phase through alloying additions of chromium and vanadium, can improve the desirable oxide skin formation. To support the above discussion, a Ti–30Al–12Cr–15V (wt.%) alloy was used for the oxidation studies. It was found that this alloy forms an alumina scale in air at 1673 K (1400°C) and has a parabolic rate constant that is the same as for NiAl, which is often used for high-temperature oxidation-resistant coatings. At lower temperatures of 1073 K (800°C), however, the rate constant is several orders of magnitude higher, partly due to the lower amount of β phase that is retained at lower temperatures.

Compared to Ti_3Al and TiAl, the lower melting point $D0_{22}$ compound $TiAl_3$ has received very little attention. The report by Yamaguchi et al. [71] on the $TiAl_3$ compound concluded that the alloy development of this material has not reached a point where it can be used as a structural compound. However, its oxidation resistance and low density make it an attractive material that could be used as a coating material on TiAl-based alloys. It is suggested that V, Hf, Zr, and Li additions can improve the ductility of the compound.

Iron Aluminides

Intermetallic compounds based on iron aluminides are expected to be an excellent and inexpensive replacement for a number of stainless steels and moderate temperature-resistant materials. Presently, the majority of the heat-resistant alloys are either nickel- or cobalt-based or high nickel- and chromium-containing steels. These conventional heat-resistant alloys often suffer from chromium evaporation at high temperatures, exhibit very poor hot corrosion properties in sulfur-containing environments, and usually contain a number of expensive and strategic elements. On the other hand, intermetallic compounds near the Fe_3Al composition exhibit good oxidation and sulfidation resistance due to their ability to form a protective alumina coating at very low oxygen partial pressures [72].

1. Fe_3Al

Investigations on Fe_3Al were initiated in the mid-1950s and quite a bit of work was done during the 1960s. However, during the mid-1970s, Marcinkowski et al. [73] showed that the fully ordered iron aluminide compound had practically no ductility at around 473 K (200°C). Interest in the iron aluminides was revived in the 1980s and the properties of the compounds were re-evaluated. It was during this period that the material studied by Marcinkowski et al. [73] was found to contain around 500 ppm of carbon that formed an Fe_3AlC compound. This carbide probably caused the brittle behavior of the iron aluminide-based intermetallic compound [74].

Extensive studies have been carried out by various investigators on the Fe–Al phase diagram, and a revised phase diagram that reviews the literature through 1981 has been depicted by Kubaschewski [75]. The order–disorder boundaries are quite complex in nature and the detailed phase diagram near the Fe_3Al composition is shown in Figure 5.14 [72,76,77]. This diagram is currently considered to be the actual Fe–Al equilibrium diagram. Another phase diagram [77] that shows a similar shape and similar phase fields but disagrees on the boundary of the phase locations is expected to be of a metastable nature. This metastable phase diagram near the Fe_3Al composition is also superimposed on the equilibrium phase diagram shown in Figure 5.14. The phases of interest are the ordered $D0_3$ compound Fe_3Al, the ordered B2 compound FeAl, and the disordered α-solid solution of iron and aluminum.

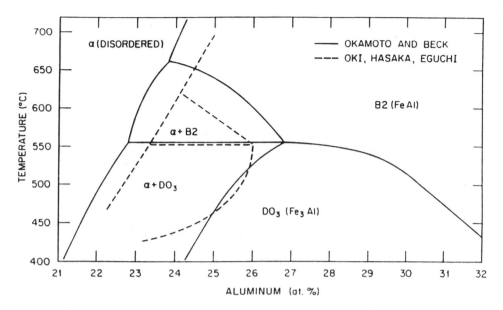

Figure 5.14: *Selected part of the iron–aluminum phase diagram (Al content between 21 and 32 at.%) showing the intermetallic compounds Fe₃Al, FeAl, and the disordered solid solution of Al in Fe. (Refs. 76,77; reprinted with permission from Ref. 72.)*

The second-order phase transformation between the ordered Fe_3Al (DO_3) and FeAl (B2) plays a significant role in determining the mechanical properties of the iron aluminides. The mechanical properties of iron aluminides ranging in composition from 23 to 40 at.% aluminum have been studied by a number of research groups [72, 78–83]. The general conclusions are extremely encouraging. The iron aluminides, especially those prepared by the powder processing route, show good room-temperature tensile elongation and strength combination, coupled with good strength, ductility, and environmental resistance at elevated temperatures.

The mechanical properties of the iron aluminide-based intermetallic compounds near the Fe_3Al composition show a strong dependence on composition, temperature, and the thermomechanical history. The yield strength increases with temperature up to a maximum at 823 K (550°C) (which is the second-order phase transformation temperature), followed by a sharp drop. According to Inouye [81], the rise in the yield strength with temperature can be attributed to the superdislocation interactions with the ordered Fe_3Al lattice. The compounds near the Fe_3Al composition can also be age-hardened by the proper selection of time and temperature. It has been shown that a Fe–24 at.%

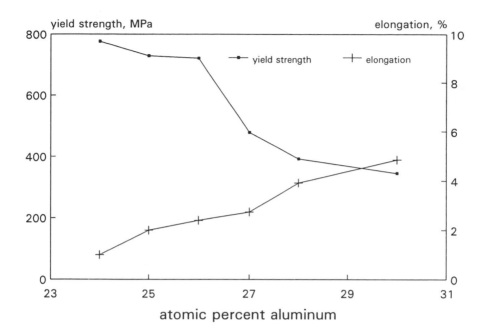

Figure 5.15: *Variation of room temperature yield strength and elongation with varying aluminum content in an iron–aluminum alloy. (Data from Ref. 72.)*

Al can be age-hardened by precipitation of the disordered α-solid solution from the ordered DO_3 phase. The variation of the yield strength with aluminum content also depends on the ordering temperature.

The effect of aluminum content on the yield strength and elongation of iron aluminides is shown in Figure 5.15 [72]. It can be observed that the yield strength was the highest for the 24–26 at.% Al, after which it dropped off sharply. The ordering temperature for the investigation was 773 K (500°C). It can be seen that the transition from the high yield strength to the low yield strength region corresponds approximately to the phase field boundary where the two-phase field region, $DO_3 + \alpha$, changes to the single-phase field region of DO_3. The tensile elongation, on the other hand, continues to increase with increasing aluminum content.

The effect of test temperature on the yield strength of Fe–24Al and a Fe–28Al (at.%) alloy is shown in Figure 5.16 [72]. For comparison, the yield strengths of 316 stainless steel and a high-temperature modified Fe–9Cr–1Mo alloy is also shown in the same figure. It can be observed that the Fe–24Al (at.%) alloy exhibits superior yield strengths over all the alloys shown in the figure up to 973 K (700°C). This is a positive reflection of the potential of this class of intermetallic compound.

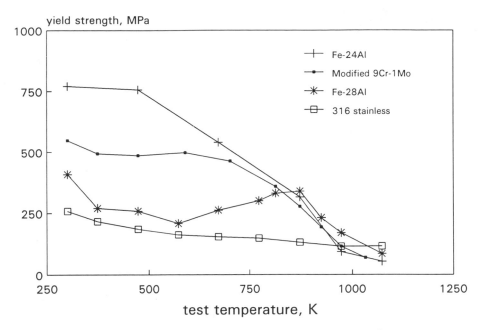

Figure 5.16: *Variation of yield strength with test temperature for Fe–9Cr–1Mo, 316 Stainless Steel, Fe–24Al (at.%), and Fe–28Al (at.%) alloy. (Data from Ref. 72.)*

The binary iron aluminides have often been alloyed with other elements in order to obtain desirable property combinations. The effects of alloying elements on the stability of the $D0_3$ phase field and their effects on the mechanical properties have been studied and reported by a number of authors [72, 83–87]. It has been observed that elements that increase the strength of iron aluminades also in turn increase the $D0_3$ to B2 transition temperature. It is also seen that the strengthening of iron aluminides is possible by both precipitation hardening and solid-solution strengthening.

An excellent review on ternary additions to iron aluminides has been given by Mendiratta et al. [88]. The ternary additions were made on the philosophy that the yield strength of the binary Fe–25Al (at.%) alloy decreased sharply at temperatures above 673 K (400°C) due to the $D0_3$ to B2 transformation, and additions that raise the transformation temperature will also improve the strength at elevated temperatures. From trials with various additives it was determined that silicon and titanium increased the transformation temperature significantly [89].

To test the effect of silicon addition, powder processing of an Fe–20Al–5Si (at.%) alloy was attempted. The consolidated of the material was achieved by extruding mixed inert-gas-atomized powders with compositions of Fe–25Al (at.%) and Fe–25Si (at.%) [90]. The consolidated intermetallic compound had a room-temperature strength of 813 MPa (118 ksi) and a ductility of 0.6% compared to a strength of 696 MPa (101 ksi) and a ductility of 8% for a Fe–25Al (at.%) binary alloy. At 873 K, however, the strength of the Fe–20Al–5Si (at.%) alloy was four times greater than the strength of the binary Fe–25Al (at.%) alloy.

Culbertson and Kortovich [85] carried out extensive investigations on alloying of the Fe_3Al composition with Ti, Nb, Cr, V, Mo, Ta, Si Cu, Ni, and Mn. Significant increases in yield strength at 873 K (600°C) were observed with the additions of Mo, Si, Nb, Cr, Ti and Ta, with a concomitant drop in the ductility with all the additions except for chromium. Similar improvement in the room-temperature ductility was also confirmed by Oak Ridge National Laboratory (ORNL) [72], where it was observed that 2% to 6% addition of chromium could produce ductilities around 7% to 10% compared to 4% for the binary alloy. The chromium and molybdenum additions also had a very strong solid-solution strengthening effect at high temperatures. Investigations carried out by ORNL [72] on a large suite of alloying additions showed that additions of Mo, Nb, Ti, B, and Zr reduced the grain size with only niobium additions showing a continuous decrease in grain sizes with increasing additions. The additions of Cr, Mn, Ce, and Y did not have much effect on the grain size. The study concluded that the formation of fine second-phase precipitates is more effective in refining grain size and increasing the recrystallization temperature than the additions that form solid solutions. It was concluded that the most promising alloys for future study were the ones containing chromium, niobium, molybdenum and boron.

The other route normally selected to enhance the properties of iron aluminides is dispersion strengthening. An interesting material for which a coherent ordered phase is precipitated in an ordered Fe_3Al matrix has been reported by Dimiduk et al. [84]. Dispersoids in the form of TiB_2 (1.5 mole) in stoichiometric Fe_3Al compound were described by Slaughter and Das [91]. Rapidly solidified powder produced by centrifugal atomization and rapid quenching in jets of helium gas was either hot isostatically pressed or forged at various temperatures. With subsequent thermomechanical treatment, the TiB_2 particles could be uniformly dispersed in the matrix and did not rapidly coarsen during the treatment. This dispersion was stable at temperatures slightly above 1273 K (1000°C). The stabilized grain size was around 1 to 2 μm.

Rapid solidification has also been used by Air Force Wright Aeronautical Laboratories of Wright Patterson Air Force Base to develop iron aluminides that contained a fine dispersion of TiB_2 [92]. Powders of the rapidly solidified material were densified by hot extrusion, which was followed by hot and warm working. One of the materials developed exhibited a room-temperature tensile elongation in the range of 15% to 20% with a tensile strength of 965 MPa and a creep rupture strength at 1255 K (982°C) that was equivalent to that of Hastealloy X. The best results were obtained from the finer grain size structures. Research was also carried out by ORNL on similar dispersoid-strengthened iron aluminide. The ORNL report [72] concluded that the TiB_2 dispersion-strengthened compounds, though good, had poor weldability.

2. FeAl

Compounds near the FeAl composition that exhibit the ordered B2 structure are also of interest for their low density and increased oxidation resistance. The equiatomic composition is extremely brittle, while the Fe–40Al (at.%) compound exhibits some amount of ductility. It has also been found that boron, which has been so successful in imparting ductility to Ni_3Al, is also effective in increasing the ductility of FeAl [82].

In one investigation, prealloyed powders of Fe–40Al (at.%) and Fe–50Al (at.%) were mixed with elemental boron powders. The powder blends were extruded twice at 1250 and 1073 K (977 and 800°C). In tension, the Fe–50Al (at.%) showed practically no ductility while the Fe–40Al showed some yielding before failure. The addition of small amounts of boron (0.05 to 0.2 wt.%) increased the ductility of the as-extruded Fe–40Al (at.%) to approximately 6% without any significant change in the yield strength. The Fe–50Al, however, remained brittle even with the boron additions. It is postulated that the boron addition improves the grain boundary cohesion in the Fe–40Al (at.%) alloy. The reason for the ineffectiveness of boron in a Fe–50Al alloy is still unclear.

Gaydosh et al. [83] described the effects of alloying a B2 intermetallic FeAl compound having 40 at.% aluminum with C, Zr, Hf, and B. The material used in the study was produced by hot extrusion of gas-atomized powders. The C, Zr, and Hf were incorporated in the prealloyed powder during gas atomization and the effects of boron addition on the alloys were studied by addition of elemental boron to the gas-atomized powders. All the ternary alloys had some ductility, though it was reduced considerably compared to the binary 40 at.% Al intermetallic. The addition of zirconium and hafnium increased strengths up to 900 K (627°C). Fe_6Al_6Zr and Fe_6Al_6Hf precipitates were identified as the principal second phase, whereas the carbon-containing alloy showed a grain boundary phase. Addition of boron to the carbon- and zirconium-containing alloys resulted in restored ductility of around 5% and an increased fraction of transgranular fracture.

Dispersion strengthening has also been tried in FeAl-based alloys. The results of the investigations carried out by Vedula and coworkers [93,94] were extremely encouraging. Their investigation focused around iron aluminides with 40 at.% aluminum. A gas-atomized FeAlZr powder was blended with submicrometer boron powder to produce a final composition of Fe–40Al–0.1Zr–0.2B (at.%). Dispersion of 1 vol.% Y_2O_3 was obtained by mechanically alloying the FeAlZrB powder with Y_2O_3. Consolidation was carried out by hot extrusion at different temperatures ranging from 1250 to 1450 K (977 to 1177°C), and a fixed extrusion ratio of 16 : 1. The oxide dispersion resulted in refined microstructures compared to the gas-atomized powders. The oxide-dispersed materials had an average grain size between 3 and 5 μm with a uniform distribution of the Y_2O_3 particles. Room-temperature tensile elongations up to 10% were obtained in the ODS alloys (among the highest ductilities observed for this class of compound) and strengths greater than 1000 MPa were also observed. The high-temperature strength of the ODS alloy was also found to be greater than that of the non-ODS counterpart. The low-temperature (1250 K; 977°C) extrusion of the ODS alloy yielded the best property combination.

Powder metallurgy techniques have been used to introduce dispersoids in a Fe–35Al (at.%) alloy. The two dispersoids studied were hafnium-rich particulates and TiB_2 [95]. The materials were consolidated by powder extrusion and the resulting dispersoid size for the hafnium-containing alloy was 0.2 μm, while for the TiB_2-containing alloy it was 0.04 μm. Both the alloys exhibited around 8% elongation and improved yield strengths up to a temperature of 873 K (600°C), above which the strength decreased rapidly.

Reactive processing of iron aluminides has been carried out by Bose et al. [96], and the results of the preliminary studies are quite encouraging. The system thermodynamics is not as favorable for the iron aluminum system as for the nickel–aluminum system, and, as a consequence, pressureless reactive sintering does not result in any significant densification. However, reactive hot isostatic pressing has been carried out successfully. In the investigation, water-atomized iron powder was mixed with elemental gas-atomized aluminum powder in the desired stoichiometric ratio. The mixed powder was cold isostatically pressed into rods. After CIP the rods were encapsulated in stainless-steel cans and hot isostatically pressed at 1473 K (1200°C) using a pressure of 82 MPa for 3600 s (1 hour). The resulting sample was approximately 90% dense, and the tensile properties (before ordering) were extremely encouraging, with a yield strength around 850 MPa, ultimate strength of 979 MPa, and a strain to failure of around 3%. The reactive processing provides an inexpensive alternative to the normal powder processing approaches where prealloyed powder is used as the starting material. Another alternative approach that has also been investigated by Bose consists of prereacting the elemental powders, milling the reacted powder cake, cold isostatic pressing into rods, canning, vacuum degassing, and hot isostatic pressing. The work is in its infancy but it also shows good potential. This process will have flexibility where additional strengthening phases like TiB_2 or Y_2O_3 particles might be added in during milling of the prereacted powder cake.

The oxidation and sulfidation behavior of the iron aluminides proved to be extremely encouraging [72,97,98]. Oxidation studies carried out by Schmidt et al. [97] with aluminum contents varying from 34 to 51 at% showed that the resistance of FeAl is far superior to 304 stainless steels. With increasing aluminum contents the oxidation rate also decreased. Oxidation tests were carried out in air at temperatures of 1173 and 1473 K (900 and 1200°C). With 3600 s (1 hour) annealing of the Fe–36Al sample the surface turned reddish, suggesting the formation of Fe_2O_3. However, for all other samples (for the Fe–36Al samples longer hold times) the oxidation tests resulted in a dark gray oxide layer, probably of alumina.

The sulfidation resistance of this alloy has also been studied [72] and found to be superior to that of the commercial high-temperature alloys presently in use. This offers a great boost to the development of this compound, especially since recent powder processing techniques [98] have provided excellent combination of ductility and strength.

The iron aluminides could be in environments where oxidation and sulfidation at elevated temperatures are major problems (for example, in fossil fuel conversion plants). The material is expected to be an inexpensive replacement for several grades of stainless steels.

Overall, the iron aluminides possess good strength at moderate temperatures of 773 K (500°C), have lower density compared to the nickel- or cobalt-based alloys and steels, and do not depend on expensive and strategic elements. Powder processing of the advanced iron aluminides offers the potential of property combinations that could be close to some of the superalloys. A case in point would be the oxide dispersion-strengthened Fe–40Al-based (at.%) alloys and the rapidly solidified TiB_2 precipitation-hardened iron aluminides. Novel processing methods such as reactive processing or prereacting of powders to serve as inexpensive starting materials should be utilized in order to enhance the process economics. It is conceivable that a combination of mechanical alloying followed by reactive sintering could produce materials with excellent properties based on iron aluminides. This processing route could also provide a tremendous economic advantage.

Silicide-based Intermetallic Compounds

Although the aluminides have been studied extensively, other classes of intermetallic compounds such as the silicides and beryllides are also being investigated. The increasing demand for high-temperature materials that can serve as hot components in advanced fossil fuel energy systems, turbines, and other sulfidizing environments is expected to stimulate and provide the impetus for the growth of the silicides. These compounds also provide a high melting point to density ratio and excellent strengths at high temperatures. Thus, it will be beneficial to discuss several other intermetallic compounds besides the aluminides described earlier.

The silicides offer a potential pool of material with extremely good high-temperature capability and good stability against hostile environments (sulfuric acid, high-temperature oxidation and sulfidation, etc.). This important class of intermetallic compounds has so far been neglected in favor of the aluminides. However, the attractive properties of the intermetallic compounds based on silicon are gradually leading to more research efforts in this area.

One of the major reasons for the great interest in the silicide-based compounds stems from the ability of these materials to form a continuous adherent layer of SiO_2 film. The commonly used high-temperature materials rely on an Al_2O_3 film or a Cr_2O_3 film for their environmental protection at high temperatures. The SiO_2 film provides comparable high-temperature oxidation resistance but far superior high-temperature sulfidation resistance when compared to the Al_2O_3 or Cr_2O_3 films. It should be remembered that the aluminides rely on the formation of an adherent alumina skin for protection against the environment. Thus, intermetallic compounds based on aluminum do not offer adequate protection against sulfidation. For example, the sulfidation resistance of nickel aluminide was found to be adequate up to 1.5 vol.% of H_2S mixed with hydrogen. With H_2S gas levels above this, nickel-rich sulfur globules formed on the scale surface, rendering the alumina scale susceptible to cracking and spallation [99]. Studies on the nickel aluminides in an H_2–H_2S environment at temperatures of 1148 K (875°C) showed that the corrosion rate is much larger than the acceptable limit [99]. Preoxidation of the advanced nickel aluminide to form the alumina skin and its subsequent exposure to a high sulfur level of 4.7 vol.% of H_2S resulted in the formation of Ni_3S_2 and HfS_2, which do not provide good protection against sulfidizing atmospheres. However, the iron aluminides do provide good resistance against sulfidation.

1. Nickel Silicides

The high-temperature oxidation and sulfidation resistance of silicides has not been studied extensively. However, from the limited research results it is clear that the oxide scale of SiO_2 exhibits very strong resistance against both oxidation and sulfidation. It has been determined that alloys of nickel or iron with 10 wt.% silicon have oxidation rates similar to or better than the commonly used moderate temperature-resistant Ni–20Cr–10Al (wt.%) alloy. The oxidation rate in air decreases as the silicon content is increased, proving that the silica (SiO_2) surface film is an excellent barrier against oxidation [100]. The sulfidation resistance of these materials proved to be much better than that of the chromium- or aluminum-containing alloys, which rely on the alumina or chromia film for their protection [101]. An excellent comparison of the sulfidation properties of a Ni–20Si intermetallic alloy with a Ni–30Cr and a Ni–18Cr–6Al alloy (wt.%) has been reported in the literature [101]. The test atmosphere was a mixture of hydrogen and hydrogen sulfide and the test was conducted at a temperature of 1223 K (950°C). The Ni–Cr–Al alloy formed a thick sulfide scale that spalled off on cooling and the material exhibited extensive attack to the bulk alloy. The Ni–30Cr (wt.%) alloy was almost entirely converted to sulfide. Even when these alloys were preoxidized to form the alumina or chromia film, the exposure to the sulfide-containing environment rapidly caused the film to be broken and resulted in catastrophic corrosion of the underlying layer. At low magnification the Ni–20Si (wt.%) alloy appeared to be unaffected, but careful high-magnification studies revealed porosity that penetrated inside the matrix. However, there was no evidence of condensed sulfide formed on the scale. When the material was preoxidized in air to form a continuous SiO_2 film, it became extremely protective against the sulfidizing atmosphere. Thus, it can be surmised that the SiO_2 film appears to offer the best protection against sulfidation as compared with the commonly employed alumina or chromia films.

The success of the intermetallic compound Ni_3Al also spurred the development of its analogous silicide compound, namely Ni_3Si. Oliver [102,103] has reported the development of intermetallic compounds based on Ni_3Si. The room-temperature ductility of this intermetallic compound is improved by small additions of boron, carbon, or beryllium and can be further enhanced by the addition of niobium. The improvement in the room-temperature ductility of Ni_3Si with the addition of carbon and boron is shown in Figure 5.17. The elevated-temperature mechanical properties of the compound, when tested in air, exhibited similar behavior to its aluminide counterpart. The strength of the compound increased, while the ductility dropped off drastically in the temperature range 773 to 873 K (500 to 600°C), as shown in Figure 5.18. The addition of chromium partially reduces the dynamic embrittlement problem when the testing is carried out in air at 873 K (600°C). Chromium also improves the hot fabricability of Ni_3Si, which is found to be better than that of Ni_3Al.

Additions of less than 1% Hf were found to improve the strength of Ni_3Si to a value similar to that of Ni_3Al. Thus, the compound Ni_3Si exhibits a number

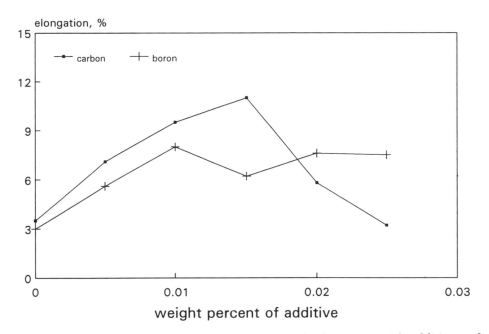

Figure 5.17: *Variation in room-temperature tensile elongation with additions of B and C to a Ni$_3$Si alloy. (Data from Ref. 103.)*

of characteristics that are similar to those of the well-studied compound Ni$_3$Al. However, it is expected to offer superior environmental protection against both oxidation and sulfidation. The intermetallic compound Ni$_3$Si also has the same L1$_2$ type of structure is Ni$_3$Al. The two compounds exhibit a pseudobinary isomorphous phase diagram, and are mutually soluble in one another over the entire composition range, forming a single-phase material that retains the L1$_2$ structure. Interestingly, it has been reported that additions of silicon to Ni$_3$Al (replacing aluminum) produce a marked rise in flow stress at all temperatures between 100 and 900 K [104]. Thus, new compounds based on Ni$_3$(Al,Si) with Cr and Hf offer the prospect of increased strength and creep resistance and better protection against oxidation/sulfidation compared to Ni$_3$Al. The use of powder processing approaches for the optimization and fabrication of such mixed aluminide–silicide compounds is reported in the literature [105].

2. Molybdenum Disilicide

Silicide-based intermetallic compounds such as molybdenum disilicide, and molybdenum disilicide alloyed with tungsten or rhenium disilicide are extensively used as high-temperature heating elements. This material can be used in

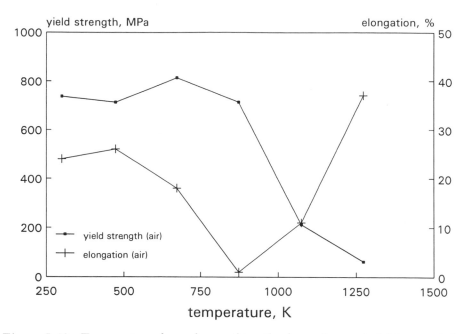

Figure 5.18: *Temperature dependence of tensile elongation and yield strength for a Ni–18.9Si (wt.%) alloy. (Data from Ref. 103.)*

air at temperatures up to 2173 K (1900°C) without undergoing any aging [106]. It can also be used at high temperatures in nitrogen, ammonia, noble gases, hydrogen, and endogas (40%N$_2$, 40%H$_2$, 20%Co nominal) atmospheres. When heated, the material quickly forms a self-generating coating of silica glass on the surface.

Monolithic molybdenum disilicide is an extremely attractive intermetallic compound having a very high melting point of 2283 K (2010°C), low density, and good oxidation resistance due to the ability to form a protective silicon dioxide layer. This film results in usable heating element life in excess of 72×10^5 s (2000 hours) at 1923 K (1650°C). This compound is currently used in high-temperature furnace heating elements (operating at air temperatures as high as 2023 K; 1750°C) but is not used as a high-temperature structural material. The extreme brittleness of this compound at room temperature has prevented its use as an elevated-temperature structural part. The material has a brittle-to-ductile transition at a temperature of 1273 K (1000°C). The resistance of the material against hot corrosion by NaCl and H$_2$S is approximately an order of magnitude better than that of the best superalloys [107]. Thus, the lure of being able to use this extremely high-temperature material for structural applications is resulting in a tremendous amount of research.

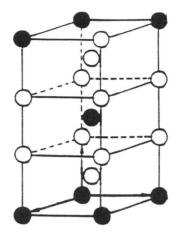

Figure 5.19: *Molybdenum disilicide crystal structure.*

Molybdenum disilicide exhibits a C11$_b$ type of ordered structure up to 2173 K (1900°C). The atomic arrangement in the C11$_b$ crystal structure is shown in Figure 5.19. It is postulated that this material may exhibit anomalous strengthening with temperature [108]. In recent publications from Los Alamos National Laboratory [109,110], toughening and strengthening of MoSi$_2$ have been achieved by SiC whisker reinforcement. In their work, MoSi$_2$ powder (particle size between 0.6 to 20 μm) or MoSi$_2$ powder dry-blended with SiC whiskers were hot pressed in graphite dies at 1898 to 1913 K (1625 to 1640°C). The dies and punches were surrounded by lamp black, which resulted in an atmosphere of carbon monoxide- and carbon dioxide-rich air. The pressing pressure was 41.4 MPa. The monolithic compound displayed significant grain growth and also the evidence of liquid phase formation. This problem was reduced when SiC whiskers were added to the material. Reinforcing a MoSi$_2$ matrix with 20 vol.% SiC whiskers resulted in a 54% increase in the fracture toughness and 100% increase in the flexural strength. The fracture toughness of 8.2 MPa m$^{1/2}$ obtained by the 20 vol.% SiC whisker-reinforced MoSi$_2$ represents a level that is still short of but approaching the acceptable regime for structural components. It serves as a pointer toward the potential of silicon carbide-reinforced molybdenum disilicide as an excellent high-temperature structural material.

In order to use the silicides as high-temperature structural components, one of the major criteria should be the enhanced creep resistance and high-temperature strength. The whisker reinforcement will certainly enhance the creep properties of the matrix. The monolithic matrix material could be further strengthened by incorporating niobium, tantalum, tungsten, or rhenium disilicides as solid solution strengtheners in the $MoSi_2$ matrix. The choice of the solid solution strengthener will have to take into consideration the density and the thermochemical compatibility of the additive and the reinforcing phases.

The other type of additive that could be used to modify the matrix will also influence the oxidation behavior of the material. Molybdenum disilicide is very attractive as a high-temperature material partly as a result of the ability to form a self-healing protective silica glass layer. This depends on the volatilization of the simultaneously formed molybdenum trioxide. Oxidation without the molybdenum trioxide volatilization causes fast intergranular oxidation with severe material damage, often referred to as the "pest". The addition of small amounts of $MoGe_2$ can reduce this problem [111].

Self-propagating high-temperature synthesis has also been successfully used to produce the molybdenum disilicide compound from a mixture of elemental powders [112]. A stoichiometric mixture of 63.1% Mo and 36.9% Si (wt.%) was loaded into a reactor in a stainless-steel can and ignited by powering a tungsten spiral. The reactor was first purged with nitrogen and then hydrogen was maintained throughout the combustion synthesis. The resultant material was fully converted to $MoSi_2$ with around 2% impurity, which was Mo_5Si_3. Reaction pressing of molybdenum disilicide and reaction bonding of molybdenum disilicide to alumina seals have also been attempted, utilizing the high exotherm of the compound formation [113]. It is therefore conceivable that a two-phase composite material can be produced by selecting the proper composition of the elemental powder mixture.

One of the most potent uses of silicide-based intermetallic compounds might be their potential to serve as the matrix material for diboride-reinforced composites that could be suitable for ultrahigh-temperature applications (temperatures greater than 1873 K; 1600°C). Investigations carried out by Vedula and Lisy [114] have demonstrated that molybdenum disilicide has the ability to protect titanium diboride against oxidation. The diborides usually have very high melting points and they also exhibit superior strengths at elevated temperatures. However, their main problem is their poor oxidation resistance, which is primarily due to the formation of a low-melting boron-based oxide. If these diborides are to be used as high-temperature reinforcements, they must be able to withstand the oxidation at ultrahigh temperatures. Oxide-based high-temperature matrices such as zirconia or alumina have proved to be ineffective in preventing oxidation at ultrahigh temperatures. The study carried out by Vedula and Lisy [114] on molybdenum disilicide matrix composite reinforced with 20 wt.% titanium diboride indicated that the composite, when exposed to air at a temperature of 1923 K (1650°C), formed a glassy protective layer. The glassy layer was not as protective as the silica layer

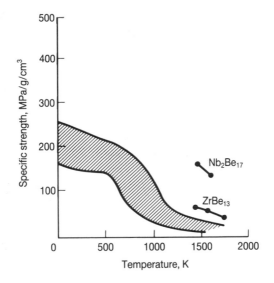

Figure 5.20: Comparison of specific strength as a function of temperature for two beryllides with various state-of-the-art materials (properties shown as a range in the shaded area.) (After Ref. 115.)

formed on the monolithic molybdenum disilicide, but it seemed to prevent any significant reaction of the diboride particles.

Other Intermetallic Compounds

Less common intermetallic compounds such as beryllides have not enjoyed the attention that has been bestowed on the aluminides and silicides. Intermetallic compounds based on beryllium could have attractive properties, since beryllium has high stiffness and low density. A comparison of the specific strength as a function of temperature for various state-of-the-art materials (shown as the shaded envelope in the figure) along with a niobium beryllide and zirconium beryllide is reproduced in Figure 5.20 [115]. The figure shows beryllides as the material of choice at temperatures above 1273 K (1000°C). Some of the materials included within the shaded zone are TiAl, Ti_3Al, NiAl, Rene 41, and rapidly solidified niobium, aluminum, and titanium alloys.

The processing problems associated with beryllium-based intermetallic compounds have inhibited their growth to a large extent. The compound Nb_2Be_{17} is perhaps the only beryllide that is on the verge of commercial exploitation. The intermetallic compound Nb_2Be_{17}, with a rhombohedral structure, has high strength at high temperatures, high specific stiffness, excellent oxidation resistance and good thermal conductivity. The ductility of the compound has been improved to 2%. The strength of this compound also increases with increasing temperature. Addition of a small amount of aluminum prevents the heavy oxide layer that forms at elevated temperatures. Powder metallurgy techniques such as vacuum hot pressing and hot isostatic pressing have been used to fabricate this intermetallic compound.

The intermetallic compounds have also been used extensively as shape memory alloys. The basis for the shape memory phenomenon is dependent on the phase transformation that occurs when a deformed material is heated above a certain temperature. To attain the shape memory effect, the material is obtained in its martensitic state and then deformed to produce apparently permanent strain. On heating, the strain is recovered when the martensite transforms back to its parent phase. A two-way shape memory effect may be produced in materials by thermomechanical processing of a shape memory alloy. This is followed by cooling through the martensite transformation region where the material undergoes a spontaneous change in shape. Heating results in an inverse shape change by the shape memory effect described earlier [116]. Due to the unique characteristics exhibited by this class of materials, they are used in number of applications such as thermally actuated electrical switches, biological sutures, and sophisticated military components.

Detailed discussion on intermetallic compounds exhibiting the shape memory effect (SME) is outside the scope of this chapter. It should be noted that a large number of intermetallic compounds exhibit SME. The most extensively studied shape memory alloy, NiTi, has been developed by Wang and coworkers [117]. The material, commercially known as Nitinol, has become an important engineering alloy. Powder processing has been used extensively to fabricate this material. The conventional powder processing techniques and some recent hot powder processing techniques have been used. Interesting work on the fabrication of NiTi intermetallic compounds from elemental powders has been described by Morris and Morris [118]. Nickel and titanium powders in equiatomic ratio were mixed and milled to various degrees of fineness, and subsequently consolidated by sintering or hot processing. The proper selection of initial milling, compactions and sintering resulted in good quality of the shape memory intermetallic alloy NiTi.

Other intermetallic compounds that are still of academic interest are compounds such as Al_3Ta, the niobium aluminides, and a number of two-phase intermetallic compounds. Reactive powder processing has been extensively used for processing of these intermetallic compounds. The compound Al_3Ta has been consolidated to full density from elemental powders by the process of reactive hot isostatic pressing (RHIP). The mixed powders were cold isostatically pressed into rods that were then inserted into stainless-steel tubes lined with tantalum foil and coated with alumina slurry. The samples were vacuum degassed for 72×10^3 s (20 hours) at a temperature of 773 K (500°C), and then vacuum sealed. The samples were then reactive hot isostatically pressed. Fully dense samples of Al_3Ta (23 at.% of Ta) were obtained using the following RHIP conditions: 1473 K (1200°C), 172 MPa, 3600 s (1 hour). The hot compressive yield strength of this compound at 1223 K (950°C) was 41 MPa. The RHIP samples of Al_3Ta obtained by the addition of 8 at.% Fe by the replacement of aluminum exhibited a hot compressive strength of 198 MPa at 1223 K (950°C) [119].

The intermetallic compound Al_3Nb, which is being investigated as a potential high-temperature structural material, especially as a matrix material for high-temperature composites, has also been consolidated to a maximum density of 95% by reactive sintering. Compact produced from the mixture of elemental 9 μm niobium powder and 30 μm aluminum powder is outgassed at 773 K (500°C) and heated to 1473 K (1200°C) at 0.25 K/s and held at the maximum temperature for 3600 s (1 hour). RHIP of this compound at 1473 K (1200°C) at 170 MPa resulted in a 98% dense material [120]. Further development in the reactive processing of this compound has been reported by Murray and German [121].

The high-temperature intermetallic compound Nb_5Si_3 is also being investigated because of its very high melting point and low density. An interesting paper [122] describes a novel method of producing this high-temperature compound by mechanically alloying stoichiometric elemental powders in a Spex mill. Interestingly, continuous milling of elemental powders produced the desired compound after approximately 18×10^3 s (300 minutes), while intermittent milling with cool down after 1800-s (30-minute) intervals achieved compound formation in only 4500 s (75 minutes). However, these exotic high-temperature intermetallic compounds are not expected to find commercial use in the near future until their important properties (such as the mechanical strengths at room and elevated temperatures, the ductility at room and elevated temperatures, the oxidation resistance at elevated temperatures, and the high-temperature creep properties) have been carefully investigated.

The Future

The lure of lightweight high-temperature materials that offer good environmental protection will continue to fuel the research and development in the area of intermetallic compounds. The major thrust for the development of these compounds is expected to come from the aerospace industry. The drive to build hypersonic-velocity commercial vehicles is expected to intensify the

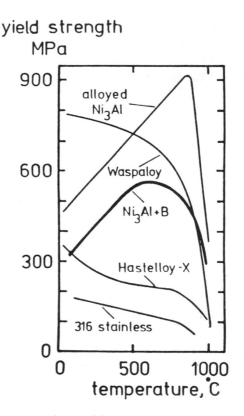

yield strength
MPa

Figure 5.21: Variation in the yield strength with temperature for several high-temperature alloys and two nickel aluminide-based intermetallic compounds. (Reprinted with permission from Ref. 124.)

search for new high-temperature materials with low specific gravity, high stiffness, and resistance to different environments. Research on intermetallic compounds is in progress in most countries around the globe.

The superalloys that have enjoyed a long and uncontested reign in the area of high-temperature materials are presently unable to retain their "super" alloy status owing to the increased performance demands. When we compare the variation with temperature of the density-compensated strength for a few of the presently used heat-resistant alloys with that for some of the long-range-ordered intermetallic compounds, the tremendous potential of the intermetallic compounds can be clearly appreciated. Some data have been collected and are graphically depicted in a publication from National Materials Advisory Board [123]. Another comparison of the superior high-temperature properties of nickel aluminides and advanced nickel aluminide with several high-temperature alloys such as 316 stainless steel, Hastelloy, and Waspaloy is shown in Figure 5.21 [124]. It should be realized that the intermetallic compounds are presently in their initial growth phase and the aluminides are definitely not the most exciting class of compound among the various intermetallics.

One of the easiest ways of getting an idea of the usefulness of the intermetallic compounds is to determine the melting point and density or specific gravity. Fleischer et al. [12] have collected the melting points of a large number of intermetallic compounds and plotted them against their specific gravity. The specific gravity and the melting point are the two intrinsic properties that are insensitive to the processing history and minor compositional variations. The data for more than 280 intermetallic compounds have been plotted by Fleischer and are redrawn in Figure 5.22 (with only a handful of intermetallic compounds). The curve that has been drawn represents a rough envelope for the data and indicates an empirical optimum for the intermetallic compounds. Unfortunately, the majority of the intermetallic compounds, especially the $L1_2$ types of compounds that offer the hope of some ductility at room temperature, lie toward the unfavorable side of the curve. A few hafnium- and rhenium-based compounds and a number of beryllides lie close to the hypothetical curve shown in Figure 5.22. Thus, at least for aerospace types of applications, the $L1_2$ type of compounds are not expected to provide any exceptional opportunities.

The detailed testing of each and every composition is really not a practical way of going about identifying real high-temperature intermetallic compounds that will be suitable for structural applications. Fleischer et al. have suggested a quick screening method that consists of measuring the variation in the microhardness with temperature. According to Fleischer, intermetallic compounds based on beryllium are extremely attractive due to their favorable position in the specific gravity–melting point diagram and their high specific stiffness. According to another study by Anton and coworkers [8], the stress is on the identification and focusing on few intermetallic systems of desirable crystal structures and adequate important engineering properties such as oxidation and corrosion resistance, high-temperature strength and creep, plasticity, and specific stiffness. The prerequisites are low density and high melting point. Other metallurgical factors such as having little or no solubility of the constitutive elements beyond the stoichiometric ratio (line compounds) and undergoing a solid-state transformation are expected to be detrimental.

It is the author's belief that only when designers and engineers are in a position to use materials with very limited ductility at ambient temperatures will the intermetallic compounds attain some commercial use in the aerospace industry. However, the valuable knowledge that is gained from the development of these materials for the aerospace industry can be, and indeed is being,

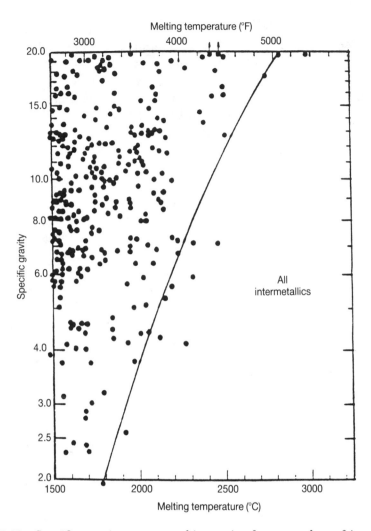

Figure 5.22: *Specific gravity versus melting point for a number of intermetallic compounds. The curve is an approximate envelope to the data. Intermetallic compounds lying to the right of the curve will be the ones with greater potential. (Reproduced, with permission, from* Annual Review of Materials Science, *vol. 19, © 1989 by Annual Reviews Inc., Ref. 12.)*

successfully used in a number of other noncritical applications. This trend is expected to continue, and by the turn of the century intermetallic compounds are expected to become a well-used material, with the nonaerospace applications outnumbering the critical and demanding aerospace applications by a wide margin. This is already becoming evident as the extensively studied compound Ni_3Al+B is being commercially exploited by the auto industry for diesel engine components such as turbocharge rotors, pistons, and valves. Other uses of intermetallic compounds in the near future include aircraft fasteners such as landing gear bolting (being developed by Valley-Todeco, California), electrical resistance heating wires, sheets and rods (Hoskins Research, Michigan), hot roll guides (being fabricated by Metallamics Inc., Michigan), and tools and dies for the glass industry [125].

The future of intermetallic compounds is bright, and their potential as high-temperature components has only recently been commercially recognized. However, it is expected that the nonaerospace applications will be the first commercial area where these intermetallic compounds will be used successfully. Powder processing is expected to lead the way in the processing and fabrication of intermetallic compounds. For very high-melting-point intermetallic compounds, powder processing will not be just an alternate processing route but the only logical choice for fabricating bulk shapes from these materials.

References

1. C.C. Koch, C.T. Liu, and N.S. Stoloff, eds., *High Temperature Ordered Intermetallic Alloys*, Materials Research Society, Pittsburgh, PA, vol. 39, 1985.
2. N.S. Stoloff, *International Metal Review*, vol. 29, p.123, 1984.
3. K. Aoki and O. Izumi, *Journal of the Japan Metallurgical Society*, vol. 43, p.358, 1979.
4. W.M. Schulson, *International Journal of Powder Metallurgy*, vol. 23, p.25, 1987.
5. K. Vedula and J.R. Stephens, *Powder Metallurgy 1986: State of the Art*, ed. W.J. Huppmann, W.A. Kaysser, and G. Petzow, Verlag Schmid, Freiburg, Germany, p.205, 1986.
6. M. Hansen and K. Anderko, *Constitution of Binary Alloys*, 2d ed., Mc-Graw Hill, New York, 1958.
7. I.M. Robertson and C.M. Wayman, *Metallography*, vol. 17, p.43, 1984.
8. D.L. Anton, D.M. Shah, D.N. Duhl, and A.F. Giamei, *Journal of Metals*, vol. 41, p.12, 1989.
9. M.J. Marcinkowski, *Electron Microscopy and Strength of Crystals*, ed. G. Thomas and J. Washburn, Wiley Interscience, New York, p.333, 1963.
10. J.M. Larsen, K.A. Williams, S.J. Balsone, and M.A. Stucke, *Proceedings of the 1989 Symposium on High Temperature Aluminides and Intermetallics*, ed. S. H. Whang, C. T. Liu, D. P. Pope, and J. O. Stiegler, The Minerals, Metals, and Materials Society, Warrendale, PA, pp. 521–556, 1990.
11. T.M.F. Ronald, *Advanced Materials and Processes*, vol. 135, no. 5, p.29, 1989.
12. R.I. Fleischer, D.M. Dimiduk, and H.A. Lipsitt, *Annual Reviews in Material Science*, vol. 19, p.231, 1989.
13. D.M. Wee, O. Noguchi, Y. Oya, and T. Suzuki, *Transactions of the Japan Institute of Metals*, vol.21, p.237, 1980.

14. I. Jung, M. Rudy and G. Sauthoff, *High-Temperature Ordered Intermetallic Alloys II*, ed. N.S. Stoloff, C.C. Koch, C.T. Liu, and O. Izumi, Materials Research Society, Pittsburgh, PA, vol. 81, p.263, 1987.
15. C.T. Liu, C.L. White, and J.A. Horton, *Acta Metallurgia*, vol. 33, p.213, 1985.
16. C.T. Liu and V.K. Sikka, *Journal of Metals*, vol. 38, p.19, 1986.
17. C.L. White, R.A. Padgett, C.T. Liu, and S.M. Yalisove, *Scripta Metallurgia*, vol. 18, p.1417, 1984.
18. A. Choudhury, C.L. White, and C.R. Brooks, *Scripta Metallurgia*, vol. 20, p.1061, 1986.
19. C.L. White and A. Choudhury, *High Temperature Ordered Intermetallic Alloys II*, ed. N.S. Stoloff, C.C. Koch, C.T. Liu, and O. Izumi, Materials Research Society, Pittsburgh, PA, vol. 81, p.427, 1987.
20. C.T. Liu, C.L. White, C.C. Koch, and E.H. Lee, *High Temperature Materials Chemistry II*, ed. Z.A. Munir and D. Cubicchiotti, p.32, 1983.
21. A.I. Taub, S.C. Huang, K.M. Chang, *Metallurgical Transactions*, vol. 15A, p.399, 1984.
22. I. Baker, E.M. Schulson, J.R. Michael, *Philosophical Magazine*, vol. B57, p.379, 1988.
23. R.A.D. Mackenzie and S.L. Sass, *Scripta Metallurgia*, vol. 22, p.1807, 1988.
24. C.C. Koch, J.A. Horton, C.T. Liu, O.B. Gavin, and J.O. Scarbrough, *Rapid Solidification Processing*, National Bureau of Standards, Washington, D.C., p.264, 1983.
25. T. Takasugi, O. Izumi, and N. Masahashi, *Acta Metallurgia*, vol. 33, p.1259, 1985.
26. D.M. Dimiduk, V.L. Weddington, and H.A. Lipsitt, *High Temperature Ordered Intermetallic Alloys II*, ed. N.S. Stoloff, C.C. Koch, C.T. Liu, and O. Izumi, Materials Research Society, Pittsburgh, PA, vol. 81, p.221, 1987.
27. C.T. Liu, C.L. White, and E.H. Lee, *Scripta Metallurgia*, vol. 19, p.1247, 1985.
28. S. Hanada, S. Watanabe, and O. Izumi, *Journal of Material Science*, vol. 21, p.203, 1985.
29. B.H. Kear and H.G.F. Wilsdorf, *Transactions of the Metallurgical Society, AIME*, vol. 224, p.382, 1962.
30. P.H. Thornton, R.G. Davies, and T.L. Johnston, *Metallurgical Transactions*, vol. 1A, p.207, 1970.
31. R.D. Rawlings and A. Staton-BeVan, *Journal of Material Science*, vol. 10, p.505, 1975.
32. C.T. Liu and C.L. White, *High Temperature Ordered Intermetallic Alloys*, ed. C.C. Koch, C.T. Liu, and N.S. Stoloff, Materials Research Society, Pittsburgh, PA, vol. 39, p.365, 1985.
33. K.M. Chang, S.C. Huang, and A.I. Taub, *High Temperature Ordered Intermetallic Alloys II*, ed. N.S. Stoloff, C.C. Koch, C.T. Liu, and O. Izumi, Materials Research Society, Pittsburgh, PA, vol. 81, p.303, 1987.
34. R.J. Lauf and C.A. Walls, *ECUT Quarterly Report*, ORNL, July 1 to September 1, 1985.
35. J.S.C. Jang and C.C. Koch, *Scripta Metallurgia*, vol. 22, p.677, 1988.
36. R.M. German, A.Bose, and D. Sims, U.S. Patent 4,762,558, August 9, 1988.
37. A. Bose, B.H. Rabin, and R.M. German, *Powder Metallurgy International*, vol. 20, p.25, 1988.

38. A. Bose, B. Moore, R.M. German, and N.S. Stoloff, *Journal of Metals*, vol. 40, p.14, 1988.
39. E.M. Schulson, *High Temperature Ordered Intermetallic Alloys*, ed. C.C. Koch, C.T. Liu, and N.S. Stoloff, Materials Research Society, Pittsburgh, PA, vol. 39, p.193, 1985.
40. E.M. Schulson and D.R. Barker, *Scripta Metallurgia*, vol. 17, p.519, 1983.
41. D.J. Gaydosh and M.A. Crimp, *High Temperature Ordered Intermetallic Alloys*, ed. C.C. Koch, C.T. Liu, and N.S. Stoloff, Materials Research Society, Pittsburgh, PA, vol. 39, p.429, 1985.
42. K. Vedula, V. Pathare, I. Aslanidis, and R.H. Titran, *High Temperature Ordered Intermetallic Alloys*, ed. C.C. Koch, C.T. Liu, and N.S. Stoloff, Materials Research Society, Pittsburgh, PA, vol. 39, p.411, 1985.
43. V. Pathare and K. Vedula, *Modern Developments in Powder Metallurgy*, ed. E.N. Aqua and C.I. Whitman, Metal Powder Industries Federation, Princeton, NJ, vol. 16, p.695, 1984.
44. J.C. Murray, R. Laag, W.A. Kaysser, and G. Petzow, *Advances in Powder Metallurgy*, compiled by E.R. Andreotti and P.J. McGeehan, Metal Powder Industries Federation, Princeton, NJ, vol. 2, p.233, 1990.
45. J.D. Whittenberger, *Material Science and Engineering*, vol. 57, p.77, 1983.
46. E.M. Garla, *Mechanical Properties of Intermetallic Compounds*, ed. J.H. Westbrook, Wiley, New York, p.358, 1960.
47. J.R. Weir Jr., *Proceedings of the Gas Research Institute Seminar on Applications for and Designing with High Temperature Materials in Natural Gas Usage*, Chicago, April 6–7, p.229, 1989.
48. T.B. Massalski, J.L. Murray, L.H. Bennett, and H. Baker, *Binary Alloy Phase Diagrams*, ASM, Metals Park, Ohio, p.175, 1986.
49. S.M. Sastry and H.A. Lipsitt, *Metallurgical Transactions*, vol. 8A, p.1543, 1977.
50. H.A. Lipsitt, *High Temperature Ordered Intermetallic Alloys*, ed. C.C. Koch, C.T. Liu, and N.S. Stoloff, Materials Research Society, Pittsburgh, PA, vol. 39, p.351, 1985.
51. P.L. Martin, H.A. Lipsitt, N.T. Nuhfer, and J.C. Williams, *Proceedings of the 4th International Conference on Titanium*, Japan, p.1245, 1980.
52. M.J. Blackburn, D.L. Ruckle, and C.E. Bevan, AFML-TR-78-18, Wright-Patterson Airforce Base, Ohio, 1978.
53. M.J. Blackburn and M.P. Smith, AFWAL-TR-80-4175, Wright-Patterson Airforce Base, Ohio, 1980.
54. M.J. Blackburn and M.P. Smith, AFWAL-TR-82-4086, Wright-Patterson Airforce Base, Ohio, 1982.
55. S.A. Court, J.P.A. Lofvander, M.H. Loretto, and H.L. Fraser, *Philosophical Magazine*, vol. A59, p.379, 1989.
56. A.G. Jackson, K. Teal, and F.H. Froes, *High Temperature Ordered Intermetallic Alloys*, ed. C.C. Koch, C.T. Liu, and N.S. Stoloff, Materials Research Society, Pittsburgh, PA, vol. 39, p.143, 1985.
57. K. S. Chan, *Metallurgical Transactions*, vol. 24A, p. 1095, 1993.
58. S.J. Balsone, *Oxidation of High-Temperature Intermetallics*, The Minerals, Metals, and Materials Society, Warrendale, PA, p.219, 1989.
59. D.G. Konitzer and H.L. Fraser, *High Temperature Ordered Intermetallic Alloys*,

ed. C.C. Koch, C.T. Liu, and N.S. Stoloff, Materials Research Society, Pittsburgh, PA, vol. 39, p.437, 1985.

60. R.G. Rowe, J.A. Sutliff, and E.F. Koch, *Rapidly Solidified Alloys and their Mechanical and Magnetic Properties*, ed. B.C. Giessen, D.R. Polk, and A.I. Taub, Materials Research Society, Pittsburgh, PA, vol. 58, p.359, 1986.

61. J.A. Sutliff and R.G. Rowe, *Rapidly Solidified Alloys and their Mechanical and Magnetic Properties*, ed. B.C. Giessen, D.R. Polk, and A.I. Taub, Materials Research Society, Pittsburgh, PA, vol. 58, p.371, 1986.

62. R.G. Rowe, J.A. Sutliff, and E.F. Koch, *Rapid Solidification Technology for Titanium alloys*, ed. F.H. Froes, D. Eylou, and S.M.L. Sastry, Proceeding of AIME Conference, New Orleans, LA, 1986.

63. H.A. Lipsitt, D. Shechtman, and R.E. Schafrik, *Metallurgical Transactions*, vol. 6A, p.1991, 1975.

64. M.J. Blackburn and M.P. Smith, AFML-TR-78-78, Wright-Patterson Airforce Base, Ohio, 1978.

65. D.S. Shong, Y. W. Kim, C. F. Yolton, and F. H. Froes, *Advances in Powder Metallurgy*, compiled by T.G. Gasbarre and W.F. Jandeska, Metal Powder Industries Federation, Princeton, NJ, vol. 3, p. 359, 1989.

66. R.E. Schafrik, *Metallurgical Transactions*, vol. 7B, p.713, 1976.

67. R. Oddone, M.S. Thesis, RPI, Troy, NY, 1989.

68. P.L. Martin, M.G. Mendiratta, and H.A. Lipsitt, *Metallurgical Transactions*, vol. 14A, p.2170, 1983.

69. S.M. Sastry and H.A. Lipsitt, *Metallurgical Transactions*, vol. 8A, p.299, 1977.

70. R.A. Perkins, K.T. Chiang, and G.H. Meier, *Scipta Metallurgia*, vol. 21, p.1505, 1987.

71. M. Yamaguchi, Y. Umakoshi, and T. Yamane, *High Temperature Ordered Intermetallic Alloys II*, ed. N.S. Stoloff, C.C. Koch, C.T. Liu, and O. Izumi, Materials Research Society, Pittsburgh, PA, vol. 81, p.275, 1987.

72. C. G. McKamey, C. T. Liu, S.A. David, J.A. Horton, D.H. Pierce, and J.J. Campbell, *Development of Iron Aluminides for Coal Conversion Systems*, ONRL/TM-10793, July, 1988.

73. M.J. Marcinkowski, M.E. Taylor, and F.X. Kayser, *Journal of Material Science*, vol. 10, p.406, 1975.

74. W.R. Kerr, *Metallurgical Transactions*, vol. 17A, p.2298, 1986.

75. O. Kubaschewski, *Iron Binary Phase Diagrams*, Springer Verlag, New York, p.5, 1982.

76. H. Okamoto and P.A. Beck, *Metallurgical Transactions*, vol. 2, p.596, 1971.

77. K. Oki, M. Hasaka, T. Eguchi, *Japan Journal of Applied Physics*, vol. 12, p.1522, 1973.

78. D.K. Chatterjee, M.G. Mendiratta, S.K. Ehlers, and H.A. Lipsitt, *Metallurgical Transactions*, vol. 18A, p.283, 1987.

79. M.G. Mendiratta, S.K. Ehlers, D.K. Chatterjee, and H.A. Lipsitt, *Proceedings of the 3rd International Conference on Rapid Solidification Processing: Materials and Technologies*, ed. R. Mehrabian, National Bureau of Standards, Washington, D.C., p.240, 1983.

80. S.K. Ehlers and M.G. Mendiratta, AFWAL-TR-82-4089, Air Force Wright Aeronautical Laboratories, Wright-Patterson Airforce Base, Ohio, 1982.

81. H. Inouye, *High Temperature Ordered Intermetallic Alloys*, ed. C.C. Koch, C.T.

Liu, and N.S. Stoloff, Materials Research Society, Pittsburgh, PA, vol. 39, p.255, 1985.

82. M.A. Crimp, K.M. Vedula, and D.J. Gaydosh, *High Temperature Ordered Intermetallic Alloys II*, ed. N.S. Stoloff, C.C. Koch, C.T. Liu, and O. Izumi, Materials Research Society, Pittsburgh, PA, vol. 81, p.499, 1987.

83. D.J. Gaydosh, S.L. Draper, and M.V. Nathal, *Metallurgical Transactions*, vol. 20A, p.1701, 1989.

84. D.M. Dimiduk, M.G. Mendiratta, D. Banergee, and H.A. Lipsitt, *Acta Metallurgia*, vol. 36, p.2947, 1988.

85. G. Culbertson and C.S. Kortovich, *Development of Iron Aluminides*, AFWAL-TR-85-4155, Air Force Wright Aeronautical Laboratories, Wright-Patterson Airforce Base, Ohio, 1982.

86. R.S. Diehm and D.E. Mikkola, *High Temperature Ordered Intermetallic Alloys II*, ed. N.S. Stoloff, C.C. Koch, C.T. Liu, and O. Izumi, Materials Research Society, Pittsburgh, PA, vol. 81, p.329, 1987.

87. R.T. Fortnum and D.E. Mikkola, *Material Science and Engineering*, vol. 91, p.223, 1987.

88. M.G. Mendiratta, S.K. Ehlers, D.M. Dimiduk, W.R. Kerr, S. Mazdiyasni, and H.A. Lipsitt, *High-Temperature Ordered Intermetallic Alloys II*, ed. N.S. Stoloff, C.C. Koch, C.T. Liu, and O. Izumi, Materials Research Society, Pittsburgh, PA, vol. 81, p.393, 1987.

89. M.G. Mendiratta and H.A. Lipsitt, *High-Temperature Ordered Intermetallic Alloys*, ed. C.C. Koch, C.T. Liu, and N.S. Stoloff, Materials Research Society, Pittsburgh, PA, vol. 39, p.155, 1985.

90. S.K. Ehlers and M.G. Mendiratta, *Journal of Material Science*, vol. 19, p.2203, 1984.

91. E.R. Slaughter and S.K. Das, *Proceedings of the 2nd International Conference on Rapid Solidification Processing*, ed. R. Mehrabian, B.H. Kear, and M. Cohen, Claitor's Publishing Division, Baton Rouge, LA, p.354, 1980.

92. R.G. Bordeau, *Development of Iron Aluminides*, AFWAL-TR-87-4009, Air Force Wright Aeronautical Laboratories, Wright Patterson Air Force Base, Ohio, 1987.

93. S. Strothers and K. Vedula, *Progress in Powder Metallurgy*, compiled by C.L. Freeby and H. Hjort, Metal Powder Industries Federation, Princeton, NJ, vol. 43, p.597, 1987.

94. K. Vedula and J.R. Stephens, *Progress in Powder Metallurgy*, Metal Powder Industries Federation, Princeton, NJ, vol. 43, p.561, 1987.

95. M.G. Mendiratta, T. Mah, and S.K. Ehlers, Interim Technical Reports under Air Force Contract F33615-84-C-5071, April 1985, November 1985, May 1986.

96. A. Bose, R.A. Page, W. Misiolek, and R.M. German, *Advances in Powder Metallurgy*, compiled by L.F. Pease III and R.J. Sansoucy, Metal Powder Industries Federation, Princeton, NJ, vol. 6, p.131, 1991.

97. B. Schmidt, P. Nagpal, I. Baker, *High-Temperature Ordered Intermetallic Alloys III*, ed. C.T. Liu, A.I. Taub, N.S. Stoloff, and C.C. Koch, Materials Research Society, Pittsburgh, PA, vol. 133, p.755, 1989.

98. J.H. Reinshagen and V.K. Sikka, "Thermal Spraying of Selected Aluminides," *Proceedings of 4th National Thermal Spraying Conference*, ASM International, Metals Park, OH, p.307, 1992.

99. N. Natesan, *High Temperature Ordered Intermetallic Alloys II*, ed. N.S. Stoloff,

C.C. Koch, C.T. Liu, and O. Izumi, Materials Research Society, Pittsburgh, PA, vol. 81, p.459, 1987.

100. G. H. Meier, *High Temperature Ordered Intermetallic Alloys II*, ed. N.S. Stoloff, C.C. Koch, C.T. Liu, and O. Izumi, Materials Research Society, Pittsburgh, PA, vol. 81, p.443, 1987.

101. G.M. Kim, Ph.D. Thesis, University of Pittsburgh, 1986.

102. W.C. Oliver and C.L. White, *High Temperature Ordered Intermetallic Alloys II*, ed. N.S. Stoloff, C.C. Koch, C.T. Liu, and O. Izumi, Materials Research Society, vol. 81, p.241, 1987.

103. W.C. Oliver, *High Temperature Ordered Intermetallic Alloys III*, ed. C.T. Liu, A.I. Taub, N.S. Stoloff, and C.C. Koch, Materials Research Society, Pittsburgh, PA, vol. 133, p.397, 1989.

104. S. Ochiai, Y. Mishima, M. Yodogawa, and T. Suzuki, *Transactions of the Japan Institute of Metallurgy*, vol. 27, p.32, 1986.

105. A. Bose, V.K. Sikka, N.S. Stoloff, and D. Alman, *Advances in Powder Metallurgy and Particulate Materials*, compiled by J.M. Kapus and R.M. German, Metal Powder Industries Federation, Princeton, NJ, vol. 7, p.123, 1992.

106. V. Bizzarri, B. Linder, and N. Lindskog, *Ceramic Bulletin*, vol. 68, p.1834, 1989.

107. J. Schlichting, *High Temperatures – High Pressures*, vol. 10, p.241, 1978.

108. Y. Umakoshi, T. Hirano, T. Sakagami, and T. Yamane, *Scripta Metallurgia*, vol. 23, p.87, 1989.

109. F.D. Gac and J.J. Petrovic, *Journal of American Ceramic Society*, vol. 68, p.200, 1985.

110. W.S. Gibbs, J.J. Petrovic, and R.E. Honnell, *Ceramic Engineering Science Proceedings*, vol. 8, p.645, 1987.

111. E. Fitzer and W. Remmele, *Proceedings of the Fifth International Conference on Composite Materials*, Part II, ed. W.C. Harrigan Jr., J. Strife, and A.K. Dhingra, The Minerals, Metals, and Materials Society, Warrendale, PA, p.515, 1985.

112. V.G. Kayuk, M.A. Kuzenkova, S.K. Dolukhanyan, and A.R. Sarkisyan, *Soviet Powder Metallurgy and Metal Ceramic*, vol. 17, p.588, 1978.

113. R.K. Stringer, J.H. Weymouth, and L.S. Williams, *Journal of Australian Institute of Metals*, vol. 14, p.271, 1969.

114. K. Vedula and F. Lisy, *P/M 90 World Conference on Powder Metallurgy*, The Institute of Metals, London, U.K., vol. 1, p.548, 1990.

115. T.G. Nieh, J. Wadsworth, and C.T. Liu, *High Temperature Ordered Intermetallic Alloys III*, ed. C.T. Liu, A.I. Taub, N.S. Stoloff, and C.C. Koch, Materials Research Society, Pittsburgh, PA, vol. 133, p.743, 1989.

116. C.M. Wayman and J.D. Harrison, *Journal of Metals*, vol. 41, p.26, 1989.

117. F.E. Wang, W.J. Buehler and S.J. Pickart, *Journal of Applied Physics*, vol. 36, p.3232, 1965.

118. D.G. Morris and M.A. Morris, *Material Science and Engineering*, vol. A110, p.139, 1989.

119. D. Alman, A. Dibble, A. Bose, R.M. German, N.S. Stoloff, and B. Moore, *Proceedings of Ceramic and Metal Matrix Composites*, ed. H. Mostaghaci, TMS of Canadian Institute of Mining & Metallurgy, Pergamon Press, vol. 17, p.217, 1989.

120. R.M. German, A. Bose, D. Alman, J. Murray, P. Korinko, R. Oddone, and N.S. Stoloff, *P/M 90 World Conference on Powder Metallurgy*, The Institute of Metals, London, U.K., vol. 1, p.310, 1990.

121. J.C. Murray and R.M. German, *Advances in Powder Metallurgy*, compiled by E.R. Andreotti and P.J. McGeehan, Metal Powder Industries Federation, Princeton, NJ, vol. 2, p.145, 1990.
122. J. Kajuch and K. Vedula, *Advances in Powder Metallurgy*, compiled by E.R. Andreotti and P.J. McGeehan, Metal Powder Industries Federation, Princeton, NJ, vol. 2, p.187, 1990.
123. Committee on Application Potential for Ductile Ordered Alloys, National Materials Advisory Board, Publication # NMAB-419, National Academy Press, Washington, D.C., 1984.
124. A. Bose and R.M. German, *Modern Developments in Powder Metallurgy*, compiled by P.U. Gummeson and D.A. Gustafson, Metal Powder Industries Federation, Princeton, NJ, vol. 18, p.299, 1988.
125. M. Hunt, *Materials Engineering*, vol. 107, p.35, 1990.

Chapter 6

Particulate Injection Molding

Introduction

In our daily life we constantly come across complex-shaped objects that are made out of a variety of materials. A large number of these complex shapes are usually made from plastics and, generally, the complex-shaped thermoplastics, which have been around for a long time, are produced by the process of injection molding.

It must have been a materials engineer's dream to be able to process complex-shaped objects from a variety of exotic materials with only a few processing steps. The process of metal casting provided the means for processing useful engineering materials such as steel, aluminum, bronze, etc., to the desired shape complexity. The surface finish and precision of the older cast products (cast in sand molds) was poor. The thermoplastics produced by injection molding, on the other hand, have extremely good dimensional tolerances and excellent surface finish. The introduction of die casting in metallic molds and investment casting into ceramic molds, to some extent, produced complex shapes from useful engineering materials, with good surface finish and dimensional tolerances.

All the near-net shaping techniques such as plastic injection molding, die casting, or investment casting in its rudimentary form, involve melting of the material followed by either pouring or forcing the liquid material under pressure into a preshaped cavity, where it is cooled and then resolidified into the shape of the cavity. The cavity is then removed and the shaped material is ready for end use with few or no secondary operations.

As materials technology advanced, the requirement for processing exotic materials in the form of complex shapes with good surface finish and high precision became evident. There was also a trend of global miniaturization, with smaller objects continually replacing their larger counterparts without compromising their performance to a great extent. Another notable demand of

272

modern technology was high volume of productivity. Gone were the days of the hand-made or one-of-a-kind engineering product. Thus, modern technology had brought with it the demand for small, high-performance, high-precision, high-volume, and complex-shaped parts. This environment was just right for the conception and the rapid growth of the newborn technology of "metal injection molding" (MIM) and "ceramic injection molding" (CIM). Often the term "powder injection molding" (PIM) has been used to refer to both the ceramic and metal injection molding processes. The author would, however, like to call this process "particulate injection molding," or (PIM) owing to the recent introduction of technically "nonpowder" types of materials, such as whiskers and chopped fibers, in the arena of injection molding. Thus, particulate injection molding (PIM) actually covers the full range of high-performance engineering materials such as metals, ceramics, whisker- or fiber-reinforced composites, cermets, and intermetallic compounds. All these materials or combinations of them, in some form or another, are presently being processed by particulate injection molding.

Particulate injection molding has a number of steps that are similar to those of plastic injection molding. In particulate injection molding, metal, intermetallic, or ceramic powders, any combination of these three powdered materials, or a combination of powders and whiskers or chopped fibers are intimately mixed with a polymeric binder. This mixture is known as the injection molding feedstock. This feedstock is granulated and fed into an injection molding machine where it is heated to a temperature at which the feedstock can flow like a viscous fluid. Injection molding of the feedstock is achieved by pressurizing the viscous mass of particulate material and binder mixture into a cavity of the desired shape, which is oversized to account for shrinkage. The feedstock is allowed to cool and solidify in the mold cavity and the solidified shape is then ejected from the die. The oversized green injection molded shape, which has good strength for subsequent handling, is then subjected to the debinding step in which all the polymeric material is removed. This leaves behind a shaped body made out of the particulate material. This shaped body is then heated to the desired temperature at which densification can occur. Thus, a dense solid material that is a smaller-sized version of the die cavity is obtained. A view of the important steps in the PIM process is shown in Figure 6.1. Even from the simplistic steps described above, one can obtain an idea of the complex and interdisciplinary nature of this process. The characteristics of the polymeric binder, the particulate material, the binder–particulate material interactions, injection molding conditions, debinding time–temperature and process schedule, and finally the sintering of the debound material, all play an extremely important role in influencing the final properties of the material. Thus, to get a proper handle and a basic understanding of the process, it is

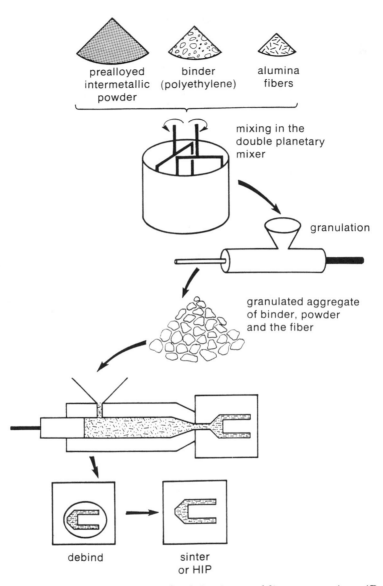

Figure 6.1: Important steps in powder injection molding processing. (Reprinted with permission from Ref. 6.)

important to have an interdisciplinary team consisting of polymer scientists for developing and understanding new binder systems, chemical engineers who can deal with the interactions of the binders and solids and develop adequate debinding cycles for the binder removal, mechanical engineers for tool design and injection molding, and finally materials engineers who can consolidate the material to the desired densities. Lack of basic understanding of the processing fundamentals was one of the prime factors responsible for the very slow growth of this material shaping process with tremendous potential. Surprisingly, sporadic accounts of the process were available from the late 1920s, and in 1949 Schwartzwalder [1] showed pictures of some of the early ceramic components that were produced by PIM.

Polymeric materials with metallic or ceramic particulate reinforcements to improve their mechanical strengths, or impart specific thermal, electrical, or magnetic properties, have been reported in the literature [2]. In this type of product, the particulate material usually occupies a low volume fraction compared to the polymeric material, and also the polymer is retained to form a part of the end product. In contrast, the PIM feedstock usually has a higher volume fraction of the particulate material, and the polymeric binder is absent in the end product. The polymeric binder in the PIM process is incorporated simply to impart a viscous flowable fluidlike characteristics to the feedstock so that it can flow under pressure to fill up the die cavity and provide sufficient handling strength to the green part after ejection from the injection molding machine. Apart from that, the polymeric binder serves no other purpose, and has to be completely removed before the material can be fully densified. Thus, the binder in the PIM process is a necessary evil that has to be tolerated and ultimately eliminated. This binder removal step, commonly known as debinding, is usually the rate-controlling step in the PIM process. The inhibited growth of PIM could to some extent be attributed to this slow rate-controlling step of debinding. However, most of the inhibiting factors are gradually being overcome and the process is poised for explosive growth.

The gestation period for PIM, which really spanned two decades, was finally over, and the latter part of the 1980s witnessed an extraordinarily rapid growth of PIM. Various factors that have hindered its growth have been discussed in great detail by Pease [3]. According to Pease, the classic problem of the long lead times that are required for a new process to gain general acceptability played a key role in stunting the growth of this technology. Among the other factors that contributed to the initial slow growth was the long processing times that used to be required to remove the binders, the difficulties associated with producing uniform feedstock, difficulties in tool design, nonuniform shrinkage due to numerous factors, the cost and availability of powders that are tailored to the PIM process, the lack of specific standards for PIM materials, and of course the lack of skilled personnel. The incubation period was extended further due to the lack of dissemination of the technology and science of the PIM process.

The industry went through its gory maturation process where a few small, undercapitalized, and short-handed companies were struggling with an enor-

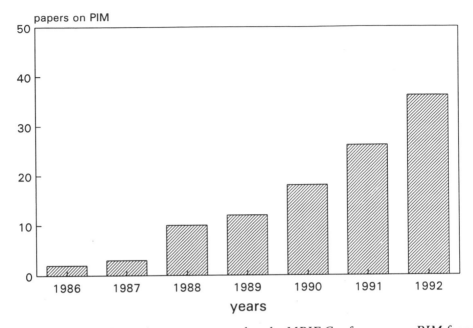

Figure 6.2: *Numbers of papers presented at the MPIF Conferences on PIM from 1986 to 1991.*

mously complex and interdisciplinary problem. The splintering process, accelerated by the basic human thought process that "I know more than the person running the organization while earning less than him," also played a major role in inhibiting growth. This splintering — people breaking off from the parent company to form a spin-off company — resulted in a host of lawsuits aimed at securing the patent rights for the process. Thus, the initial period of growth of this technology was fraught with numerous impediments. It is therefore extremely pleasing to finally see the process burgeon into a rapidly developing near-net-shape forming commercial technology. The rapid dissemination of information, the willingness of industries to finally share part of their experience in open platforms, the formation of consortiums to enhance the underlying scientific and basic principles of PIM processing, and the formation of the Metal Injection Molding Association (MIMA) to further the cause of PIM, have all contributed to the rapid growth of the technology. An interesting guide through which one can easily get a glimpse of the rapid dissemination of this technology can be obtained by following the number of presentations that have been made in the Powder Metallurgy Annual Meetings since 1985. The number of presentations on PIM from 1986 to 1992 is shown in Figure 6.2. It can be concluded that the industry, academe, and the research organizations are now working toward better fundamental understanding and process development by the exchange of ideas and experience.

Another noticeable fact is the gradual consolidation of the PIM industry, where in the past few years a number of small companies have gone bankrupt, and the surviving companies have exhibited rapid growth. This trend is expected to continue, as it is gradually realized that although the process on the surface may seem amazingly simple, it is in reality a very involved interdisciplinary technology. The process may not be extremely capital-intensive, but it is definitely technology-intensive. Companies trying to invest or diverge into this area should be cautious and have the proper technical backbone to support the business, or be prepared to take heavy losses.

In the following sections, attempts will be made to bring into focus some of the underlying scientific and technological principles that govern this rejuvenated process of PIM. It would be impossible to totally cover all the subtle processing variations and all the materials that are processed by PIM within the scope of this book. Instead, discussion of the generic techniques and their applications in the PIM industry will be the focus of this chapter. The principal aim is to provide readers with an overall view of the process and also deal with some of the new and exciting developments in the area of PIM. It is hoped that this chapter will prevent the uninitiated from being blinded by the forecasts of extremely rapid growth in PIM and jumping into this rather difficult technology; but it is also hoped that technocrats and businessmen who satisfy themselves that the team they have is adequate to overcome the technical challenges posed by this technology will be encouraged to quickly move into this area. The technology is truly poised for explosive growth and, in spite of its highly involved technical requirements, this is perhaps one of the best times to enter to the PIM market.

Factors Driving the Growth of Particulate Injection Molding

After pointing out the technological difficulties of the process, it is now time to look at the positive factors that are providing the driving force for the rapid growth of this industry. According to a number of authors, the average growth rate of the PIM industry is expected to be anywhere between 20% and 50% and should be sustained over a period of a few years, thus making it a technology whose annual sales could soon exceed the $250 million mark and could potentially have a sales volume of $1 to 2 billion by the turn of the century [3–6].

If we neglect the few companies that were producing some PIM parts in the early part of this century, we can see that the first serious emergence of this technology was spawned by Wiech, Zueger, and Rivers in the 1970s. They were successful in obtaining a number of patents covering various aspects of PIM [7–11]. PIM has come a very long way from those early days, and is gradually encroaching into areas that were the strongholds of investment casting or die casting process. The markets for PIM components have become extremely diverse and include orthodontic braces, dental and medical devices, magnets, business machines, guns and firearm components, musical instruments, household appliances, components of high-temperature ceramic engines, watches, electronic packages, drill-bit blanks, chemical machinery components, food and beverage industry components, small- and large-caliber fins for penetrators, aerospace components, automotive components, sunglasses hinges, lock parts, small ignition components, toothbrush drive shafts, and computer peripherals [6,12–15]. It is expected that this list is only partial in nature, and new applications will continue to develop as the technology expands its horizons.

Essentially any part that is small in size, complex in shape, requires a high production volume, and is made from high-performance materials is a potential candidate for PIM processing. According to Merhar [16], a candidate for particulate injection molding on volume basis is typically smaller than a tennis ball, with the process being cost-effective when the parts are smaller than a golf ball. On a weight basis the process is cost-effective in the range of 1 to 25 g per part, and a wall thickness range of 0.75 mm (0.03 inch) to 6.4 mm (0.25 inch) is considered practical. The greater the added value, either in terms of shape complexity or in terms of high-performance materials, the greater is the opportunity for PIM.

Apart from the rapid expansion of the marketplace, the other encouraging factor for this technology is its rapid spread all over the world. It is as though the whole world is gradually waking up to the realization that PIM is a technology with virtually unlimited potential. There is a great deal of interest in this technology in Japan, Taiwan, Singapore, Europe, India, Hong Kong, and China. The worldwide MIM sales volume increased from $21 million in 1988, to $43 million in 1989 and is expected to be around $200 million in 1994. Interestingly, while the sales volume in the United States increased from $16 million to $32 million, the sales in Japan jumped from $2.2 million to $6 million, an increase of over 170%, and the sales of the other countries combined jumped from $2.8 million in 1988 to $6 million in 1989, providing a growth of over 110% [13]. This increased globalization of the technology opens up new marketplaces all over the world and will serve as a major driving force for the expansion of the technology.

The ability of this process to manufacture very small parts such as orthodontic devices that are extremely difficult to process by other metalworking techniques has created a niche for this technology. Rocky Mountain Orthodontics, American Orthodontics, and 3M's Unitech Division are using this technology to produce a variety of orthodontic devices. The value-added aspect of this technology is providing companies with a larger profit margin compared to competing metalworking technologies.

As the material becomes more and more difficult to machine or as the price of the material increases, the greater is the value-added nature of the process and the greater is the profit margin. As a case in point, electronic packaging materials made from Kovar, which is a difficult material to machine, provide an ideal candidate for PIM processing. There are, however, a large number of materials that cannot easily be shaped by casting techniques, usually due to very high melting points and extremely high segregation problems. These materials, such as ceramics or refractory metals, are almost "natural" candidates for PIM. Another factor that favors the PIM process is the fact that the gates and all other extraneous parts of the green component can be recycled by using the material directly as the PIM feedstock. Thus, extremely low material wastage is one of the attributes of this process. Obviously, the higher the price of the material, the greater will be the savings.

One major technical factor that is expected to fuel the growth of this technology is the awakening of the powder manufacturers to the fact that the powder requirements for the PIM industry are different from those of the general powder metallurgy industry. In general, melt atomization, which is the most common technique of powder production, yields powders that are often considered too coarse for PIM. The yield of the fine powder fraction is too low to provide a reasonable cost for the desired powders. For a long time the PIM industry was forced to use powders that were meant for use by the classic pressing and sintering P/M industry. The realization of the tremendous market potential of the PIM technique has spurred a number of powder producers to try to produce production-scale powders catering specifically to the PIM industry. For example, Avesta is meeting part of the demand for stainless-steel powder by capturing the fines produced during their atomization process. ISP (formerly GAF) and BASF have been the leading suppliers of fine, spherical iron powder that is used extensively by the PIM industry. The installation of the Hoeganaes ultrasonic gas atomization unit to produce high yields of fine particles is expected to cater directly to the needs of the PIM industry. Ultrafine Powder Technologies' patented atomization process is expected to produce powders finer than 20 μm of various materials (several grades of stainless and high-carbon steels, superalloys, etc.) at the anticipated rate of approximately 46 g/s ($\approx$4000 kg per day). The setup of the gas atomization unit geared for processing powders for the MIM industry has also been taken up by

Anval, with capacities being in the range of 3 to 16 g/s ($\approx$250 to 1400 kg per day). Ametek has been involved in the production of precipitation-hardened stainless-steel powders and several aluminide powders for the PIM industry. The process of microatomization, which uses a plasma gun to process fine powders of refractory metals and ceramics as well as a variety of other materials, is being used by Micromaterials and GTE to produce specialty powders for the MIM industry. INCO has been producing various grades of nickel powder (by the carbonyl process) which is being extensively used by the PIM industry. Pacific Metals Company of Japan has been supplying various grades of steel powders produced by the high-pressure water atomization technique and Mitsubishi Steel Manufacturing (powder division) is also producing PIM powders by water atomization. Currently several other companies such as Starck (Germany), Osprey (UK), and Kawasaki Steel (Japan) have shown keen interest in supplying suitable powders to the PIM industry. This concerted effort of powder manufacturers is expected to go a very long way in supporting the phenomenal growth of the PIM industry.

The other important development that would really advance the cause of PIM is the involvement of the large furnace manufacturers. The proper design of furnaces to suit the PIM process is gradually entering the marketplace. The technology is presently attempting to make the transition from being a batch process to being a high-volume continuous production process. This demand for continuous furnaces is being addressed by several large furnace manufacturers. Faster binder removal and even the development of a batch furnace in which the same cycle can remove the binder and also carry out the sintering, have recently emerged. This furnace can be used for stainless steels and other oxidation-sensitive materials [3]. The PIM industry should also take note of the development in the sinter+HIP furnaces that could be used to process materials (which are first sintered and subjected to HIP) to almost full density using the same furnace.

Giant strides have been taken in the area of binder development, which was among the most closely guarded secrets in the PIM industry, and which was undoubtedly the reason for the extremely long gestation period for this industry. Early PIM technology generally required around 3 to 4 days of slow thermal removal of the binder at around 450 K (177°C) in air, which was then followed by sintering. Thus, a small complex-shaped steel part would be produced in 2 to 3 days. It was surprising that, in spite of that, PIM was considered viable for small and extremely complex shapes. As understanding of the requirements of the binders increased, and their combined rheological behavior became better known (mainly due to the joint efforts of the industry and university), research activity in this area increased to a great extent. This

Table 6.1. Some of the binders commonly used for processing PIM feedstock

1. Paraffin wax 77% Low molecular-weight polypropylene 22.2% Stearic acid 0.8%	5. Poly(butyl methacrylate) copolymer with ethylene–vinyl acetate 45% Atactic polypropylene 23% Dibutyl Phthalate 9% Wax 23%
2. Water 98% PVA 2%	6. Polystyrene 20% Acetanilide used with SA-coated Fe 80%
3. High-density polyethylene 20% Carnauba wax 10% Paraffin wax 69% Stearic acid 1%	7. Water 60% Methylcellulose 25% Glycerine 15%
4. Epoxy resin 60% Paraffin wax 30% Butyl stearate 10%	8. Vegetable oil 55% High-density polyethylene 40% Polystyrene 5%

has resulted in the development of a variety of binders, a collection of which is given in Table 6.1 and which was first published in greater detail by German in his book *Powder Injection Molding* [6].

The Rivers process is a well-known technique in which a short debinding time has been achieved. Short debinding time in the Rivers process has been demonstrated in a 3-inch-sided cubic stellite part. Another notable development in the fast debinding cycles has been in the process of solvent extraction. In this process, a part of the binder is removed by soaking the green part in an organic solvent bath. Since a major part of the binder has been removed, the remaining binder can quickly be eliminated by thermal debinding inside the sintering furnace. Variations of this process are being used by several PIM companies to process a wide variety of materials. Other developments that may be of significance in the future include the freeze drying techniques, which are dependent on a water-based binder, and the claims by one company that it has developed a technique of binder-free PIM processing. One trend that is extremely encouraging is the attempt by research groups to produce binders based on the fundamental requirements of the PIM industry. In the past the practice has been to try to modify slightly an existing binder that is already in use with the aim of improving the binder characteristics. Recent efforts by MPS (France) and BASF (Germany), however, have been to develop a new binder that will reduce the debinding time and reduce the part warpage during the debinding process [17]. It is hoped that this form of endeavor will continue to boost research and development activities in the area of new binder development.

The automation that exists in plastic injection molding processes has not yet been incorporated by the PIM manufacturers. It is expected that as the PIM industry grows, the assimilation of automation into processing will become routine. The PIM process, due to its very nature, will be amenable to the process of automation, and this is gradually being adopted by some of the successful PIM companies.

Understanding of the science of mold filling, analysis of the flow behavior of injection-molded feedstock, and optimization of gate location are all leading to the better use and faster growth of this process. The use of statistical packages such as Orthogonal Array Methods to solve the problem of numerous variables that are encountered during the molding process is also accelerating the pace of growth and helping in the isolation of the critical variables. For example, investigations into the optimization of the molding process by Rensselaer Polytechnic Institute (RPI) using the Orthogonal Array Method have resulted in a total of 18 tests instead of the 39 366 tests that would have been required for process optimization. The development by the Cornell Consortium of sophisticated software for the tool design stage is also expected to aid the PIM industry. The use of analytical or computer simulation models for studying the various process optimizations has led to the development of the temperature and density profiles that are generated within the mold after it has been filled. This, in turn, can predict the proper gate locations for obtaining an optimized green part [18]. The flow analysis software, which has now developed into a readily accessible tool for any mold designer, is also capable of working on mainframe computers as well as on personal computers. The software can graphically display the intricate filling process to the designer. Armed with this software, one can visualize the material flow and its solidification within the mold cavity. Use of this technique can spot design problems even before the mold is fabricated [19]. Current investigations at the Pennsylvanya State University on mixing, molding, debinding, and sintering of PIM parts is expected to go a long way toward providing fundamental understanding of the science and technology of this process. All these advances will help the PIM industry to become a much more cost-effective process and will provide the impetus for rapid growth of this technology.

A number of other factors, though nontechnical in nature, have contributed and are expected to contribute rapidly in the growth of this technology. The inflow of large amounts of capital into reasonably successful PIM companies has provided a great deal of confidence in the industry. Some examples are the purchase of Remington Arms by a venture capital group, the acquisition of

Omark by Blount IMMP, and the relatively recent acquisition of Advanced Forming Technology by Precision Castparts Corporation. There has also been the acquisition of the PIM technology by various large companies in the form of licensing agreements between the large companies and smaller well-established PIM companies. For example, Mitsubishi and Juki of Japan and IMPAC (Division of Vallourec) in France have recently entered into licensing agreement with Parmatech Corporation in California for using Parmatech's powder injection molding process, which is based on a patented solvent extraction binder removal technique. Similarly, Hitachi of Japan has acquired a license from Advanced Forming Technology. All this is indeed a very good sign for the industry and the fact that it has gained the confidence of large investors and corporations is more than likely to induce other large investors to invest in this technology.

The infusion of new technical talents in this area has been the outcome of a multiclient industrial consortium that was set up at Rensselaer Polytechnic Institute (RPI), Troy, New York. Several universities and institutes (such as Penn State University, RPI, University of Alabama, etc.) have also developed active interest in the area of PIM and they have been primarily responsible for providing a pool of graduate students who are well trained in the highly interdisciplinary process of PIM. Also, the rapid development of this technology is helped by the realization of various companies involved in the PIM business that an open-door philosophy is not only laudable but also makes good business sense as it provides them the opportunity of promoting their capabilities to a very wide cross-section of end users. The recognition that was given to the PIM industry for the first time in the form of two Awards of Distinction by the Metal Powder Industries Federation (MPIF), one for a screw seal for use in the airline industry and the other for a niobium alloy thrust chamber and injector for a rocket engine (both won by Parmatech Corporation of California), boosted the image of PIM as a high-performance metalforming technology. This recognition has been further improved by the formation of the Metal Injection Molding Association (MIMA), and the establishment of a separate award of distinction category for powder injection molding. All these important nontechnical factors are expected to foster a large expansion to this industry.

The Particulate Injection Molding Process

In general, all the PIM processes can be divided into four broad categories. The four steps are

1. Feedstock preparation

2. Injection molding
3. Debinding
4. Sintering

In some cases additional steps such as presintering may precede the sintering step, or a heat-treatment step may follow the final sintering step. In some cases it may be cost effective to carry out some secondary operations such as drilling, deburring, or tapping on the green part. Several secondary operations such as repressing (to create flat surfaces), coining, final grinding, plating, etc., are also carried out on the sintered PIM part. This chapter will, however, only present detailed discussion of the four areas mentioned above, namely feedstock preparation, injection molding, debinding, and sintering. Due to the large volume of literature that is presently available, it would be impossible to cover all the aspects except in terms of the generic scientific and technological factors underlying each of the four key processing steps. For excellent and detailed coverage, readers are referred to the book by German on powder injection molding [6] and some of the papers that will be cited during the course of the discussions.

1. Feedstock Preparation

During feedstock preparation, fine particulate material, preferably in the range of 10 to 25 μm or less and having a predominantly spherical shape (with usually a small quantity of irregular powder), is intimately mixed with a binder, which is usually a polymer-based material. The mixing is carried out at a temperature at which the binder is in a liquid state and the binder–particulate mix forms a viscous fluid that has the consistency of a dough mix. The binder serves as a temporary agent and imparts the desired flow characteristics such that a uniform particulate–binder mixture may easily take up the shape of the mold cavity.

To get a better understanding of the feedstock preparation and its characteristics, it is logical to discuss the constituents that form the feedstock. There are essentially two components of the feedstock: one is the binder and the other is the particulate constituent (or the solid). The mixture of the solid and the binder, usually in the granulated form, is known as the feedstock.

A. Binders
The binder compositions and the techniques applied for their removal are usually the basis for the different PIM processes. The binder is an essential vehicle for imparting the flow characteristics to the feedstock, and it also helps in retaining the overall shape of the part until some green strength is attained

due to the effect of sinter bonding of the particles. However, the binder exerts a great influence on the PIM processing steps, and also has a large bearing on the final properties of the part, though it does not form a part of the final material. The binder, like the powders, influences the feedstock by exerting an influence on volume fraction of solid loading, feedstock mixing, feedstock rheology, injection molding, and debinding, and even influencing the distortion and defects that are observed in the final part.

There are a number of attributes that a PIM binder should possess. The binders used for the PIM process are almost invariably multicomponent binders, as no single binder material would satisfy the diverse requirements for forming the PIM feedstock. Consideration of the binder selection should also involve the binder removal step, as this too is heavily influenced by the type of binder.

One of the primary requirements of the binder is that it does not react with the particulate material with which it is mixed. This would totally defeat the purpose of PIM, and lead to extreme difficulties in the process control stage. Thus, the binder needs to be essentially passive. The binder also has to have good adhesion to the powders, wet the powder surface, and impart its fluidity to the feedstock for better packing and filling up of the mold cavity.

The effect of adhesion of the binder to the powders has been demonstrated in an experiment where in one case silicone oil was mixed with iron powder and in the other carnauba wax was mixed with the same iron powder. For the same volume fraction of particulate loading, and using mixing temperatures at which both the binders have similar viscosities, it has been observed that the mixture with silicone oil is extremely nonuniform and stiff while the one with carnauba wax is very fluid [20]. This behavior is attributed to the fact that silicone oil has no adhesion with iron while the high ester content in the carnauba wax provides some affinity with the iron powder.

Usually, small amounts of coupling agents (surfactants, additive, etc.) are used to decrease the viscosity of the feedstock. One terminal of the molecule (polar head group) of the coupling agent provides strong chemical interaction with the particulate constituent of the feedstock, while the other terminal of the molecule (nonpolar side) results in a higher affinity with the other major constituent, namely, the binder. Addition of extremely small surface-modifying agents has been known to decrease the viscosity of the feedstock to a large extent by creating an interfacial bridge between the constituents of the feedstock. Additives such as stearates, silanes, and phosphates have been shown to effectively decrease the viscosity of the feedstock. This form of addition to influence the viscosity is routinely employed in the case of ceramic injection molding due to the very fine particle sizes used. The effect of adding extremely minute quantities of sodium pyrophosphate to a 0.35 volume

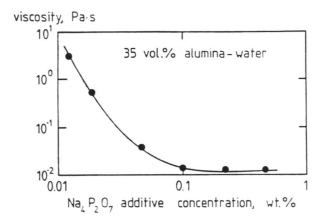

Figure 6.3: *Effect of sodium pyrophosphate on the viscosity of an alumina–water mixture. (Reprinted with permission from Ref. 6.)*

fraction of alumina powder loading in water is shown in Figure 6.3 [6, 21]. It can be observed that the addition of only 0.1 wt.% of the sodium pyrophosphate additive lowers the viscosity of the binder–particulate mixture by two orders of magnitude. Some of the titanates have been reported to be effective with alumina, but a similar titanate did not result in any improvement to the polyethylene wax (PEW)–iron mixture. This illustrates that there is no such thing as a universal coupling agent. This is to be expected, as the coupling agent acts as a bridge to provide the affinity between the two major feedstock constituents, and changing one of the major constituents changes the nature of the adhesion. Thus, an effective coupling agent has to be able to bond with both constituents of the PIM feedstock.

Another interesting observation was that coating iron powders with stearic acid reduced the viscosity of the plain polyethylene wax – iron mixture. However, the effectiveness of stearic acid, which acts as the coupling agent, is significantly decreased when the feedstock is repeatedly mixed [20]. This happens as the stearic acid is dissolved into the PEW and is gradually moved away from the surface of the iron. Thus, stearic acid will not be a good coupling agent where PEW is a major constituent in the binder phase.

There is usually a viscosity range that the binder should possess at the molding temperature such that the optimum feedstock viscosity can be attained. Low viscosities will result in the separation of the powder and the binder, while too high viscosities will create defects and produce an extremely difficult feedstock to mold. Generally, a viscosity below 10 Pa s is required for binders that can be used in the PIM process. The effect of binder viscosity is clearly demonstrated in the experiments of Chung and his coworkers [20].

PEW at 408 K (135°C) had a viscosity of around 0.42 Pa s. The mixing of the binder with iron powder was extremely difficult. In contrast, paraffin wax, which had a viscosity of around 0.009 Pa s at 373 K (100°C), could easily be mixed with the iron powder.

The viscosity of a mixture of binder and particulates with greater than 60 vol.% of loading of the particulate material can be determined from the following equation developed by Maron and Pierce [22]:

$$\eta_r = \frac{\eta}{\eta_0} = \left(1 - \frac{\phi}{\phi_m}\right)^{-2} \tag{1}$$

where

η_r is the relative viscosity;
η is the mixture viscosity (Pa.s);
η_0 is the binder viscosity (Pa.s);
ϕ is the volume fraction of solid loading;
ϕ_m is the maximum powder volume fraction at which the mixture viscosity becomes infinite.

Thus, polymers with lower molecular weights such as polypropylene, polystyrene, and polyethylene, and waxes such as paraffin, carnauba, and beeswax, which also have low viscosities in the molten state, provide an ideal pool of materials to choose from. Multicomponent binders can effectively use one or more of these materials as the ingredients making up the final binder composition.

It is preferable to have a binder whose viscosity does not change to a great extent with temperature during the molding process, but also one that has a very rapid increase in viscosity during the cooling stage so as to hold the compact shape. If the fluctuation of viscosity is extremely high in the molding temperature range, process control becomes extremely difficult as small fluctuations in temperature during molding can lead to drastic changes in the flow behavior of the particulate feedstock. Since the binders usually have a much higher thermal shrinkage and heat capacity than the particulates, it is preferable to use binders with low melting points so that the processing can be carried out at low temperatures and thereby avoid excessive thermal shrinkages.

Another important attribute of the binder material is that, ideally, it should not exhibit any molecular orientation with flow. This may result in nonisotropic green PIM parts. The problem of molecular orientation becomes acute with the long-chain polymers and polymers with various functional groups. This again points toward the low-molecular-weight polymeric materials (oligomers) as being the choice for the binders in PIM applications. The lower the molecular orientation, the more uniform will be the part shrinkage.

The binder plays an important role in determining the properties of the green injection-molded parts, which are subjected to quite a bit of handling; in certain cases the green parts are also subjected to machining. The binders should have good mechanical strength to withstand this subsequent handling and machining. The green strength is, of course, dependent on the room-temperature strength of the binder and also on the capability of the binder to adhere properly to the particulates. Measurement of the flexural strengths of green bars produced from the PIM feedstock provides a first-hand clue for predicting whether the particulate–binder combination will have adequate strength. The strength of the polymers usually forming the primary constituents of binder is briefly discussed in a later part of this chapter.

It should be underlined that the green strength is not simply a function of the binder strength at room temperature, but also reflects the adhesion between the powder and the binder. Thus, simply choosing a binder that has good mechanical strength does not necessarily guarantee that the binder will provide adequate green strength when used as a PIM feedstock component; however, a binder that does not have good mechanical strength will never produce a part with sufficient green strength. Figure 6.4 shows the variation in yield strength for two different feedstocks produced by loading 62 vol.% iron powder in a polyethylene binder in one case, and a paraffin wax-based binder in the second case. Also included in the figure are the yield strengths of the pure binders without any solid loading. It can be seen from the figure that though the strength of the paraffin-based binder is lower than that of the polyethylene binder, the strength of the mixture of the paraffin-based binder and iron powder is greater than that of the mixture of the polyethylene binder and iron powder (same volume fraction of solid loading in both cases). This is due to the ability of the paraffin-based binder system to adhere to the iron powder particles better than does polyethylene. The data were provided by B. Rhee [23].

The effect of binder properties becomes extremely important during the crucial step of debinding or binder removal. In fact, the choice of the different constituents of a multicomponent binder is often dictated by the debinding process that is to used to remove the binders. At this juncture, it is worth while discussing the characteristics and the types of materials that are commonly used as binder materials.

The binders that are commonly used in the PIM process can be categorized as thermoplastics, thermosets, water-based systems, and polymer systems with

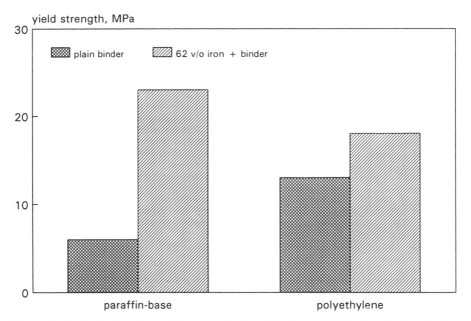

Figure 6.4: Variation in yield strength for PE–Fe and PW–Fe feedstock along with the yield strength of pure PE and PW. (Data from B. Rhea, personal communication.)

sol–gel properties. Various combinations of these are used to suit specific processing routes. The thermoplastic polymers are the most popular.

Thermoplastic compounds can be hardened and softened and, thus, their viscosity can be changed by changing the temperature. Some common examples of thermoplastic binders used in PIM are polyethylene, polypropylene, polystyrene, poly(vinyl chloride), poly(vinyl alcohol), etc. Polymers that are linear or weakly branched and not cross-linked are generally termed thermoplastic polymers. They can be dissolved in solvents and they melt and flow when heated. In their usual form, their application temperatures lie below their melting temperatures (in the case of crystalline polymers) or below their glass transition temperature (in the case of highly amorphous polymers). On heating above those characteristic temperatures, they can be transformed to an easily deformable "plastic" state. This plastic state can be termed liquid from the viewpoint of molecular order and viscoelastic with respect to rheological properties. Cooling below the characteristic temperatures results in the polymers adopting the shaped form. The shape of parts made from thermoplastics can be changed and/or modified upon heating above the melting point followed by subsequent cooling. This heating and cooling process can be repeated a number of times before a significant amount of polymer decomposition occurs. This advantage is often the reason for using thermoplastics as the feedstock material, and the feedstock can be recycled a number of times during PIM process.

The thermosetting polymers, in contrast, attain permanent rigidity by forming irreversible cross-links upon heating or during the synthesis of the material. Thermosets are prepared by cross-linking low-molar-mass compounds. These low-molar-mass compounds can be monomers and oligomers (prepolymers) whose molecules are irreversibly cross-linked via covalent bonds. Due to cross-linking or the formation of compact three-dimensional structures, the polymer chains lose their ability to flow past one another and the material exhibits a considerable degree of strengthening at elevated temperatures. Thermosetting polymers, in general, will not melt after they have cross-linked to form a rigid structure. To manufacture useful products from pure thermosetting polymers, it is necessary to perform the cross-linking reaction (during processing) or temporarily disrupt the cross-linking (also during processing) to allow the flow of the macromolecules. These polymers are insoluble in most polar organic solvents because the cross-linking reaction causes a tremendous increase in the molecular weight. However, under certain conditions, thermosets are known to swell in the presence of solvents. The high cross-link density restricts molecular segment crystallization and at the same time increases the activation energy for segmental motion. As a result, thermoset polymers are mostly noncrystalline in nature. In addition, they exhibit an increased glass transition temperature or none at all. In contrast to thermoplastics, the thermosetting polymers do not soften on reheating because of irreversible chemical cross-linking. However, like most polymers, they can and do gradually decompose at high temperatures. Examples of thermosetting material include bakelite, epoxy resins, urea–formaldehyde resin, and phenol–formaldehyde resin.

The gelation process in the case of thermosets involves the formation of a single molecule that extends via covalent bonding. The viscosity of this massive molecule increases rapidly as the gel reaction progresses. The gel reaction can be initiated by either heating or cooling depending on the type of gel reaction. These macromolecules can often associate with each other or with other small molecules, for example, with water. This process, known as associative behavior, gives gels of various types. Examples of such gels are cellulose, inorganic glass-gels like sodium silicate solutions, gums of hydrated polysaccharides, and agar. Water is also used as a major binder constituent, especially in freeze drying processes. Water has been used with cellulose, aniline, agar, poly(vinyl alcohol), etc., to form the desired binder.

The morphology or the supermolecular structure of a polymeric system plays an important role in determining the binder properties. The state of polymeric systems can extend from the completely random, known as amorphous state, to a very orderly state known as the crystalline state. The physical structures are not only controlled by the polymeric framework and the micro- or macro-conformation caused by it, but also by the experimental conditions. It is accepted that these physical structures do not necessarily correspond to a state of equilibrium. Unlike small compounds, polymers are not found in completely crystalline state, primarily due to their chemical structure. In general, crystallinity is favored by symmetrical chain structures that allow close packing of polymer units to the maximum advantage of secondary forces. Crystallinity is also favored by strong interchain interactions that include hydrogen bonding, London forces, and Van der Waals' interactions. Crystallinity can be induced in a number of ways such as evaporation of a polymer solution, cooling of molten polymer, or heating of a polymer under vacuum at a specified temperature.

Almost fully amorphous polymers, for example atactic polystyrene and poly(methyl methyacrylate), are assumed to consist of randomly coiled and entangled chains. An analogy of this at a molecular level is that the polymer system looks like a bucket of worms. In semicrystalline systems, the crystalline order exists in crystalline domains known as crystallites or lamellae, which are usually surrounded by an amorphous matrix. It should be emphasized that the crystallites do not have the shape of most organic crystals. They are small in size, typically between 1×10^{-8} and 2×10^{-8} m, and have a large number of imperfections. Normally a macromolecule is over several hundred nanometers long and the chains are aligned normal to the crystalline surfaces. Thus, a polymer molecule can remain in the crystalline state for only 1×10^{-7} m before it reaches the surface of the crystal, where it usually folds back and reenters the crystal at some other point. Some chains do not reenter the crystal. It should be emphasized that there is normally no sharp boundary between amorphous and crystalline regions, and it is therefore useful to consider them as one phase system with varying degrees of order. Since the crystalline and amorphous regions are continuous, the crystallites increase the cohesiveness of the bulk polymer. In reality, they are very much like covalent cross-links that are stable with respect to time but not with respect to temperature.

Semicrystalline polymers are tougher, stiffer, and more opaque, have higher densities, and are more resistant to solvents than their amorphous counterparts. For a polymer to have a high amount of crystallinity it must be able to assume a stereoregular conformation. In other words, the three-dimensional arrangement of the substituents must be regular. For example, polyethylene crystallizes in a fully extended, planar zig-zag conformation. Polymers with short, regularly spaced bulky groups tend to assume a helical conformation that allows the substituents to pack closely together without any appreciable distortion of the chain. Table 6.2 lists polymers according to their tendency to

Table 6.2. Tendency of some polymers to form crystallites

Low	Moderate	High
PMMA	Polystyrene	Polypropylene
Atactic polymers	PVC	Poly(ethylene oxide)
Random copolymers	*cis*-Polyisoprene	Stereoregular PP
Thermosets	PVC	High-density PE
	Cellulose derivatives	Nylon
		Cellulose
		PVDC

form crystallites. In general, yield stress, strength, stiffness, elastic modulus, and hardness increase with an increase of crystallinity.

The two common classes of polymers, the thermoplastic and the thermosetting, differ significantly in their mechanical properties. Since thermoset polymers are highly cross-linked systems, the elongation at break is low compared to thermoplastics. The tensile strengths are about the same. This is interesting because thermosets have covalent intermolecular bonds of high bond energy compared to the intermolecular physical bonds of thermoplastics. However, cross-linking in a thermoset is in most cases inhomogeneous; thus, only a few cross-links have to take the full load. Thus, the tensile strength of a thermoset is the result of a few effective bonds of high bond energy and is about the same as that of the thermoplastics whose strength is the result of many intermolecular interactions of low bond energies.

A polymeric system can fracture in several different ways depending on its type, stress, and condition. The type of fracture can be either brittle or ductile. It is important to appreciate that any specific mechanical or physical properties of polymers can be modified by the use of additives or by forming a "polymeric alloy." Table 6.3 gives the mechanical properties of some of the commonly used polymers and Table 6.4 gives the common names and the basic polymer repeating units of some popular polymeric materials used in PIM processing. It should be realized that in order to produce a multicomponent binder system, various different types of polymeric materials, oils, water, etc., may be mixed together.

The general philosophy of multicomponent binders is the gradual removal of a part of the binder while the other part is retained to provide some rigidity to the green body. The key element in the debinding step is the total removal of all the binders in a reasonable amount of time without leaving behind any binder residue (unless it is intentionally desired). Another extremely important

Table 6.3. Mechanical properties of some commonly used polymers

Polymer	Tensile strength (MPa)	Tensile elongation (%)	Modulus (MPa)	Compressive strength (MPa)	Flexural strength (MPa)	Impact strength (N/cm)
LDPE	9–30	100–600	170–250	–	–	No break
HDPE	20–30	10–1200	1070–10990	20–25	–	0.23–2.3
PVC	40–50	40–80	2400–4100	55–90	69–110	0.23–1.3
PP	30–40	100–600	1150–1700	38–55	40–55	0.23–0.57
PS	35–52	1.2–2.5	2280–3280	83–90	89–101	0.20–0.26
PMMA	48–75	2–10	2240–3240	72–124	72–131	0.17–0.34

Table 6.4. Common names and the polymer repeating units of some polymers used as binders in PIM

Common names	Polymer repeating unit
Polyethylene	$-[-CH_2-CH_2-]-$
Polypropylene	$-[-CH_2-CHCH_3-]-$
Polystyrene	$-[-CH_2-CHC_6H_5-]-$
Poly(ethylene glycol)	$-[-CH_2-CH_2O-]-$
Poly(vinyl chloride)	$-[-CH_2-CHCl-]-$
Poly(vinyl alcohol)	$-[-CH_2-CHOH-]-$
Poly(vinyl acetate)	$-[-CH_2-CHOCOCH_3-]-$
Poly(methyl acrylate)	$-[-CH_2-CHCOOCH_3-]-$

consideration is that the binder should be removable without causing distortion or cracking of the injection-molded part.

In the early days of powder injection molding, the debinding was carried out over a period of 3 to 4 days because it usually depended on total thermal debinding. However, the need for rapid debinding cycles spurred the development of various multicomponent binders because a single-component binder was definitely unsuitable since it was unable to cater to the diverse requirements of a good PIM binder.

A multicomponent binder is usually made up of a primary binder that occupies the major volume fraction of the total binder and other secondary binders that serve various secondary functions. The primary binder is first removed during the sequential steps of debinding. This removal leaves behind a large amount of pore channels that facilitate the secondary binder removal during the later stages. The softening or the total liquefaction of the volume of binder within the green part can cause slumping, distortion, or cracking. Thus, a small part of the binder that remains rigid under the conditions in which the primary binder is removed serves the purpose of retaining the green strength and the rigidity of the green part. Other secondary binder constituents may have been employed to improve the adhesion and wetting, and to produce a lowering of the feedstock viscosity. The major part of the secondary binder should be of the type that can be quickly pyrolyzed at temperatures high enough to promote some amount of particulate–particulate bonding (presintering) to hold the net shape of the green compact after all the binder has been removed.

The removal of the primary binder constituent may be accomplished by various means. In the case of a binder with constituents that can be pyrolyzed at different temperatures, a progressive thermal debinding schedule may be quite appropriate. A typical example of such a binder could have a composition of 75% paraffin wax, 24% low-molecular-weight polypropylene, and 1% stearic acid. In this case, the lower-melting-point materials such as paraffin wax and stearic acid can be removed at temperatures below 573 K (300°C), while the low-molecular-weight polypropylene is burnt out at temperatures around 673 K (400°C). A wicking approach may also be used to remove the primary binder by embedding the green body in a wick material and heating to a temperature at which the major constituent of the binder melts and is rapidly wicked away from the green body into the wicking material. A third approach may involve different forms of solvent extraction where the major constituent of the binder is removed by immersing the green shape in an organic solvent, or by exposing the green parts to the solvent vapor, or a combination of both. The solvent extraction process dissolves the major constituent of the binder; the minor constituent in this case needs to have higher strength and rigidity and should not be taken into solution by the solvent used to dissolve the primary binder phase. The higher rigidity is needed as the green part is usually handled after the solvent-extraction stage.

Thus, a combination of thermoplastic materials with oligomeric low-molecular-weight materials such as waxes can provide the two important binder properties. The time for the hardening of thermosetting polymers is usually quite long. A combination of thermoplastic and thermosetting type polymers could provide a reasonable compromise, where the thermoplastic material provides the low-temperature strength and the thermosetting compound that can form the three-dimensional cross-linked network at high temperatures provides the high-temperature rigidity and thus shape retention. It should be cautioned that the possibility of phase separation between thermosetting and thermoplastic polymers could lead to tremendous problems during the mixing.

Another form of binder development work carried out at Rensselaer Polytechnic Institute has resulted in the concept of solid polymer solution (SPS) type of binders [24]. This is also a form of multicomponent binder in which a polymer is dissolved in a solvent that is a solid at room temperature. The solid solvent can be selectively leached out with a solvent, leaving behind the rigid polymer, which is later removed by thermal pyrolysis. An example of the SPS binder system is the mixture of acetanilide and polystyrene (1000 to 10 000 units in a chain).

With some other forms of binder, especially in the case of water-based binders, the technique of freeze drying is often employed. In this technique, the major constituent of the binder is extracted directly from the solid state by the process of sublimation. Examples of such processes include water and aniline or water and PVA [25,26]. The process of sublimation creates very little distortion of the green part.

The gel reaction has also been utilized to produce behavior similar to that of thermoplastic binders. Agar, at low concentrations in water, has low viscosity at high temperatures and high viscosity on cooling. The gel entraps the lower-melting-temperature material such as water, which can be gradually removed while leaving behind the rigid gel. The gel is then removed at high temperatures while the major part of the binder (water) has already been removed and has left behind pore channels through which the gel decomposition products can easily escape. Alternately, some gelation reaction processes produce hardening with temperature and the process can also trap other molecules such as water, which can subsequently be removed. The cellulose–water-based binder is an example of such a system.

Recent efforts in binder development have generally tried to address the issues of powder–binder separation during injection molding, reducing debinding times, and also reducing distortion and warpage during debinding. In recent binder development work reported by BASF, polyacetal chemistry has been applied to solve these problems [17]. In this case, the removal of the binder is made to occur homogeneously from the outside surface to the inside core by reacting the polyacetal binder with gaseous acids at elevated temperatures of 393 to 423 K (120 to 150°C) to produce a reacted flue gas of formaldehyde. The binder residue of this reaction is almost zero and since the gas exchange is limited only to the outer volume, which is porous, the debinding process does not lead to any distortion. The reaction is controlled by the concentration of the acid catalyst. A small quantity of residual binder component provides rigidity to the green part and can later be removed by pyrolyzing the material using heating rates in excess of 1 K/s.

A large variety and number of binder systems based on the general principles of multicomponent binders as outlined in the previous paragraphs are presently in use. This part of the technology is considered to be proprietary in nature, and a number of companies are in the business of selling proprietary feedstocks to individual customers. The large variety of binders in existence is probably due to the efforts of companies to have their own binders that do not infringe on the patent rights of binders developed by other companies. However, the underlying principles as described in this chapter remain essentially the same.

Based on this discussion, it should not come as any surprise that there are probably hundreds of binder compositions that have been used for particulate injection molding. The constituent chemicals used in the binder systems are equally numerous and vary from the mundane to the exotic. Some of the constituents used as PIM binders include polyethylene (PE), polystyrene (PS), polypropylene (PP), poly(vinyl chloride) (PVC), poly(vinyl alcohol) (PVA), poly(vinyl acetate) (PVAc), paraffin wax (PW), a number of copolymers, beeswax (BW), carnauba wax (CW), cellulose, some ketones, stearic acid, oleic acid, epoxy resin, aniline, glycerine, vegetable oils, fish oils, peanut oil, octyl acid phosphate, sodium pyrophosphate, acetanilide, polyacetals, and agar.

The rheological properties of the feedstock are an extremely important parameter which is dependent on the characteristics of both the binder and the particulates in the feedstock mixture. It is therefore appropriate to dwell briefly on the rheological characteristics of the binder itself.

In the case of very low molecular weight liquids, such as water, the viscosity depends almost entirely on temperature and pressure, and this type of liquid is known as Newtonian. Most of the polymers used in the PIM process, however, show changes in viscosity not only with pressure and temperature, but also with shear rate. A basic understanding of shear rate and its relation to viscosity can be understood by considering two thin, parallel plates of material within a liquid, separated by a thickness t. If one of the plates is assumed to be stationary with respect to the other, and the second plate is being acted on by a shear stress (force per unit area) denoted by τ, then there will be a relative displacement between the two layers. The relative velocity between the two parallel plates, denoted by V, is related to the shear strain rate or the shear rate, $\dot{\gamma}$, by

$$V = \dot{\gamma}t \tag{2}$$

The viscosity, η, which is a measure of the resistance of the fluid to shear, is related to the shear stress and the shear strain by

$$\eta = \frac{\tau}{\dot{\gamma}} \tag{3}$$

The Newtonian concept of fluids is not valid for the general PIM feedstock, which shows considerable variation in viscosity with shear rate. In a Newtonian fluid, the variation of the shear stress with shear rate can be represented by a straight line. In case of a pseudoplastic fluid, the rate of increase of shear stress gradually decreases with increasing shear strain, while in case of a dilatant fluid, the rate of increase in shear stress increases with shear rate. Considering the empirical equation

$$\tau = K\dot{\gamma}^m \tag{4}$$

The value of m will be less than, equal to, or greater than unity when the fluid exhibits pseudoplastic, Newtonian, or dilatant behavior, respectively. The three different types of fluids also show varying behavior when their viscosities are measured with respect to shear rates. In the case of Newtonian fluids, the viscosity essentially remains constant with shear rate; in pseudoplastic fluids, the viscosity decreases with shear rate (shear thinning); and in dilatant fluids, the viscosity increases with shear rate (shear thickening). Preferred binders used in PIM are usually Newtonian or pseudoplastic in nature, as binders that exhibit shear thickening are undesirable because this increases the tendency of separation between the particles and the binder [27]. Schematic representations of the variation of shear stress with shear rate, and of the shear rate with viscosity, for fluids exhibiting Newtonian, pseudoplastic, and dilatant types of behavior are depicted in Figures 6.5a through 6.5f. It should be pointed out that there exist fluids with other subtle variations in their flow characteristics (Bingham fluids and St. Venant fluids), but a detailed discussion of those will be outside the scope of this chapter.

The viscosity of the binders also depends on temperature and pressure. Generally, for PIM applications, the viscosity of the binders decreases with increasing temperature. In contrast, with increasing pressures, the viscosity of the fluid generally increases. The effect of temperature is, however, more prominent than that of pressure. It should also be noted that with the high shear rate conditions that are often encountered during the injection molding of a PIM feedstock there could be an increase in the binder temperature.

A multicomponent binder usually consists of a mixture of a number of constituents that have different viscosities at the same temperature and shear rate. A rough approximation of the overall multicomponent binder viscosity, designated η_b, can be derived from the following equation [28]:

$$\ln(\eta_b) = W_1 \ln(\eta_1) + W_2 \ln(\eta_2) + \ldots W_n \ln(\eta_n) \tag{5}$$

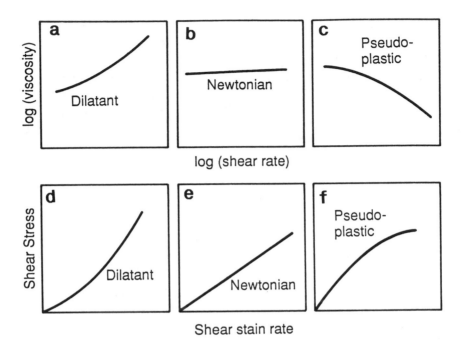

log (shear rate)

Shear stain rate

Figure 6.5: *(a) to (c) Schematic variation of viscosity with shear strain rate for (a) a dilatant type of material, (b) a Newtonian type of material, (c) a pseudoplastic type of material. (d) to (f) Schematic variation of shear stress with shear strain rate for (d) a dilatant type of material, (e) a Newtonian type of material, (f) a pseudoplastic type of material.*

where

$$\eta_b = \text{viscosity of the multicomponent binder (Pa.s)};$$

$W_1, W_2, \ldots, W_n$ = weight fractions of binder components 1, 2, $\ldots$, n, respectively;

$\eta_1, \eta_2, \ldots, \eta_n$ = viscosities of binder components 1, 2, $\ldots$, n, respectively (Pa.s);

n = total number of binder components.

A number of other factors, such as the binder composition, molecular weight, and glass transition temperature, are all expected to play a role in determining the final viscosity of a multicomponent binder. As a general guideline, the viscosity of the binder at the molding temperature should be below 10 Pa s, and the binder should exhibit Newtonian or pseudoplastic behavior.

A detailed listing of some of the commercially used binders can be obtained from the book by German [6].

B. Particulates

It should be realized that the particulates that are generally suited for conventional press-and-sinter P/M techniques are usually not the powders that are desirable for particulate injection molding. This part of the chapter will briefly discuss the particle characteristics that are desirable in the PIM process. Let us assume for the moment that the particles are not agglomerated and that they represent the smallest physical unit that cannot be further subdivided unless some mechanical energy is applied to further comminute the particles.

In the case of conventional press-and-sinter P/M techniques, the applied pressure is responsible for the high powder packing and high green densities. In PIM, the pressure applied is not sufficient to produce high green densities and, thus, the process has to rely on high powder packing to attain good green densities. Spherical-shaped powders provide the maximum packing density and are therefore the natural powders of choice. However, the use of purely spherical powders could result in slumping of the shape when the binder is removed, so it is common for the powder in the feedstock to have a small fraction of irregular-shaped powders.

Irregular powders, without substantial pressurization, have poor packing density but can provide good shape retention due to mechanical interlocking. The shape retention by mechanical interlocking becomes critical in the conventional P/M process of pressing and sintering as this process does not use substantial additions of binders to hold the green part together. However, in PIM, the green part actually contains a high volume fraction (around 0.3 to 0.5) of the binder, which acts as a glue to provide the desired shape retention capability. The poor packing of the irregular powders would mean a lower green density, which would translate into a higher shrinkage requirement to fully densify the desired part. High shrinkages often lead to greater distortion. Thus, the preferred powder shape in the PIM process is generally spherical. The particle shape can be determined by simple microscopic observation of the particles.

Having established that spherical powders (or teardrop-shaped powders) with a small fraction of irregular powders provide the optimum packing characteristics and strength after binder removal, we need to turn our attention to the particle size. Though most of the modeling on packing and volume fraction of solid loading have been carried out on monosized spherical powders, in reality, the commercially available powders usually have a particle size distribution.

Various particle size analysis techniques are available from which one can

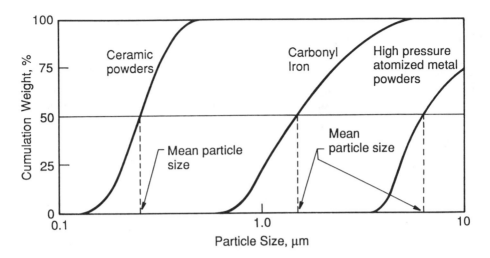

Figure 6.6: Cumulative plots of three powder sizes.

obtain the particle size distribution of the batch of powder from which the PIM feedstock is to be prepared. Most typical are cumulative particle size distribution plots, which graphically depict the cumulative percentage of particles that are below or above a certain particle size. The median particle size from a cumulative particle size distribution plot can be obtained from the particle size corresponding to the 50% value. The standard deviation can be determined by the usual methods and corresponds to the particle sizes at roughly 85% and 15%. A graphical representation of cumulative plots of three different types of powders with their mean sizes is shown in Figure 6.6.

The particle shape and size of the powders determine the actual surface area of the powders. In turn, the surface area of the powder to some extent determines the interparticle friction of the powders, which plays an important role in PIM processing. The surface area of the powder is connected to the density and particle size (for spherical particles) by

$$\text{Surface area} = \frac{6}{\text{Density} \times \text{Particle size}} \tag{6}$$

The surface area of the same material will be increased if the particle size is finer. Deviations from sphericity will also increase the surface area for the same volume of material. Thus, spheres, which have the least surface area, will also have the lowest interparticle friction. The interparticle friction increases as the surface area of the powders increases or the powder particle size decreases. The higher the interparticle friction, the lower is the flow of the powders and

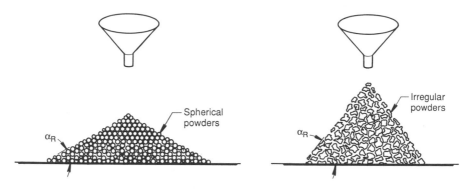

Figure 6.7: Effect of powder particle shape on the angle of repose, α_R.

the packing density. The best measure for the interparticle friction of a powder mass is the angle of repose. The angle of repose of a powder can be obtained by allowing a mass of powder to flow freely though an orifice and allowing the powder to form a cone on the bottom resting surface (ground). The angle that is formed between the ground and the side of the powder cone is the angle of repose. The smaller the angle of repose, the lower is the interparticle friction and the powder mass tends to spread out much more on the ground surface. This is shown schematically in Figure 6.7, where a spherical powder and an irregular powder after free-falling from a funnel placed at the same height from the ground exhibit distinctly different angles of repose. The spherical powder is seen to spread out much more flatly on the ground than the irregular powder. It can be visualized that coarse spherical powders will have the least amount of interparticle friction while the fine irregular powders have the highest degree of interparticle friction. The greater the interparticle friction, the greater will be the resistance of the injection-molded part against distortion and slumping as the binder is removed. Thus, coarse spherical powders, though good for providing high packing densities, would not be useful for PIM, as the part made from that powder would slump and distort as the binder was removed.

The role of interparticle friction is demonstrated with coarse spherical intermetallic powders [29] based on Ni_3Al. When this powder was mixed with a binder, compression molded into small cylindrical pellets, and debound within a wicking medium of fine alumina powder, the top surface of the pellet invariably collapsed. Even with volume fractions of solid loading as high as 0.7, this top surface cave-in was clearly visible. The phenomenon is discussed in greater detail in the debinding section of this chapter. Debinding without the wicking medium resulted in extensive flow and total loss of green shape. It was like forming a shape from spherical glass beads (intermetallic powder) that are held together by a glue (binder). It can be visualized that, once the glue is removed, the glass beads will collapse due to the extremely low interparticle friction.

The interparticle friction could theoretically be increased by the addition of fine particulates with high surface areas. The addition of only 5 vol.% of a fine spiky nickel powder with an average particle size of 5 μm to the powder feedstock (replacing the same volume of intermetallic powder) resulted in a green pellet that retained its shape after debinding. The top surface of the material did not cave in as it did when only coarse spherical powders of the intermetallic compound were used. Interestingly, it was found that the addition of 5 vol.% of chopped alumina fibers was also successful in significantly reducing the collapse of the top surface. SEM photomicrographs of the coarse spherical powder of the advanced nickel aluminide, the fine spiky nickel powder, and the alumina fiber used in the above experiments are shown in Figures 6.8a through 6.8c, respectively.

The importance of interparticle friction in influencing the final shape retention capability and warpage during debinding a shaped PIM feedstock is clearly demonstrated. A more detailed discussion on this subject has been carried out by Khipput and German [30]. Thus, although coarse spherical powders are quite inexpensive and easily produced by the conventional gas atomization techniques and would seem to be the likely choice of the PIM process due to their high packing density and the ease of powder availability, they cannot be used alone due to the severe distortion problems that occur during the debinding step. Fine spherical powders with an average particle size less than 10 μm could provide sufficient interparticle friction and allow shape retention during the debinding process. The other alternative is to use slightly coarse spherical powders (less than 25 to 30 μm) with a sufficient amount of irregular powder to increase the interparticle friction. This could provide an inexpensive alternative to the fine spherical powders that are being sought by PIM manufacturers. Another way of increasing the interparticle friction is to use short fibers or whiskers along with the coarse spherical powders.

Most particulates that are suitable for use in PIM should have an angle of repose of at least 55° or more. Among the alternatives discussed above, the use of spherical powders in the range 20 to 30 μm with some amount of irregular powder to increase the interparticle friction seems quite attractive from an economic standpoint. The irregular powders are usually far less expensive than the gas-atomized spherical powders, and conventional gas atomization with some modifications can produce high yields of powders that are in the 20 to 30 μm size range. However, with the recent demand for fine spherical powders, the powder manufacturers are also gearing up to try to meet the demand. As the productivity of the fine spherical powders increases, their price is expected to drop.

Figure 6.8: *(a) SEM of IC-218; (b) SEM of nickel powder; (c) SEM of alumina fiber. (Reprinted with permission from Ref. 29.)*

Powders with large amounts of interconnected porosity are not desirable for use in particulate injection molding. Even though the external shape of the powders may not be extremely irregular (their packing densities may be poor), the interconnected porosity will result in the binders being drawn into the fine internal pores due to capillary action. Binders that are drawn into the fine pores will be difficult to eliminate during the debinding step. This essentially eliminates a large segment of inexpensive powders that are generally produced by the process of oxide reduction. Subjecting these porous oxide-reduced powders to an additional grinding step to break up the powders into smaller particles so that they no longer have interconnected porosity could be one possible way to utilize these inexpensive powders. However, the powders are still expected to be irregular in shape.

Another form of porosity that is unwanted in powders for PIM is closed or sealed-off porosity within a powder particle. Due to various powder producing conditions, it is possible to have gases entrapped within a powder particle, resulting in a sealed off porosity that is not connected to the surface of the powder particle. If the measured powder density is lower than the actual theoretical density of the powder, it is quite likely that the powder has closed porosity. Visual observation of these pores can be made by observing the polished cross-section of a number of powder particles under a microscope. Determination of the density of the particulates by a pycnometer should also provide some clue to the amount of such closed porosity that is present in the mass of particulate material. It is extremely difficult to remove these pores during the final sintering stage, and they are generally retained in the bulk material as flaws. Expensive techniques such as post-sintering hot isostatic pressing would probably remove those pores, but would also lower the productivity and increase the expense of the PIM part. Thus, the powder should be carefully chosen and should preferably have very few particles with such entrapped porosity.

The chemistry of the powder particles will also play an important role in the PIM process. It is clear that the rheological properties of the particulate–binder mixture are influenced greatly by the wetting and adhesion characteristics of the binder to the particles. These phenomena are usually surface related. Powders, usually due to their fine nature, could have adsorbed gases or thin oxide layers on the surface. These could result in a change in the wetting and adhesion characteristics of the binder, which in turn could influence the volume fraction of solid loading, flow, and the feedstock green strength properties. Proper control of the powders, especially during storage, is an important consideration. Some powders, such as fine copper, if left on the shelf in a moist environment, could form a thin layer of oxide on the surface. Thus, to the surprise of many unsuspecting PIM manufacturers, the same batch of copper powder taken from the shelf after a period of time (maybe 3 months) can show entirely different mixing and flow properties. Therefore, care should be taken to prevent the storage of the powders in a warm and moist atmosphere. In some materials, such as aluminum, titanium, hafnium, etc., the oxides can create a great deal of difficulty during the sintering stage, resulting in poorly densified materials.

The question of powder chemistry also arises when a choice is to be made between prealloyed powders versus elemental powder mixes. Both types of powders are often used in PIM processing. The elemental powder blends often rely on diffusion and homogenization during the final sintering stage, while densification is the key concern during sintering of parts produced by injection molding prealloyed powders. The prealloyed powders are often harder than the elemental powder blend and also the powders tend to be more spherical in nature. This makes the prealloyed powders difficult to press in conventional dies, and it also results in a faster wear of the dies. Thus, in conventional P/M processes, the elemental powder blends are often favored over the prealloyed powders. Since the green shape in the PIM process is retained due to the presence of binders, it is much easier to use prealloyed powders in this process compared to the conventional P/M process of pressing and sintering. It should be remembered that the characteristics of the two types of material will be entirely different. There are, however, situations in PIM where the elemental powder blends would be more favorable. For example, when attempts are being made to directionally align fibers in a matrix material, it is important to have fine powders of the matrix. In certain cases, such as in NiAl–matrix composites, it is extremely difficult to obtain NiAl powder in the 5 μm size range, a size adequate for aligning the 20 μm diameter chopped fibers. Coarse powders will not provide the desired fiber alignment [31]. Elemental nickel and aluminum powders are commercially available in powder sizes of 3 μm. Blends of elemental powder mix can be reacted to form the desired compound, and with the fine starting powders it is possible to have excellent fiber alignment in the matrix.

The role of tailored powder particle sizes for higher powder packing density and improved sinterability has also become an extremely important research area. It is possible to produce extremely high powder packing densities based on suitable mixtures of powder particle sizes. However, the role of tailored powder size mixes in the overall sintering behavior as well as the debinding and molding characteristics is still not entirely clear. It is generally felt that the lower part of the curve in the cumulative weight percentage curves, as shown in Figure 6.6, has a very important part in determining the suitability of the powder for PIM processing. A more detailed discussion of this will be taken up in the next section of the chapter dealing with binder–particulate mixtures.

C. Binder–Particulate Material Mixture

One of the key steps in forming a good PIM feedstock is the proper mixing of the particulate material with the organic-based binder. The feedstock has to satisfy a number of conflicting requirements, and mixing of the two totally different materials, which have entirely different purposes in the feedstock, poses a real technological challenge. Some of the requirements and concerns are outlined below.

1. What is the optimum amount of particulate material loading in the binder?
2. Should tailored powder particle size distributions be used to try to increase the volume fraction of solid loading to very high levels?
3. How should the feedstock be mixed, i.e., in what kind of equipment and under what processing conditions?
4. How is one to assess the mixed feedstock?

This section will dwell briefly on some of the key concerns that have been outlined above, and try to illustrate the underlying principles. It will be best to start the discussion with the process of mixing and then gradually move into the area of producing good feedstock by optimizing the volume of solid loading in the mixture. The discussion will concentrate on the preparation of the extensively used form of PIM feedstock in which the multicomponent binder material is a mixture made of thermoplastics, waxes, oils, and surface-modifying agents. The binder needs to be heated to some elevated temperature so as to liquefy it. However, numerous other variations in the binders are possible, and they usually require different mixing regimes. It would not be possible to cover all the variations in the mixing regimes due to entirely different binder characteristics.

The process of mixing in PIM usually involves melting of the polymeric binder phase into which the particulate material is gradually added. The two components, one in a liquefied state and one in the solid state, are then agitated inside a single vessel for sufficient length of time to produce a homogeneous mix. The mixing usually has to be carried out at elevated temperatures to ensure that the polymeric binder component is in the liquefied state, but in case of some binders, such as water-based binders, where the binder component is already in the liquid state, mixing may be carried out at room temperatures. Once the mixing is completed, the mixture is usually cooled to produce a solid chunk of feedstock material; this is then removed from the vessel (usually called a tank), and the process of mixing is started all over again. This form of mixing is a batch-type mixing. When a new material is used in the feedstock, the whole chamber (tank) and the agitating arms have to be thoroughly cleaned before introducing the components of the new feedstock into the mixing chamber.

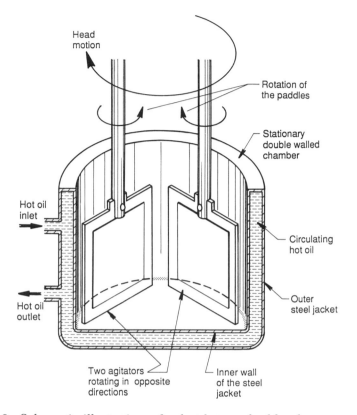

Figure 6.9: Schematic illustration of a batch-type double planetary mixer.

The agitation, which provides the mixing, is an extremely important step that determines the homogeneity of the feedstock. The pseudoplastic nature of the feedstock requires that all regions of the mixture be equally sheared. Thus, PIM mixers usually are designed to provide a high degree of shear.

To better understand the process of mixing, let us consider a common form of PIM mixer, namely the batch-type double planetary mixer, illustrated schematically in Figure 6.9. The double planetary mixer consists of a stationary double-walled steel chamber (tank) in which the batch mixing is carried out. The chamber has provision for circulation of oil that can be separately heated and pumped into the double-wall cavity. The temperature of the circulating oil can be controlled precisely, allowing heating of the mixing tank to the desired temperature.

The two agitator arms, commonly known as impellers, rotate in opposite directions. They are actually attached to the top lid of the chamber (not shown in the figure), which can be closed in order to render the system airtight. The speed of rotation of the two impellers can be controlled to provide high levels of shear during the final stages of mixing. The equipment has provision for a vacuum atmosphere to be maintained within the tank while the mixing is in progress. There are also some port holes toward the top of the mixing chamber that provide the operator the opportunity to look into the mixer and actually witness the process of mixing.

The process of mixing is usually initiated by introducing the individual components of the multicomponent binder into the mixing tank, which is preheated to the desired temperature by the circulating hot oil. The highest-melting-point component of the binder is the one that is liquefied first, and this is followed by the addition of the lower-melting-point components in sequence. The sequential addition of the lower-melting binder constituents is usually accompanied by decreasing the circulating oil temperature. This precaution is necessary so that the lower-melting-point binders are not evaporated or degraded by high temperatures in the mixing chamber. The temperature of the tank is then maintained at a point where the multicomponent binder in the liquid state has the desired viscosity. Once the components of the binders have been introduced into the preheated tank, a very short mixing of the binder components may be carried out by rotating the impeller arms. This is usually done to attain a homogeneous multicomponent binder before the solid particulates are introduced into the tank. To this liquefied binder, the total volume fraction of particulate material (solid) that needs to be loaded, is gradually added. The amount of particulate material to be added to the binder is determined by prior experimentation. A more detailed discussion on the optimum volume fraction of solid loading will be taken up at a later stage.

Initially around one-third of the total amount of particulate material is introduced into the tank with the binder. The impeller blades are rotated at a slow speed and the gradual progress of the mixing is observed through the port holes. The impeller blades should be rotated slowly at first as the addition of the particulate solid material produces in a rapid decrease in the mix temperature, resulting in a very high initial mix viscosity that could cause damage to the mixer if the impellers were rotating at high speeds. However, the heated oil circulating around the tank soon restores the temperature to its desired value. Usually within a very short time the solid particulate material is engulfed by the liquefied binder and the total mixture resembles a pot of thick creamy soup. At this juncture the rotation of the impeller arms is stopped, the top lid is opened, and another one-third of the solid is gradually added. The mixing process is carried out as before, and the sequence of steps is repeated until the addition of total volume of particulate material is completed. After the last amount of solid has been added to the mix, the top lid is closed, the impellers are activated. At this point the tank is usually evacuated and the mixing is carried out under vacuum to provide proper degassing of the mix, which ensures that the mix is free from tiny voids formed by attachment of air bubbles to the solid particles. The speed of the twin impeller blades is gradually increased to achieve high-shear mixing within the tank. The progress of mixing can again be observed though the port holes. The mixing is continued until the total mixture has been kneaded to the consistency of a thin dough. At this point the impeller blades are gradually slowed and finally stopped. Air is released into the system, the top lid is opened, and the temperature of the circulating hot oil is reduced to ambient temperature. The whole mass of the doughlike mix rapidly cools, contracts away from the side walls of the tank (due to shrinkage), and then solidifies into a block. It is advisable to introduce the coiled end of a thin steel rod into the doughlike feedstock before it has solidified. This enables the easy removal of the solidified chunk of feedstock from the tank by simply pulling on the top part of the steel rod. The tank bottom and the side walls must be smooth so as to allow easy removal of the solidified feedstock, and to prevent cross-contamination between different feedstocks. The feedstock stuck to the impeller arms is also scraped off. The feedstock is then either subjected to a secondary high shearing extrusion for better mix homogenization or directly pelletized into tiny fragments suitable for particulate injection molding.

An important issue that has to be addressed by the PIM community is the provision of scaled-up productivity. This has to include the conversion from batch-type mixers to continuous mixers. Especially in operations involving the use of very high volumes of a single feedstock, it is imperative that continuous mixing is employed. The design of the continuous mixer should be such that the mixture constituents can be subjected to high-shear mixing and continuous removal from the mixing vessel. The ribbon blender, and single-screw and twin-screw extruders can be used for continuous mixing. Generally shear is generated between the screw and the vessel walls, and the material is usually extruded out in the form of a cylindrical rodlike product. These devices usually have high shearing but poor blending capacity. The mixing time for the continuous mixers is usually much smaller than that of batch mixers. Thus, care needs to be taken to ensure that the feedstock is subjected to a proper amount of shearing that will result in the formation of a uniform mixture.

In general, for small volumes of different feedstocks the batch-type mixer is preferable, while large volumes of a single feedstock would be more economically processed by continuous mixers. The batch mixer is far less expensive and much easier to clean. The continuous mixer avoids batch-to-batch variations and is necessary for very high-productivity situations. Various forms of agitation devices are available for producing the PIM feedstock. All of them basically apply high shear during the mixing step. Some of the typical mixers

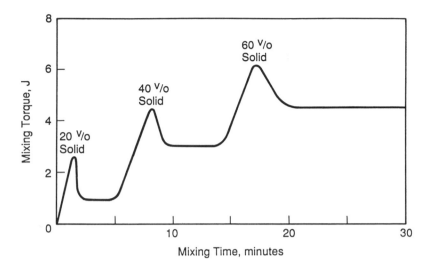

Figure 6.10: *The approximate variation in the mixing torque for a sequence of particulate solid additions in steps of approximately 20 vol.% (the curve is smoothed out).*

include double planetary mixers, single-screw extruders, twin-screw extruder, plunger extruders, sigma-blade or z-blade mixers, ribbon blenders, and twin-cam blenders. The double planetary blender is one of the most popular blenders, as it probably provides the best balance between reasonably good shear mixing, cost, easy scaleup, and ease of cleaning of the tank and the impeller arms. However, for continuous mixing, other forms of mixers have to be employed. A detailed discussion of various mixing equipment has been given by Snider [32].

It is important to know the homogeneity of the PIM feedstock after it has been mixed. Ways of determining it include observation of small portions of feedstock selected from random areas of the mixture and observing them under a microscope; measuring the density variations of samples taken from different areas of the feedstock and observing the variation in density from the calculated theoretical density of the feedstock; or continually monitoring the mixing torque. For a particular solid loading, the mixing torque should attain a steady state after which it should remain nearly constant with mixing time. Once the steady state has been achieved, the mixture can be said to be homogeneous. The approximate variation in the mixing torque for the sequence of particulate solid additions discussed earlier is depicted in Figure 6.10. Let us assume that the optimum solid loading is around 60 vol.%, and the solid is added in three equal steps. The addition of the particulate material to the binder will cause an immediate rise in the mixing torque, depicted as the first hump in the figure. A steady state is very quickly reached as the volume fraction of solid loading is quite low. The addition of the next amount of solid is responsible for the next hump in the figure. Here the attainment of steady state requires a little longer mixing time, and the steady-state torque is considerably higher compared to the 20 vol.% solid addition. The addition of the final amount of particulate material to attain the 60 vol.% solid loading is responsible for the last hump in the figure. In this case the attainment of the steady-state torque takes longer and the steady state torque has the highest value. The mixture with 60 vol.% solid loading has attained homogeneity within approximately 1200 s (20 minutes), which is verified by monitoring the mixing torque for 1800 s (30 minutes). The attainment of steady-state torque is therefore a quick and simple means of assuring that a homogeneous feedstock has been obtained after mixing. It should be pointed out that the shear rate has been assumed to be constant during the mixing; a change in the shear rate will result in a change in the torque since the feedstock is pseudoplastic in nature.

A number of other measurements to determine the homogeneity of the feedstock have been proposed: the measurement of the density of samples taken from various areas of the feedstock; measuring the variation in viscosity with shear rate; monitoring of the particulate concentration in samples obtained from various sections of the feedstock. Care should be exercised to obtain a nonbiased sample from the feedstock. Large deviations from a mean value often indicate inhomogeneous feedstocks. The PIM industry needs some continuous monitoring system that can provide a measure of the mix homogeneity during the mixing step, to avoid unnecessarily long mixing times that can produce nonuniform feedstock. The incorporation of a torque-measuring device in the mixer itself could provide the opportunity of monitoring the mix homogeneity as the mixing is in progress.

The ultimate goal of mixing is to provide a uniform coating of the binder on each individual particle, and impart to the mixture the desired flow characteristics during the molding stage. Uniform coating of the powders implies the breaking up of the powder agglomerates and dispersion of the binder components between the particles. The mixing temperature should be high enough that the mixture yield-point phenomenon is avoided. The shear induced in the mixer initially breaks up the particle clusters and the liquefied binder gradually seeps into the interparticle pores to provide a uniform coating on the particles.

A major problem associated with PIM feedstock is the separation of the binder from the particulate material, causing zones of low solid content or even small solid-denuded areas. Very low viscosity of the binder can result in segregation of the feedstock components. The density difference between the individual powders (in an elemental powder blend) can also lead to uneven packing density and ultimately result in greater degree of distortion in the final part. Commercially available powders are usually not monosized, and difference in the powder particle sizes can also lead to nonuniform packing, which in turn can result in higher distortion in the sintered condition. The larger the size difference, the greater is the problem of segregation, with the spherical-shaped powders tending to separate more than the irregular powders. Thus, from the point of view of segregation, small and irregular-shaped powders would seem to be the best. The positive factors of irregular powders are lowered segregation, stronger compact, and lowered compact slumping or distortion due to high interparticle friction. However, homogeneous mixing of fine irregular powders with the binder is also very difficult. Thus, for powders that show a tendency for segregation, possible solutions include the use of finer powder sizes, reduction of the binder viscosity, or addition of small quantities of irregular-shaped powders to the mix.

The volume fraction of solid loading is one of the key questions that must be resolved during the mixing stage. It controls the net dimensional shrinkage and influences the final distortion of the part. The volume fraction of solid loading can be manipulated to maintain the dimensional tolerances. When a solid particulate material is mixed with a liquefied binder, the volume of solid in the binder can be too little, too much, or optimum. With too little solid loading, the presence of excessive binder in the feedstock that has to be eliminated will result in excessive dimensional shrinkage, distortion, and defects like compact slumping. With too high a solid loading, the binder is unable to coat all the particles uniformly, resulting in voids in the feedstock. Also, the viscosity of the feedstock increases to very high levels, which creates difficulties during the injection molding stage. Optimal solid loading in a PIM feedstock should allow uniform coating of the particles with the binder. A schematic of the three situations of solid loading is shown in Figures 6.11a through 6.11c. The optimum volume fraction of solid loading should also maintain a level of viscosity at which the feedstock can be easily molded and the molded sample will not slump once the total binder has been removed.

In theory, for randomly packed spherical powders, each particle should have 4.75 contacts in order to retain the shape [33], which translates to a fractional packing density of 0.35. However, in practice, a packing density of approximately 0.6 is required to maintain geometrical stability [6]. For coarse spherical powders, even a powder packing density of 0.7 was found to be inadequate for retaining the geometrical stability [34]. Thus, surface area and surface roughness, which influence the interparticle friction, are also necessary considerations for processing a feedstock mixture of good quality.

The maximum packing density for monosized spheres is around 0.64. It is

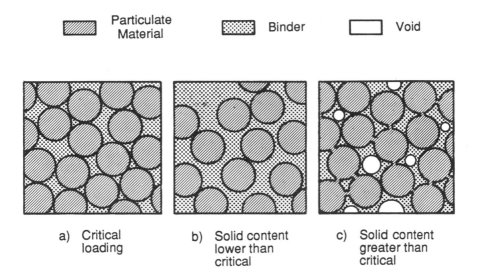

Figure 6.11: Schematic of the three loading conditions of a PIM feedstock, showing (a) critical volume fraction of solid loading; (b) low volume fraction of solid loading; and (c) too high a volume fraction of solid loading.

possible to increase the powder packing density by choosing a bimodal or trimodal distribution, or a wide range of powder particle size distributions. Thus, there exists the possibility of having tailored powder particle size distributions that can be used to attain high packing densities. Interestingly, very high powder packing densities are possible by adjusting the powder size ratio and the composition in a trimodal powder size distribution. A trimodal powder in which the particle sizes are in the ratio 25 : 5 : 1 will provide a maximum packing density of 0.85 when the compositions are approximately 69% large powder (25 μm), 9% medium size powder (5 μm), and 22% small powder (1 μm) [35]. For a trimodal mixture, the maximum density possible at infinite particle size ratio is estimated to be 0.95 [36]. It would be of great interest to to produce a PIM feedstock using a trimodal powder particle size distribution and verify experimentally whether it is possible to obtain volume fractions of solid loading as high as 0.90. However, it is expected that the mixture would be too stiff and too viscous and, therefore, unsuitable for PIM feedstock.

Theoretically, the density of the mix increases as the solid particulate material is gradually added to the binder. This is on the assumption that the particulate material has a higher density than the binder (which is almost always true). A few important parameters are necessary for generating a loading curve from which the optimum volume fraction of solid loading can be estimated. Usually, the weight of each component of the feedstock is carefully monitored. During the mixing process, incremental additions of the particulate material are made to a fixed amount of binder. It is important to be able to find the theoretical volume fraction of solid loading from the weight and density of the two individual components of the mixture. The densities of the binder and the solid are usually known quantities, and their respective weight fractions can be easily measured. The volume fraction of solid loading can be calculated from

$$\Phi = \left(1 - \frac{\rho_s W_b}{\rho_b W_s} \right)^{-1} \tag{7}$$

where

Φ is the volume fraction of solid particle loading in the binder;

ρ_s and ρ_b are the theoretical densities of the solid particulates and the binder, respectively (g/cm^3);

W_s and W_b are the weight fractions of the solid particulates and the binder, respectively.

It is also important to know the theoretical density of the binder–particulate material mixture for a particular volume fraction of solid loading. The theoretical density of the mixture is connected to the volume fraction of solid loading by

$$\rho_{th} = \rho_b + \Phi(\rho_s - \rho_b) \tag{8}$$

where

ρ_{th} is the theoretical density of the binder–solid particulate mixture (g/cm^3) and the rest of the terms identify the same quantities described in equation (7).

The theoretical part of the loading curve can be generated by plotting the variation in the theoretical density of the binder–particulate material mixture with increasing volume fraction of solid (particulate material) loading. The plot should represent a straight line, where one end represents the theoretical density of the pure binder, and the other end is the theoretical density of the pure solid (or mixture of solids).

To provide a better understanding of how the loading curve can be experimentally determined, and how the characteristics of the particulates affect the loading curve, consider two actual experiments [31]. In the first case a coarse spherical powder was mixed with a polyethylene binder and in another case the same spherical powder mixed with 5 vol.% of a chopped fiber was mixed with the same polyethylene binder.

For the first experiment, the advanced nickel aluminide powder, produced by gas atomization, was obtained from Homogeneous Metals, Inc. The particles were spherical in nature and the average particle size of the powder was approximately 70 μm (determined by laser light scattering). The composition of the powder was very close to that of the alloy developed by Oak Ridge National Laboratory, known as IC-218. The composition by weight of the various elements is 7.67Cr–8.18Al–0.80Zr–0.02B–rest Ni. In the following discussion, the compound will be referred to as IC-218. For the second experiment, chopped alumina fiber (around 20 μm in diameter and 3 to 10 mm long) obtained from Du Pont was used along with the IC-218 powder. Scanning electron micrographs of the IC-218 powder and the alumina fiber have been shown earlier in Figures 6.8a and 6.8c, respectively.

The binder chosen for both experiments was polyethylene wax (PE). The solid preweighed binder was first poured into the tank of a double planetary mixer, which has been described earlier. The hot oil circulation maintained the temperature of the tank between 398 and 408 K (125 and 135°C). For the first experiment, the polyethylene was liquefied and to it the desired amount of IC-218 powder was added to obtain an initial solid loading of 0.4. The volume fraction of the powder and the theoretical density of the aggregate were calculated from the standard formulas. The theoretical loading curve was first constructed by plotting several points of the calculated theoretical density of the binder–particulate material mixture versus the calculated volume fraction of solid loading. This should be a straight line and when extrapolated to the two extremes (i.e. volume fraction of solid loading 0 and 1) the two extremities should exhibit the density equivalents of the pure binder and the pure particulate material. This is shown as a solid dark line in the loading curve represented in Figure 6.12.

The 0.4 volume fraction of solid loading was first mixed in the double planetary mixer for 600 s (10 minutes) in a vacuum atmosphere. The mixing was stopped after 600 s (10 minutes), and small samples were taken from three different positions of the tank. The samples were obtained when the mixture was still in the liquid state. The densities of the samples were measured by water immersion technique and the average density of the three samples was determined. This was then compared to the calculated theoretical density for that particular volume fraction of solid, which was 0.4 in this case. The measured density was found to be 95% of the theoretical. The mixing was continued for another 600 s (10 minutes). Samples were again obtained and

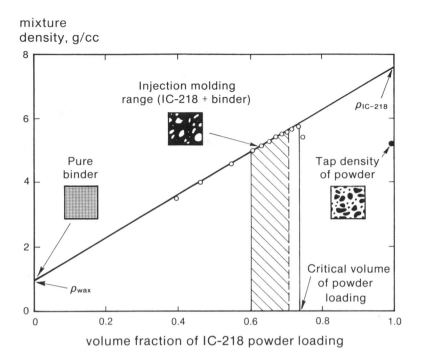

Figure 6.12: *Loading curve for IC-218 with polyethylene binder. The figure shows the calculated theoretical density variation with the volume fraction of solid addition and the critical volume fraction of solid loading. See text for experimental details. (Reprinted with permission from Ref. 29.)*

their densities measured in the manner described earlier. It was found that a total mixing time of 1200 s (20 minutes) brought the measured density of the mixture up to 99% of the theoretical. This has been plotted in Figure 6.12 as the first density point, which is shown by an open circle corresponding to the 0.4 volume fraction of solid loading. To the 0.4 volume fraction solid mix, more IC-218 powder was added to increase the volume fraction of solid. After each addition, the above described process of calculating the volume fraction of solid loading, mixing, determining the density of the mixture (by pulling out at least three samples), and comparing it with the theoretical density of the mixture, was performed. Each of the points is shown in Figure 6.12 by the open circles. Note the excellent agreement between the measured densities (open circles) and the theoretical densities (the solid line) for various volume fractions of solid loadings. This process of powder addition was continued up to a point where the polymer/metal mixture was no longer a viscous fluid but became discrete porous blobs. The experimentally determined density shows a large decrease compared to the theoretical density, showing that the critical volume fraction of solid loading has been crossed. This point is also shown in Figure 6.12 as the last open circle and the only point that has a measured density that is significantly below the theoretical density line. PE was again added to the tank to slightly lower the volume fraction of solid loading to a point where the aggregate once again just became a viscous fluid. The last point up to which the measured density of the mixture is nearly equal to the theoretical density of the mixture can be termed the critical volume of solid (or particulate material) loading. This point of inflection where the measured density substantially drops off from the theoretical density line is the critical volume fraction of solid loading, which means that any further solid loading will result in nonhomogeneous porous lumps that are unsuitable for injection molding. This is the point where the particles are in their best packing orientation with just the right amount of binder to produce a thin coating on the particulates and also fill the interstices between the particulates.

As mentioned before, the particulate characteristics will also influence the volume fraction of solid loading. The shape, size, size distribution, and tap density of the particulate material will influence the critical solid loading. This has been demonstrated from the second set of experiments, which develops the loading curve of a composite made from 95 vol.% of the IC-218 powder with 5 vol.% of chopped alumina fiber.

For the second set of experiments, the IC-218 powder and the chopped alumina fiber (around 30 mm long, 20 μm diameter) were mixed with the same polyethylene wax used in the first experiment. As before, the polyethylene wax was first poured into the tank of the double planetary mixer and melted at a temperature between 398 and 408 K (125 and 135°C). To the molten PE the desired amount of IC-218 powder and chopped alumina fiber was added such that the mixture had 5 vol.% fiber with respect to the IC-218 powder volume. The volume fraction of the solid mixture and the theoretical density of the aggregate were calculated from the standard formulas. The actual theoretical density of a mixture of solids, denoted TD_{mix}, which is composed of n individual solid components, where the theoretical density of each component is denoted as TD_a, TD_b, . . ., TD_n and the weight fraction of each component is denoted as W_a, W_b, . . ., W_n, is determined from

$$\frac{1}{TD_{\mathrm{mix}}} = \frac{W_a}{TD_a} + \frac{W_b}{TD_b} + \frac{W_c}{TD_c} + \ldots + \frac{W_n}{TD_n} \tag{9}$$

The initial particulate loading was around 0.63. On completion of the mixing of the composite with the polyethylene (mixed for 1200 s in vacuum), small samples were taken from three different positions of the basket. The densities

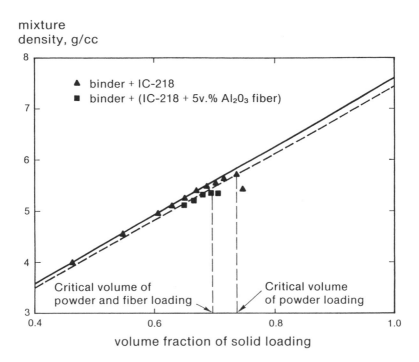

Figure 6.13: Loading curve for the IC-218 + 5 vol.% alumina fiber/polyethylene mix (broken line and squares). Also included is the loading curve for the IC-218/polyethylene mix (solid line and triangles). (Reprinted with permission from Ref. 29.)

of the samples were measured and compared to the calculated theoretical density for that particular volume fraction of solid. The density was found to be 99% of the theoretical. To the starting volume fraction of solid mix, more IC-218 and alumina fiber was added (maintaining the 5 vol.% ratio of fiber) to increase the volume fraction of solid. After each addition, the above process of calculating the theoretical density of the mixture, mixing, determining the density of the mixture, and comparing it with the theoretical density was repeated up to the point where the density of the mixture deviated from the theoretical density. The loading curve for the IC-218 + 5 vol.% alumina fiber/polyethylene mix is shown in Figure 6.13. The theoretical density variation with the volume fraction of composite loading is shown in the figure by the dashed line and the experimentally obtained densities at various volume fractions of solid are shown in the same figure as squares. The very close proximity between the theoretical density line and the experimental density points shows the mixing efficiency. After 0.68 volume fraction of solid loading, the mix density no longer follows the theoretical density line, and this can therefore be considered as the critical volume fraction of the composite. For comparison, the loading curve for the IC-218 powder has also been drawn in the figure, where the theoretical density variation is shown as a solid line and the measured densities are shown as triangles. Note that the theoretical density line for the composite (dashed line) is always lower than the theoretical density line for the plain IC-218 powder (shown as a solid line). This difference is due to the lowering of the calculated density of the composite, which consists of IC-218 + 5 vol.% chopped alumina fiber, compared to the plain IC-218 powder. It can also be observed from the figure that the critical volume fraction of solid loading that is possible with the plain spherical IC-218 powder is higher than the critical volume fraction solid loading of the IC-218 + 5 vol.% chopped alumina fiber. These two experiments demonstrate the sensitivity of the critical volume fraction of solid loading to the particulate shape. Similar experiments can be used to determine the effects of other particulate characteristics.

The optimum solid loading should be within a range of volume fraction of solid that is slightly lower than the critical volume fraction of solid loading. This is necessary since a composition very close to the critical volume of solid loading could result in an unstable PIM feedstock owing to very small errors in the weighing process. It is also necessary to absorb the small variations in the batch-to-batch characteristics of the particulates or the binders. Moving to a slightly lower volume fraction of solid loading is often necessary owing to the high mixture stiffness or viscosity at the high volume fraction of solid loading (very close to the critical). A very small error in the powder loading can push the viscosity to a level approaching infinity. Often the inability of the mixer to handle such a high-viscosity mix can limit the volume fraction of solid addition that can be made in the feedstock. The binder viscosity is seen to increase very rapidly at high volume fraction of solid loadings. The viscosity rise is often dependent on the powder particle size, shape, size distribution, and the critical volume fraction of solid loading that is possible. This is shown in Figure 6.14 where the same polyethylene binder has been used to mix two different powders, namely, the IC-218 powder and a commonly used injection molding grade of iron powder (manufactured by BASF). For both the materials, the viscosity measurements were taken at three different volume fractions of solid loading. It can be observed that the' IC-218 powder, which exhibits a higher critical volume fraction of solid loading than the BASF iron powder (0.76 versus 0.65), has a much lower viscosity level for similar volume fraction of solid additions. Alternately, it is observed that for the same relative viscosity, higher volume fraction of solid loading is possible for the IC-218 powder. This is a reflection of the higher critical volume fraction of powder loading possible for IC-218, which in turn is due to the coarse spherical nature of the powder.

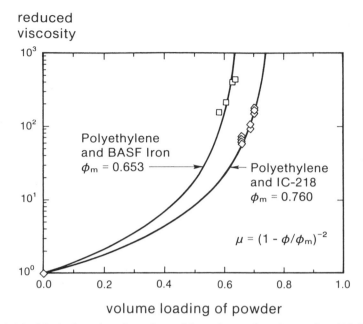

reduced
viscosity

Polyethylene
and BASF Iron
$\phi_m = 0.653$ ———————→   ←— Polyethylene
and IC-218
$\phi_m = 0.760$

$\mu = (1 - \phi/\phi_m)^{-2}$

volume loading of powder

Figure 6.14: *Variation in viscosity with volume fraction of solid loading of IC-218 and BASF iron powders mixed with the same polyethylene binder. (Reprinted with permission from Ref. 29.)*

Also, from a practical standpoint, it would make more sense to have a single tool set in which thousands of pieces are injection molded. Given the batch-to-batch fluctuations in the raw materials, it is conceivable that some adjustments in the solid loading would have to be made to maintain dimensional accuracy of the parts. Let us assume that the first batch of feedstock is prepared using a critical volume fraction of solid loading. Consider the problems associated in producing a second batch of feedstock that requires a slightly higher volume fraction of solid than that used in the first batch to maintain proper dimensional control. Mixing and molding of the second batch would be nearly impossible due to the viscosity constraint and void nucleation within the feedstock, as the volume fraction required is slightly above the critical solid loading. On the other hand, using the same volume fraction as the first batch of feedstock would result in parts outside the tolerance limit. This would leave the production engineers with the unacceptable option of changing the tool set to offset the shrinkage differential between the first and second batches of feedstock.

Various practical problems can be seen to arise if an attempt is made to produce a powder feedstock loaded with particulates to a point that is extremely close to the critical volume fraction of solid loading. The solid loading of feedstocks for particulate injection molding purposes is often kept at a level that is at least 5 vol.% lower than the critical solid loading. Figure 6.12 shows a range of volume fractions (hatched area) where a PIM feedstock prepared from a mixture of IC-218 and polyethylene binder will be suitable for powder injection molding.

Figure 6.12 also shows the tap density of the IC-218 powder (indicated by the solid circle). The tap density is the density of a powder or particulate material after it has been vibrated under standard conditions [37] and it provides an indication of the inherent packing density of the particulate material [38]. The tap density indicates the highest possible packing density of particulates in the pure state. As a general rule, particulate materials with lower tap densities will exhibit a lower critical volume fraction of solid loading. However, particulate materials like whiskers or chopped fibers, which may exhibit extremely low tap densities, can still have high critical volume fraction of solid loading when compared to some fine irregular powders with comparatively higher tap densities. The critical volume percentage of solid loading is usually 10% to 20% higher than that predicted from the tap density of the powder [6]. This is also shown in Figure 6.12.

In general, factors that tend to lower the particulate tap density will also lower the critical volume fraction of solid loading. For monosized particles, the departure from sphericity tends to decrease the critical solid loading; the smaller the particle size, especially for submicrometer powders, the lower will be the critical solid loading as more binder will be required to coat the surface of the powders; particulates with a broader size distribution, where the smaller particles could fill in the interstices of the larger particles, will allow higher critical solid loading compared to monosized particles; wetting agents that in minute quantities can alter the wetting characteristics can substantially increase the critical volume fraction of solid loading that is possible, especially in submicrometer-sized powders.

One way of rapidly measuring the critical solid loading for a particular binder–particulate material combination is to track the steady-state mixing torque. Markhoff et al. have shown that the steady-state mixing torque is at its maximum when the critical volume fraction of solid loading is attained [39]. The introduction of various torque-measuring devices can provide a rapid tool for determining the critical solid loading that is possible.

Once the critical volume fraction of solid loading for a particular binder–particulate combination has been established, the required quantity of feedstock is prepared for injection molding. The feedstock just after mixing is usually present in the form of a large chunk of material that cannot directly be introduced into the injection molding machine. The as-mixed material has to undergo an additional step before it can be ready for injection molding.

D. Pelletizing

The optimum feedstock is usually pelletized (or granulated) to form small pieces that can be suitably stored and also easily introduced into the injection molding machine. Each piece of the pelletized feedstock should ideally have the same composition of particulate material and binder. Often the sprues, runners, and rejected green PIM parts are recycled by mixing them with the freshly formed feedstock and then pelletizing the materials together. The feedstock, which is brittle at room temperature, can be granulated to smaller pieces by devices consisting of knives attached to a rotating shaft. The pelletized pieces are then classified by passing them through a coarse sieve.

An alternate method of pelletizing might include the formation of thin extruded ligaments of the feedstock (which could be a combination of freshly formed feedstock and recycled material) and chopping these into smaller pieces as it exits from the extruder. To facilitate the chopping action, the rodlike feedstock exiting from the extruder can be cooled rapidly and then chopped by rotating knife edges.

Extreme care should be taken to prevent cross-contamination between different feedstocks. Thus, if an iron-based feedstock has been pelletized and is to be followed by a nickel-based superalloy, the whole pelletizing equipment and the sieves need to be thoroughly cleaned. If a PIM company is involved in the processing of primarily two different materials in large volumes, it is definitely advisable to have two separate pelletizers dedicated to the two different materials. In spite of the initial investment, over the long run this would prove to be beneficial because of the cumulative time consumed in cleaning the pelletizing machine.

Once the pelletizing is complete, the granulated feedstock is ready for injection molding. These granulated feedstock materials are designated as the particulate injection molding feedstock since they are actually ready to be directly introduced into the injection molding machine. Interestingly, there are some small low-pressure injection molding machines in which the mixed feedstock, without any solidification, is directly injection molded. These machines do not require the use of pelletized feedstock. However, for the vast majority of cases, the process of particulate injection molding is separate from the mixing operation and pelletizing is a necessary step to produce feedstock in the size range suitable for particulate injection molding.

This concludes the discussions of the feedstock preparation for particulate injection molding. A few notes of caution should, however, be mentioned at this point. Since the particulates used in the PIM feedstock are usually very fine, special care should be taken to prevent the accumulation of too high levels of the particulates in the work environment. This is especially the case for materials that are known to be carcinogenic. Special care should also be taken during the mixing of fine particulates of highly reactive materials to prevent a potential explosion hazard. Care is also necessary in any mixing that involves the use of fine whiskers, which, due to their shape and bioinertness, are especially harmful if inhaled even in small quantities. However, with the declining interest in whisker-reinforced composites it is unlikely that general PIM manufacturers will have the occasion to produce PIM feedstocks involving fine whiskers. The OSHA guidelines in determining the level of particulates that can be tolerated in the work environment should be very scrupulously adhered to by PIM manufacturers.

2. *Injection Molding*

The process of particulate injection molding is essentially similar to that of plastic injection molding. During this step, green parts (containing both the particulate material and the binder) with considerable shape complexity and intricate detail can be molded in a shaped die cavity. However, due to the mixture of two materials of widely varying characteristics, the process of particulate injection molding is perhaps one of the most intricate and involved processes among all the metal-shaping or net-shaping techniques. There are innumerable variables that can contribute toward the optimization of the particulate injection molding process.

This section will briefly provide the reader with an idea of the complex process of PIM. Again the discussion will primarily concentrate on binder–particulate material feedstock where the binder is a multicomponent mixture mainly consisting of a thermoplastic material that melts at elevated temperatures to provide overall flowability to the whole feedstock. It should again be pointed out that there are numerous other variations of the feedstock that lead to totally different injection molding sequences. It will be outside the scope of this chapter to deal with all those variations in the PIM process.

A. *Injection Molding Equipment*

The machines used for particulate injection molding are similar in many respects to the equipment used for plastic injection molding. The key features of the molding machine along with the tool holder and the tool set and the operational sequence of the machine are shown in Figure 6.15.

There are, however, major differences between the tool set that is used in plastic injection molding and the tool set used for PIM. The plastic injection molding tools have dimensions that are somewhat closer to the final dimensions of the actual finished part. In PIM, the final part usually shrinks around 15% to 20% from its injection-molded green counterpart. In some cases, depending on the packing level that is possible, the shrinkage may be as high as 25% to 30%.

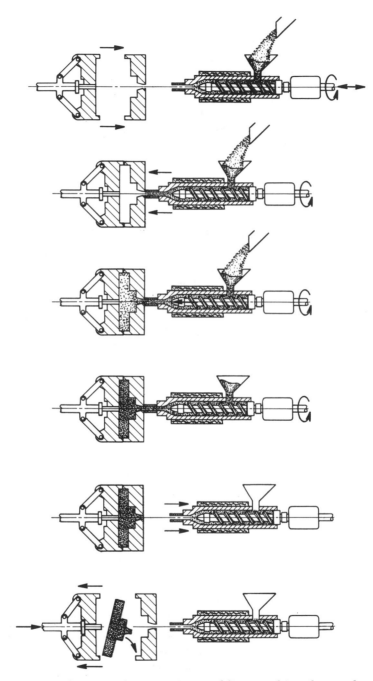

Figure 6.15: *Key features of an injection molding machine shown along with the tool holder and the tool set. The figure also shows the injection molding sequence.*

This high level of shrinkage makes it necessary to design tool sets that are oversized compared to the final dimensions of the parts.

Owing to the extremely complex nature of the PIM process design, the injection molding machine needs to utilize sophisticated and advanced computer-aided controls. In general, for the thermoplastic binders that are commonly used for PIM processes, a reciprocating screw-type of injection molding machine is commonly used. The reciprocating screw with a helical pitch operates inside a cylinder or barrel that in turn is surrounded by heaters. The typical screw diameter is around 25.4 mm (1 inch). The reciprocating screw, driven by hydraulic motors that allow easy torque adjustment, provides the necessary plasticization of the PIM feedstock and also helps in guiding the feedstock to the front of the barrel. The screw is normally rotated at speeds of around 40 rpm. The screw is attached to a plunger mechanism for forcing the feedstock into the mold cavity.

The reciprocating screw and the inner lining of the barrel are always subjected to high levels of abrasion, especially when very hard materials, such as tungsten carbides or ceramics, are being injection molded. It is therefore always advantageous to coat or clad the surface of the barrel and the screw with a hard and stable material that has considerable abrasion resistance. Apart from the harder barrel and screw faces, the other major hardware used for particulate injection molding is nearly the same as that of a conventional plastic injection molding machine. However, the machine settings are quite different as particulate injection molding conditions differ from those for plastic injection molding.

The temperature requirement of the PIM process is usually dictated by the necessity to obtain a temperature that will result in a feedstock viscosity lower than 100 Pa.s (or approximately 1000 poise). The temperature in the barrel is carefully controlled by heaters that can be independently controlled, allowing the temperature in the barrel to be regulated in different zones. The nozzle temperature can also be separately controlled. The time the feedstock is present within the barrel is dictated by the time required to obtain the desired uniform temperature and viscosity level in the feedstock. For most systems, the operating temperature is below 473 K (200°C). Molding pressures are typically in the range 15 to 20 MPa. The packing pressure can be as low as 5 MPa, with the packing times being typically around 5 s or longer depending on conditions such as the part size, feedstock temperature, die temperature, etc.

Frequently the specification of the molding machine is given in terms of the quantity of feedstock that can be delivered to the mold cavity in each stroke, or the clamping force on the die, or the maximum injection pressure. As a rule of thumb, the actual volume of the part that will be injection molded should be less than three-quarters of the volume of the actual shot size.

The design of advanced injection molding machines relies on the identification of key machine variables. Orthogonal arrays have been used to find the optimum injection molding conditions. Included in the numerous variables are the machine controls, which include four to five regions of barrel and nozzle temperature control, injection speed and profile, packing pressure and profile, packing time, cooling time, and mold temperature. These experiments can successfully account for the role of various machine settings for a particular feedstock in order to obtain a part within the tolerance limits [40].

Computer simulations and critical experiments have now identified the key control points where sophisticated controls are necessary for designing a good injection molding machine. For example, feedstock pressure inside the die is of greater importance than the hydraulic system pressure that is commonly measured; during die filling, the injection speed is of greater importance; during ejection, the temperature of the die is more important than the cooling cycle time. Modern injection molding machines are gradually incorporating closed-loop feedback control. The advanced molding machine is controlled by microprocessors that sequentially adjust all the major molding variables. The molding machines can be programmed to simulate the desired molding cycle. Microprocessor-controlled machines can also be used as process control equipment that provides information about potential problems. It is expected that with better understanding of the key controlling steps in the particulate injection molding process, better and more responsive machines will be introduced to the market.

Among the equipment used for particulate injection molding, the tool set is perhaps one of the most expensive consumable items. Thus, very careful design of the tool set is an important part of success in PIM. The proper design of the tool set includes not only the design of the die cavity but also the design and proper placement of the sprue, runner, gate, parting line, and ejector pins.

To aid the proper design of a die cavity with the optimum location of various other segments, computer simulations have often been used. Various softwares have been developed to analyze the flow behavior of the feedstock within the mold cavity. A finite element model of the mold has to be first constructed in a computer. The rheological and thermodynamic data of the feedstock are provided to model the flow behavior. The input also consists of injection molding parameters such as melt temperature, mold temperature, and injection molding time. During this stage, the designer can also vary the molding conditions, gate locations and number of gates, as well as modify of the part design. Since all the trial-and-error type of modifications are carried out on the computer, a large number of iterations is possible. The iterations are continued until the desirable molding conditions such as controlled flow pattern and balanced flow, proper and uniform cooling, maintenance of uniform pressure profiles, etc., can be attained in the computer simulations. The melt front advancement plots for the molding of a business machine part simulated on a computer are shown in Figure 6.16a, and the actual short shot (where the feedstock is injected into the die in small amounts so as not to fill the die completely) series of the same part is shown in Figure 6.16b [19]. The injection molding parameters for the short shots were carefully chosen to match the input parameters used for the computer simulations. The accuracy of the computer simulation in predicting the actual mold filling sequence is clearly demonstrated from the two pictures.

One of the great advantages of the computer flow simulation is that it enables designers to modify their part design and find an ideal gate location for the part that has to be injection molded. The flow analysis provides an important tool for the tool designers in designing the proper gate location without physically producing the tool itself. An interesting prediction that is obtained from the computer simulation and has been verified by the actual business machine part is that the thick section is filled first, then followed by the filling of the thin section. This provides an indication of the ideal gate location. If the gate is located too close to the thin section, it could result in interruption of flow, causing the area to freeze and create short shots; while overpacking may result in the thick section if the gate is located too far from the thin section [19]. The computer simulation can also provide temperature gradient plots for different gate locations. It was observed in a medical forceps blank that different gate designs gave rise to different temperature gradient plots. The actual distortion of the part closely matched that of the thermal gradient plot that was generated by the computer simulation. Thus, the flow analysis provides valuable information to the designers, and helps in preventing the need for expensive tool alterations by alerting them to possible design flaws.

The location of the parting line is also of great importance in designing a successful tool set for powder injection molding. Due to the packing pressure, elastic deflection may occur in the tool set between the movable sections of the die. There is often a faint line on the compact surface where the split had occurred, and thus the parting line should be in a noncritical location of the part. A similar noncritical location is also desirable for the gates and also for the section where the ejection pin is used to push the compact out. Often the ejection pin leaves a mark on the compact surface. A schematic of a tool set showing the parting line, sprue, runner, gate, and die cavity is shown in Figure 6.17.

A question that is often considered is whether to use a single-cavity tool set or a multi-cavity set. A number of factors should be considered before taking a decision on selecting either. The key consideration should be the desired

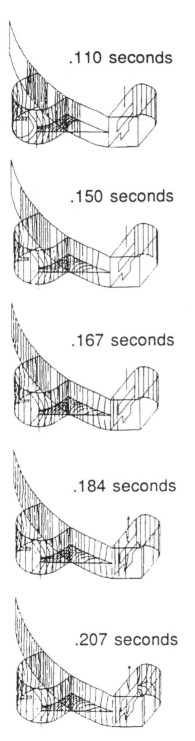

.110 seconds

.150 seconds

.167 seconds

.184 seconds

.207 seconds

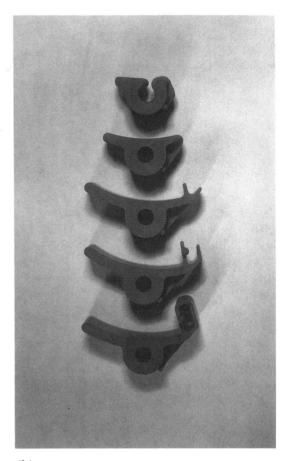

(b)

Figure 6.16: *(a) Melt front advancement plots for the molding of a business-machine part simulated by computer. (b) The actual short shot series of the part that is shown in (a). (Reprinted with permission from Ref. 19.)*

(a)

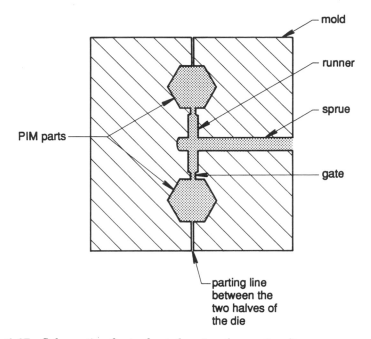

Figure 6.17: *Schematic of a tool set showing the parting line, sprue, runner, gate, and die cavity.*

production quantity that is required. If low production quantities are required, it is better to use a single-cavity mold, since multiple-cavity molds cannot be justified. However, if very high production quantities are required, it is advisable to select a multiple-cavity die set. It should be realized that the total volume of the combined cavities including the sprue, runners, and gates should be lower than the shot capacity of the injection molding machine. Sufficient clamping force should also be available. And of course the choice of a multiple-cavity tool set would greatly increase the cost of the tool set. Some common multiple-cavity geometries are shown in Figure 6.18.

The tool is often made of tool steels or other hardenable steels. Other tool materials such as stainless steels, low-carbon steels, maraging steels, etc., have also been used. For prototyping, where only a few pieces of sample need to be injection molded, softer and more easily machinable materials such as aluminum could conceivably be used, but it should be realized that the softer material will scratch and wear much faster than the harder materials. Hard materials such as cemented carbides have also been used for tools, but their lack of toughness makes them a low-priority material. Easily hardenable and machinable materials such as tool steels or die steels are usually the most popular material of choice. A heat-treated hardness of around HRC 50 is

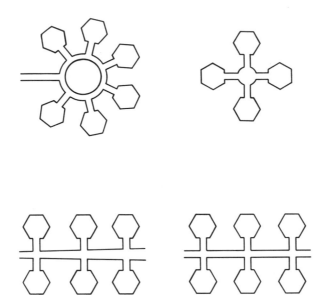

Figure 6.18: Some of the common multiple-cavity geometries.

usually desirable. Given the abrasion that a tool surface has to undergo, a typical tool can probably produce around a quarter of a million parts. This number can be considerably increased by providing a good surface-hardening treatment. As the tool is one of most expensive disposable items, it is advisable to invest the extra amount and have the die cavity coated with a high-hardness material.

There is always the requirement for producing dies that can be used for rapid prototyping. The TAFA process has made a significant contribution in this respect [41]. The TAFA process starts with a model that is exactly the size and shape of the desired tool cavity. The model can be made of virtually any material including wood, wax, plastic, or plaster. A thin shell is then formed on the model by spraying a low-melting-point zinc alloy. This arc spraying process is accomplished using wire as the feedstock. The electrically charged wire is melted at the gun's head and carried to the model surface by atomizing gases. No distortion or overheating of the model occurs during the spraying process. The process is also quite rapid, as a 1.6-mm ($\approx \frac{1}{16}$ inch)-thick shell has been developed on a 9.29×10^{-2} m^2 (1 ft^2) model within 900 s (15 minutes). According to Zanchuk and Grant [42], this spray process can produce tools with virtually no size limitations. The process has been used to produce tools about the size of a dime and also tools for plastics that have a surface area of approximately 6.5 m^2.

The Kirksite type of zinc alloy, which has a melting point around 700 K (427°C), is about three times harder than pure zinc. This material has been used as a convenient shell material, and small injection molding production runs have been made in such molds. Two insulated wires of the desired spraying material are gradually fed into the gun, where they form an arc and melt. The atomizing gas shears off the molten part of the wire as fine particles and they impinge directly on the substrate material to form a dense coating. The atomizing gas serves the dual purpose of propelling the fine particles toward the substrate material as well as keeping the substrate material cool even though it is being impinged by the molten zinc alloy. The particles solidify on impact after forming a splat with excellent interlocking. The shell temperature never reaches 328 K (55°C) when proper spraying conditions are utilized. The hand-held spray gun can be easily manipulated. A reasonable rate of spray material consumption is around 1.5 to 3 g/s (12 to 24 lb/hour).

Once a thin shell has been formed, the shell is removed from the model. All the surface imperfections are very well reflected by the shell. The model is first sprayed with a high-temperature barrier coat and a parting agent such as a special kind of PVA. The parting agent serves the very important purpose of providing excellent metal spray adherence to the pattern. Interestingly, the spray will not be retained on the model without the use of the parting agent. Once the shell has been formed, the parting agent helps in the easy stripping of the shell from the master model.

The thin shells cannot be directly mounted on molding machines, and without any backing they would not be able to withstand the pressures produced inside the cavity during the injection molding operation. Before backing up the material, a frame of aluminum is usually constructed around the shell to hold the backing material. The most common backup materials for the shell are methyl methacrylate resin filled with alumina powders or a bismuth-based alloy.

The process offers tremendous advantages in terms of net savings in price as well as the required turnaround times for rapid prototyping jobs. Tools produced by this process have been used to produce hundreds of different PIM parts.

B. Injection Molding Steps

This is a suitable point for a brief discussion of the steps involved in the injection molding process. During PIM, the solid granulated feedstock is fed into the injection molding machine barrel through a feed hopper. Heaters bring the barrel or cylinder of the injection molding machine to the desired temperature. The granulated feedstock enters the barrel, where it is melted, metered, and packed by the action of the reciprocating screw. The tool holder that holds the die and the attached ejector pin is advanced to a point where the nozzle of the injection molding machine leads directly into the sprue through which the plasticized PIM feedstock can be injected into the mold cavity. The actual die filling is obtained when the reciprocating screw is thrust forward within the barrel to push the feedstock through the sprue, runner, and gate, and then into the die cavity. The mold filling rate is an extremely important factor that, unless carefully controlled, can lead to variety of defects in the green molded part.

The mold filling consist of three different steps: filling, packing, and holding [19]. The filling step is initiated when the ram pushes the feedstock into the die cavity. This step is completed when the die cavity is completely filled. During this stage the ram is the only moving section as the screw is not rotating. It is best if the filling part is accomplished under a fixed programmed speed control profile.

The packing step takes over the instant the die cavity is completely filled. The transition can be measured by having pressure transducers in the die cavity. According to Hens and coworkers [43], it is best if the injection molding machine at this point switches over to a pressure-versus-time profile. Here the ram continues to push material into the mold until a point at which hydrostatic equilibrium is obtained. To obtain a clearer picture of the packing step, readers should remember that the feedstock material that has taken up the shape of the die cavity starts to cool and shrink. Pressure needs to be maintained to continue packing the die with material in order to offset the shrinkage. If the die is just filled and then allowed to cool, the material ejected from the die will have no dimensional accuracy, as it will not conform to the dimensions of the tooling. Also, when the material is cooling in the die, its viscosity tends to increase rapidly. Thus, to maintain the dimensional stability, the packing pressure needs to be increased with progress of time.

During the holding stage, the ram is allowed to creep forward slowly. This also compensates for shrinkage of the material. However, the pressure is not increased during the holding stage but is essentially maintained at the peak level. Once the gate has cooled, pressure should be adjusted to properly pack the sprue and runner.

Depressurization and cooling are carried out after this point. This allows the material in the die to cool sufficiently so that it can be ejected without problems. However, too long cooling times can create problems of part cracking, especially around intricate cores. For decompression, the screw is withdrawn without rotation, preventing the melt from dribbling out though the nozzle. The tool set and the injection molding machine carriage are then retracted and the mold is separated. The ejector pin pushes out the solidified part, and the total injection molding cycle is repeated again. The general sequence of the molding steps is shown in Figure 6.15.

The process of injection molding looks deceptively simple. However, once some analysis has been made of the problems associated with the various stages of injection molding, and a general assessment has been made of the variables that can cause imperfect green parts, readers will be in a position to better appreciate the problems associated with this extremely complex step.

C. Defects Generated during Molding

If the molding variables are not well balanced, this can lead to a host of problems such as short shot, jetting, sink marks, density gradients, flash formation, void entrapment, cracks, particle–binder separation, sticking of the green part to the mold wall, etc. Various molding conditions can be the cause of these defects. It is worth while to dwell on the defects and their origin during the process of molding; we shall consider the possible defects that may arise as we proceed with the injection molding cycle.

The viscosity of the PIM feedstock is lower than that of regular injection-moldable plastics; however, its thermal conductivity is much higher. As a result, during the process of mold filling, while the feedstock is flowing into the mold to fill the die cavity, the part of the feedstock that comes in contact with the die wall is quickly solidified. Restricted flow still continues to occur because the material slightly away from the wall is still molten and under pressure. The material near the frozen wall is highly oriented and is subjected to very high shear stress, while the stress in the center is much lower with much lower orientation effects. The orientation effects become critical when long-chain polymers are used as binders. Also, large nonisotropic residual stresses may be generated within the green shape due to the fast-freezing nature of the feedstock when in contact with the cold die wall. This could eventually lead to warpage of the part. The highly stressed region very near to the wall can cause particle bridging. This would create a situation where the particles form something akin to a filter through which the low-viscosity binder may be extruded. This could result in particle–binder segregation, which ultimately causes large distortions and flow lines that can cause lamellar cracks.

Solutions to the above problems include the use of high molding temperatures, short filling times, and the use of particulate loading close to the critical volume fraction of solid loading. The use of binders with low molecular weights can go a long way in decreasing the problems associated with the orientation effect. The use of a multicomponent binder that includes oil as one the components can help in reducing the cracking and residual stresses, as the oil never hardens and helps in relieving the residual stresses.

Jetting and short shot problems also occur during the mold filling step. When the speed of mold filling is slow, the contact of the feedstock with the cold die wall can lead to rapid freezing of the material after it has partially filled the die cavity. Short shots can also occur due to hesitation during the mold filling step, which in essence means that the melt flow should not be interrupted before the die cavity has been completely filled. Short shot results in parts that are incomplete in nature. The chances of producing a short shot are very high if slow injection speeds are used in conjunction with a low molding temperature and high feedstock viscosity.

If the speed with which the feedstock is injected into the die cavity is too fast, jetting problems can often occur. Conventional jetting involves the squirting out of a thin ligament of molten feedstock (especially when the molten feedstock is forced through a thin section like a gate into a much larger cross section of the mold cavity), which impinges directly on the opposite wall. Low-viscosity feedstocks when injection molded into the die cavity at high speed can often result in this form of jetting behavior. In this case, the filling of the mold starts to occur from the wall opposite to the gate, instead of the desired fill direction. Air vents are usually placed at the section of the die cavity where the last part of the mold filling is expected to occur. This is done to ensure that the air inside the mold is gradually expelled as the mold is filled with the feedstock in the desired manner. With jetting, the material first hits and solidifies in the section that most likely contains the vents. This is expected to result in air entrapment, which will result in void formation. Even without the jetting problem, improper placement of the vents can lead to air entrapment in the green part. This can be prevented by using lower injection speeds during the initial stage of die filling so that a larger front has been built up in the larger cross section of the die cavity [43]. Care should be taken to increase the speed once the larger front has been produced within the die. If this is not done, it may result in the freezing of the feedstock before the die has been totally filled. Another alternative is to try to produce low pressure or a vacuum within the die cavity just before the die filling actually starts.

Another form of jetting may occur under almost the opposite sort of circumstances. If the speed of die filling and the melt temperature are low, multiple jetting can occur within the cavity. A short shot sample where multiple jetting has occurred will look more like a cactus tree with various branches. This happens as a stream of metal filling into a die cavity freezes up at the front. However, there may be zones within the stream where the feedstock may still be molten. Once the leading tip of the flow front has frozen, but pressure is still being maintained, new flow fronts will be established by the molten feedstock squirting out in various directions, emanating from different areas of the original melt stream whose front tip was frozen. This form of flow condition and short multiple jetting may occur until all the feedstock in the mold is frozen up.

Improper mold filling conditions, such as too high a feedstock viscosity or low feedstock temperature, can cause weld lines to occur when the feedstock has to flow around an internal core rod. The flow of the feedstock filling the die from one region, when encountering a core rod, has to flow around the core rod. This could create a weld joint along the line where the flow rejoins again. Unless the temperature of the feedstock is high enough to give low viscosity of the melt, the weld lines formed due to the rejoining of the flow paths cannot be eliminated. These weld lines can be the cause of potential defects in the PIM shape. Schematic views of the ideal die filling condition, where a stable broad front is gradually moving forward; of conventional jetting, where a stream of material hits the opposite wall and the mold filling occurs backward; of multiple jetting, where short shot occurs; of the melt front when the gate placement is at the side of the die cavity, which can create better die filling conditions compared to a gate placed at the center; of the use of a fan gate to provide uniform filling of the die cavity; and of the possible creation of weld joints when the melt flows around an internal core are shown in Figures 6.19a through 6.19f.

Possible solutions to the problems discussed above may include proper control of feedstock temperature, which in turn will control the feedstock viscosity; proper positioning of the gate; proper size and cross section of the gate; optimum speed of die filling, which may have to be varied during the die filling operation; and the use of molds heated to a temperature below the softening point of the feedstock.

If a proper packing pressure is not maintained to offset the shrinkage, sink marks may form on the compact surface. This happens due to the insufficient amount of feedstock in the mold. However, it should be remembered that overpacking can also lead to distortion and crack formation in the compacts. The use of too high a pressure during the packing stage, with a high temperature of the feedstock, will result in the separation in the die and consequently the formation of flashes where the binder or the feedstock is forced into the gap between the die. Maintaining optimum temperature and pressure during the packing and the holding step is the key in obtaining defect-free green parts.

The wrong location of the gates also can lead to a host of problems such as improper die filling, uneven density distribution, and undesirable temperature gradients during cooling that could ultimately lead to part distortion.

Once the molding cycle is completed, the part is cooled in the die cavity and ejected. A number of problems can arise during this stage also. The problem of

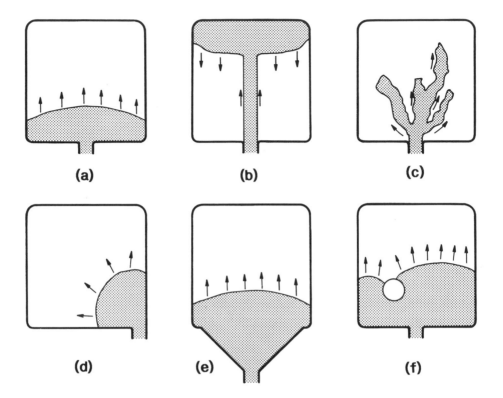

Figure 6.19: *A schematic view of the melt front: (a) ideal die-filling conditions; (b) conventional jetting; (c) multiple jetting; (d) gate placement at the side of the die cavity; (e) use of a fan gate; (f) possible creation of weld joints when the melt flows around an internal core.*

compact sticking to the die wall can occur for several reasons. One of the simplest conditions that can result in the compact sticking is the presence of high surface roughness of the die wall or the presence of scratches on the die surface. If the mold is opened at a time when proper cooling of the green compact has not been attained, the likelihood of the compact sticking to the mold is quite high. Improper cooling can also result in samples that are too "soft" in nature, and the ejection forces can create distortion of the green parts. At the other end, too long a cooling cycle can result in the tendency of the green part to adhere strongly around an internal core, resulting in cracks on ejection.

Some solutions to the above problems include the use of dies with highly polished surfaces, and the proper maintenance of the dies to prevent any deep surface scratches. Although the use of mold release agents to prevent sticking of the green compacts with the die wall automatically suggests itself, careful consideration must be given to the fact that the mold release agent could react with the components of the feedstock, and its uniform application to the die wall may create such problems as a slow molding cycle, which can hurt productivity. However, the best solution to the problem of mold sticking is the design of an optimum cooling cycle and the opening of the mold at the desired temperature and pressure.

The cooled part is usually ejected from the die by the action of the ejector pins. Unless the compact has cooled sufficiently, the ejector pins can leave a deep impression on the surface of the green molded part. The removal of the gates should also be carried out with care so as not to damage the part itself. Lastly, it is important to carefully collect the green parts after ejection from the die cavity. If the parts are allowed to free-fall over a great distance, and if the solidified feedstock is not quite strong, the part may be chipped or even broken. Thus, careful handling of the green PIM part is recommended. It is also recommended that at least a quick visual inspection of the green parts be carried out before the parts are released for further processing. Rejected parts can be reground and mixed with freshly prepared feedstock and used for molding.

D. Moldability

It is important to know how well a particular feedstock can be molded under a certain set of molding conditions such as temperature and pressure. The materials as well as the molding parameters tend to control the ease with which a feedstock can be molded. The major controlling factor is usually the viscosity of the feedstock under the molding conditions.

One of the most popular measures of moldability is the "spiral flow test." In this test, the feedstock is forced to flow through a mold that has an essentially spiral shape, shown in Figure 6.20. The length of the spiral that can be filled up by a particular feedstock under particular molding conditions is considered to be a measure of the moldability. The spiral fill length will increase when the solid loading in the binder is low, the molding temperature is high, the molding pressure is high, and the feedstock viscosity is low. However, if the solid loading is quite low the spiral fill length will be quite high, and there may be considerable separation between the solid and the binder along the length of the spiral [44]. The effect of both solid loading and molding temperature for an alumina-based solid is shown in Figure 6.21 [45]. It can be observed that with increasing volume fraction of solid loading the spiral fill length decreases, while with increasing molding temperature the spiral fill length increases. A similar influence of solid loading for a silicon carbide-based material has also been observed by Willeermet and coworkers [46].

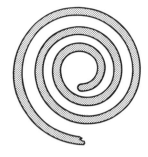

Figure 6.20: *Schematic of the spiral flow test.*

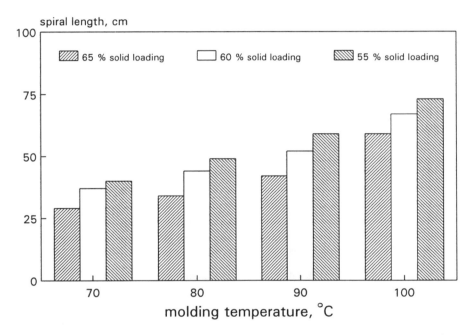

Figure 6.21: *The effect of both solid loading and molding temperature for an alumina-based solid in the spiral mold test. (Data from Ref. 45.)*

Another test in which the moldability of the feedstock can be measured constitutes a zig-zag form of die insert. In this configuration, a change of direction of flow is encountered by the feedstock. This is a more tortuous test in which the flow reversal causes an increase in the flow resistance and also results in separation of powder and binder. A view of this test configuration is shown in Figure 6.22 [43]. Experiments have shown that at low molding temperatures the feedstock, instead of filling the straight part completely, starts

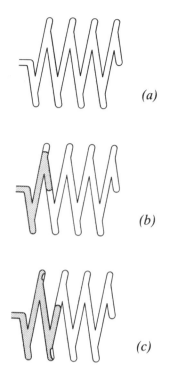

Figure 6.22: (a) A view of the zig-zag flow test configuration. (b) Filling behavior of an iron/40 vol.% binder mixture injection molded at 368 K (95°C). (c) Filling behavior of a iron/40 vol.% binder mixture injection molded at 388 K (115°C). (After Ref. 43.)

to fill the cavity after the branching. Even in higher-temperature tests, though the first straight section was fully filled, the second straight section only was partially filled. This test shows that major problems can arise when the flow direction within the mold has to suddenly change. More investigations are needed to determine the conditions that cause the freeze-up of the flow front when the flow reversal occurs. In designing the tool set, care should be taken to avoid sudden changes in the direction of the flow front in order to ensure proper die fill.

This completes the discussion on injection molding. The key operating steps, some of the problems that could arise during the process of injection molding, means of avoiding design problems by computer simulations, and ways of measuring the moldability, have been briefly discussed. However, a tremendous number of subtle variations exist in this crucial processing that would be impossible to cover within the scope of this chapter.

3. Debinding

Debinding, as the name suggests, is a process by which the binder material that is present in the green PIM part is removed. Generally it is the aim of the debinding process to remove all the binder in a manner that leaves behind no residue whatsoever. This step usually follows directly after the injection molding step, unless the green part needs some additional secondary operations before debinding.

Debinding is one of the most crucial steps in the powder injection molding process and is usually the rate-controlling step in the total PIM process. Numerous variations in the binders used by different companies are responsible for the extremely large variety of debinding processes that are used in particulate injection molding. This is one of the most delicate steps in the total sequence, in which a significant volume fraction of the green part has to be removed without creating any distortion of the part. Common sense predicts that debinding has to be a very slow process that proceeds in small steps and allows small amounts of binder to be removed over different time intervals. It would not be possible to remove all the binder in one instant, as this would cause swelling, distortion, and cracking of the green part.

The strength of the green compact is derived from the binder that essentially glues together the otherwise loose particles. There is a major difference between the green parts produced by die pressing of powders and green parts produced in the PIM process. The magnitude of the pressure that is applied during the powder pressing stage is much larger than that used in the PIM process and, consequently, during die pressing the powders undergo extensive plastic deformation and shape modification. As a result, good bonding develops between the particles and this, coupled with good mechanical interlocking between the powder particles, generates sufficient green strength in the as-pressed powder samples. In the case of green PIM parts, apart from the interparticle friction of the powders and the glueing effect of the binder, there is practically nothing else that holds the shape together.

Most debinding processes rely on heating the compact in some form or another, and during that step the organic binders usually soften or melt. The green part in this condition would be unable to withstand any internal stresses that are generated. Thus, the shape retention has to rely totally on the interparticle friction characteristics of the powder, at least during the initial stage of debinding. It is possible that during the later stages of debinding, especially with fine powders, some form of contact necks may develop between the individual particles and lend more shape rigidity to the green part. It is therefore not difficult to appreciate the extremely important role that the interparticle friction characteristics play in the process of debinding. That is one reason why finer powders are preferred for PIM processing. On the other hand, the use of irregular powders, which have higher interparticle friction characteristics, is restricted in the PIM process due to the poor powder packing characteristics exhibited by this type of powder. Thus, during the debinding step, the interparticle friction needs to be quite high, and the extraction of the binder should occur smoothly either as a liquid or a vapor.

The success of debinding is dependent on a number of interdependent factors such as the debinding time–temperature cycle, debinding atmosphere, binder chemistry, particulate–binder interactions, the type and size of pore structure, which in essence depend on the particulate characteristics, and the interparticle friction. The earlier binders, which relied on one or two components and pure thermal debinding for binder removal, often had debinding times as high as several days. Under modern manufacturing environments, such long time frames are totally unacceptable. As the understanding of the complex process of debinding progressed, it was realized that a multicomponent binder system was almost mandatory. The components of the multicomponent binder system should be so chosen that parts of the binder can be removed in gradual steps while leaving behind the other binder constituents for shape retention. It is ideal if a very small fraction of the binder component is retained to a sufficiently high temperature level where particle–particle bonding has just started to occur, whereby the part can maintain its shape rigidity even when the last trace of binder has been removed.

The practical goal of debinding is to remove all the binder (generally organic molecules), in the shortest time possible, using the least amount of energy, without leaving behind any binder residue, and without creating any problems in the green part. Among all these goals, the retention of part geometry and desired characteristics is primary. All the other steps should be optimized bearing in mind that, without the retention of the proper debound part geometry, it is useless even if all the other goals such as short debinding times, no binder residue, etc., can be achieved. Of course, if the binder residue reacts with the particulate material, then the purpose of the PIM processing is defeated.

The process of debinding can be divided into two major categories, namely, thermal debinding and solvent debinding. Solvent debinding uses chemical solvents to remove part of the binder present in a green part. The process of solvent debinding always uses a combination of some form of chemical extraction with a final thermal burnout. Thermal debinding, on the other hand, is a process of binder removal in which all the binder material is removed by the application of heat alone. Some of the gel types of binders such as methylcellulose, glycerine, and boric acid and water mixtures, also use thermal debinding in the sense that the water is removed by evaporation and then the remaining binder is removed by thermal degradation. Even the process of freeze drying, where the water is directly removed from the solid state in vacuum, can be considered to be a form of thermal debinding step. The process of wicking, where the binder is converted into a liquid of low viscosity and allowed to interpenetrate into the pores of an inert sacrificial material, is also thermally activated debinding. Thus, thermal debinding is still much more prevalent than the solvent extraction type of debinding. This section will discuss some of the debinding processes that have been used in practice. To obtain a more detailed discussion about the fundamentals of debinding, readers are referred to the chapter on debinding fundamentals in the book by German [6].

A. Thermal Debinding

All the debinding processes use thermal debinding as the final step for removing all of the binder present in the green part. However, some of the processes use chemical solvents to partially remove some of the binders during the initial debinding stage. These will be considered as a separate segment, which will be under the heading of solvent debinding. This section will only deal with debinding processes that utilize thermal energy from the initial to the final stage of debinding.

Thermal debinding usually depends on the thermally induced degradation of the binder material or the evaporation of the binder constituents. This degradation process of a particular polymeric compound usually depends on the temperature and the atmosphere in which the material is heated. The molecular chain length of the polymeric material is also an important factor that determines the rate of thermal degradation of a particular binder.

Figure 6.23 shows the degradation behavior with temperature for a polypropylene (PP) copolymer in an oxidizing atmosphere and a reducing atmosphere. It can be observed that the oxidizing atmosphere results in faster degradation of the binder, as the complete weight loss is attained by 648 K (375°C) compared to 798 K (525°C) required for the same material in a reducing atmosphere. It can also be observed that debinding in an oxidizing atmosphere results in the polymer degrading over a wider range of temperature than in degradation of the same material in a reducing atmosphere. The fact that different polymers will degrade at different rates in the same atmosphere is illustrated in Figure 6.24 [47]. This figure shows the thermogravimetric analysis (TGA) curves for stearic acid, diphenyl carbonate, and polystyrene. The role of an air-based oxidizing atmosphere versus a reasonably inert atmosphere like nitrogen for a polyethylene-based binder is shown in Figure 6.25. In this case also the air-based atmosphere (which contains oxygen) results in complete degradation of the binder at a lower temperature. Also the

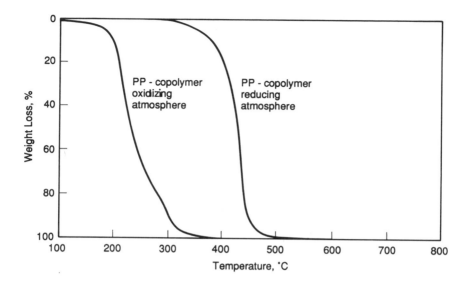

Figure 6.23: Degradation behavior with temperature of a polypropylene copolymer in oxidizing and reducing atmospheres.

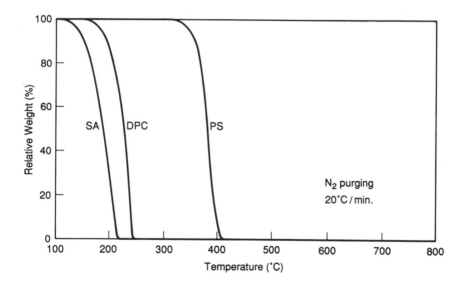

Figure 6.24: TGA curves of polystyrene (PS), diphenyl carbonate (DPC), and stearic acid (SA), at a heating rate of 0.333 K/s (20°C/min) in a nitrogen atmosphere. (Reprinted with permission from Ref. 47.)

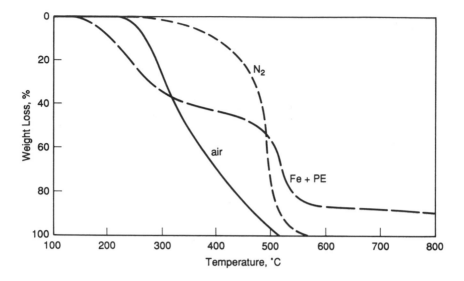

Figure 6.25: *Degradation behavior with temperature for polyethylene (PE) in air and in nitrogen. This figure also shows the degradation of PE+Fe.*

degradation behavior of the particulate-loaded polymer is quite different from that of a pure polymer. This is also shown in Figure 6.25 for a plain polyethylene binder versus a polyethylene binder loaded with iron powder. It can be observed that the weight loss of the binder is retarded as the iron powder is added to the pure polyethylene [48]. As the powder particle size becomes smaller, which essentially makes the pore sizes smaller, the retardation effect will become more pronounced. A larger thickness of the part will also require a longer time for total evaporation of the binder.

It would seem that the answer to the slow debinding step is to use an oxidizing atmosphere and a high debinding temperature. However, there are obviously problems associated with both these approaches. The oxidizing atmosphere definitely aids in the faster degradation of the polymer molecules, but in the case of some materials that are prone to severe oxidation the use of an oxidizing atmosphere could result in the formation of a surface oxide skin that would hinder the process of sintering and would often require the use of long pre-reduction cycles before the actual sintering step. Debining at too high a temperature can cause excessive vapor pressure inside the compact, resulting in a host of problems such as cracking, swelling, and distortion. The other major problem associated with very high debinding temperature is the degradation of the polymer into carbon instead of the desirable volatile species that can be easily removed. This carbon residue is a major source of contamination in both ferrous and nonferrous parts. Even in parts where small controlled amounts of carbon are beneficial, the carbon residue left behind by the binders makes carbon control an exceedingly difficult task. Thus, the aim during the thermal debinding step is to remove the binders at a temperature that will result in the gradual degradation of one component at a time in a multicomponent binder, with the lower-melting-point species being removed first. It is also the aim to remove the degraded products in the form of various hydrocarbon vapors such as methane or other small molecules like ethylene, propylene, carbon dioxide, and carbon monoxide gases. The binder should never be degraded so rapidly as to produce pure carbon as one of the degradation products. A flowing atmosphere or a vacuum atmosphere will also help in removing the volatile products away from the debound parts. This would also result in a faster debinding efficiency.

As mentioned earlier, for faster debinding it is mandatory to have a multicomponent binder. The multicomponent binder will usually have one component that accounts for the majority of the volume fraction of the total binder phase. A number of secondary binders may be used and they may serve various functions such as improving the wettability and adhesion of the binders to the particulate materials, increasing the mechanical strength of the green part, and reducing the viscosity of the overall feedstock. However, the primary requirement for the presence of a secondary binder phase in a multicomponent binder is to remain in the green part up to a temperature where most of the primary binder constituent has been evaporated out. The purpose of this secondary binder is to offer rigidity of the green part to a sufficiently high temperature at which some degree of presintering or particle–particle bond formation occurs. This secondary binder can be removed fairly quickly since, by then, most of the primary binder that constituted a major part of the binder volume fraction has been removed, leaving behind open pore channels through which the gaseous pyrolysis product of the last secondary binder can be rapidly removed. Figure 6.26 shows the TGA curve for a paraffin wax-based multicomponent binder system made up of the following constituents: 77% paraffin wax, 22.2% low-molecular-weight polypropylene, and 0.8% stearic acid. The pyrolysis was carried out in a nitrogen atmosphere. The primary binder, which is the paraffin wax, and one of the secondary binders — stearic acid — are both totally removed by pyrolysis at a temperature slightly below 573 K (300°C). However, the other secondary binder, the low-molecular-weight polypropylene, is still retained and is removed in the temperature range 653 to 703 K (380 to 430°C) [47].

Ultimately, in the case of thermal debinding, all the binder material is gradually burned out as gases and removed from the compact by diffusion-controlled or permeation-controlled mechanisms. Which mechanism is opera-

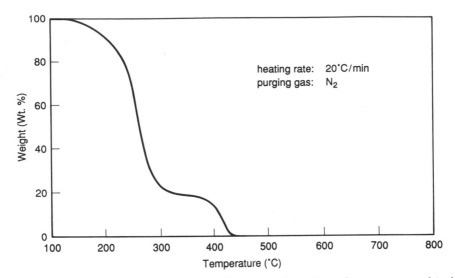

Figure 6.26: TGA curve for a paraffin wax- based multicomponent binder system made up of 77% paraffin wax (PW), 22.2% low-molecular-weight polypropylene, and 0.8% stearic acid. (Reprinted with permission from Ref. 20.)

tive will depend on the mean free path of the gas species, which decreases with gas pressure and increases with the molecular mass of the gas. When the mean free path of the vapor species is an order of magnitude greater than the mean pore radius, the mechanism is said to be diffusion controlled [49]. At small gas pressures and small particle sizes, the diffusion-controlled mechanism will dominate. This process is not dependent on the fluid viscosity. When the mean free path is smaller and collision between the gas molecules rather than collision of the gas molecules with the pore walls (i.e., particles) is the controlling mechanism, permeation is the operating mechanism. Permeation-controlled debinding will occur in compacts with large pore sizes and higher vapor pressures (where collisions between gas molecules increase). In permeation-controlled mechanism, the fluid viscosity plays an important role.

The debinding time, t (in seconds), for the diffusion- and the permeation-controlled mechanisms has been developed by German [50]. The two equations that relate to the times are outlined below.

$$t = \frac{H^2(MkT)^{1/2}}{[2D(P - P_0)E^2 U]} \qquad \text{(diffusion-controlled)} \qquad (10)$$

$$t = \frac{22.5H^2 G(1 - E)^2 P}{[E^3 D^2 F(P^2 - P_0^2)]} \qquad \text{(permeation-controlled)} \qquad (11)$$

where

H = compact thickness (m);
M = molecular weight of vapor (kg/mol);
k = Boltzmann's constant (8.32 kg m^2/s^2 mol K);
T = absolute temperature (K);
D = particle diameter (m);
P = pressure at the binder-vapor interface (kg/s^2 m);
P_0 = ambient pressure (kg/s^2 m);
E = fractional porosity (dimensionless);
U = molecular volume of the vapor (m^3/mol);
G = viscosity (kg/ms);
F = binder volume/vapor volume (dimensionless).

Some of the conclusions that emerge from the proposed models [50] are quite significant. It can be concluded that the general thermal debinding practice can result in binder removal whose primary mechanism can be either diffusion- or permeation-controlled. For a fixed gas species, lower particle sizes, lower ambient pressures, and higher porosities will tend to favor the diffusion-controlled mechanism. In both the cases the debinding times rapidly increase with part thickness. Smaller particle sizes will decrease the pore radius and could lead to longer debinding cycles. Higher porosity in the compact will be beneficial, and lower ambient pressures will result in faster debinding times. Thus, vacuum debinding will tend to decrease the debinding times irrespective of whether the process is diffusion-controlled or permeation-controlled.

Thermal debinding in a vacuum atmosphere is used in practice. The advantages of vacuum debinding are that the cracking and specimen swelling due to the vapor pressure buildup within the part are almost eliminated and there is no impediment due to permeation. The rate of debinding is also faster in vacuum atmospheres. However, in a vacuum atmosphere there is no gas species such as oxygen that can be used to cleave the binder molecules. Temperature control during vacuum debinding is also much more difficult; thus, the process of vacuum debinding has to be carried out slowly using extremely slow heating rates around 0.0008 K/s. One possible way to overcome this difficulty is to flush the system with oxygen or to use a very poor vacuum and retain sufficient pressure of oxygen in the system to yield faster cleavage of the polymeric binders. Even partial pressures of reactive gases can be beneficial.

The process of vacuum debinding is extremely suitable for debinding of binders based on simple molecules such as water. Water-based binders can be used in the form of gels like methylcellulose or in the form of water that is frozen in the mold and can then be removed at elevated temperatures in a vacuum atmosphere. The latter process, known as freeze drying, is ideally debound in vacuum furnaces, where the process of sublimation can occur and the binder can be removed directly from the solid state without going through the intermediate liquid stage. This process is discussed in greater detail in a later part of this chapter. In the case of gels such as water and methylcellulose, the first stage of thermally assisted binder removal can be achieved by the removal of water (or alcohol) from the gel at slightly elevated temperatures. The second stage of thermal debinding constitutes the removal of the actual gel, such as methylcellulose, by heating the part to high temperatures. Thus, in this case also, the underlying fundamental principle is to remove a large volume of the binder (water or alcohol) during the initial evaporation cycle, which then creates a large volume of porosity. The gel component of the binder, which degrades at higher temperature, can be removed later by a rapid thermal debinding process due to the large volume of open pore channels.

B. Wicking

To speed up the debinding kinetics, thermal debinding using a wicking medium has also been employed. The wicking process can serve the twin purpose of allowing additional support for the green compact during debinding, and thereby preventing to some extent the occurrence of slumping and shape distortion, while the fine porosity in the wicking medium can rapidly absorb the molten binder material and thereby open up channels for the total binder removal by thermal degradation. The wicking medium is essentially a fine powder of materials that do not react with the binder or the compact to affect the properties of the compact. Examples of wicking materials are alumina, zirconia, silica, clay, and graphite. Graphite wick will not be suitable for ferrous materials or materials that have a strong tendency to form stable carbides, but it could be suitable for carbide-based systems. The key here is to have a wicking powder that will result in a much smaller pore size than that of the pore size present in the compact. The capillary pressure, P, predicted for a porous structure made by packing spherical powders of diameter D is given in equation (12) [51].

$$P = \frac{10\tau \cos(\Theta)}{D} \tag{12}$$

where

τ = binder–vapor surface energy;
Θ = contact angle.

The capillary action is responsible for the wicking, and it can be observed that smaller particle diameter of the wicking material will result in a greater capillary pressure and better wicking. The wicking medium can rapidly absorb the molten binder up to the irreducible saturation limit. The binder removal time, t, in the wicking process as developed by German [50] is given in equation (13).

$$t = \frac{4.5(1 - E_c)^2 H^2 D_w G}{[E_c^3 W D_c (D_c - D_w)]} \tag{13}$$

where E_c is the fractional porosity of the compact, and D_c and D_w are the diameters of particles forming the compact and the wick material (in meters), respectively; the rest of the terms have the same connotation as described earlier.

It can be predicted from the equation that the time will increase rapidly with section thickness, and can be decreased by increasing the difference between the diameter of the particles of the compact and that of the wicking material. Higher temperature is expected to decrease the wicking time, as that will decrease the viscosity of the binder.

Problems associated with the wicking process include the removal of the sample from the wicking material. It should be realized that not all the binder will be removed in the wicking process and one has ultimately to resort to a pure thermal debinding process to remove the remaining binder. If an attempt is made to remove the samples after partial debinding by wicking, parts that are fragile can break easily. It is often necessary to rub off the wicking material after partial debinding, especially for materials that will be liquid phase sintered. This process can also lead to handling problems. An alternate way is to remove the samples from the wick after total debinding and sufficient presintering has been accomplished so that the compacts may have good handling strengths. It should also be pointed out that though the wicking medium can also act as a support for the compact and prevent shape distortion, powders that have very low interparticle friction will still slump even when embedded in a wicking medium. This has been demonstrated using coarse spherical nickel aluminide powder mixed with low-density polyethylene binder that has been debound in an alumina wick. Samples formed by compression-molding small cylindrical pellets were embedded in fine alumina wicking powder and then heated to a temperature at which the binder material would melt and flow. In all cases, the top face of the sample invariably caved in. A view of the compression-molded specimen and a view of the gravity-induced slumping are shown in Figures 6.27 and 6.28, respectively [31]. Without the wicking support, however, the material simply melted and flowed, forming a small puddle, with absolutely no semblance of shape retention.

10 mm
green sample.
66 vol. % solid.

Figure 6.27: *A compression-molded specimen of IC-218 powder in polyethylene.*
(Reprinted with permission from Ref. 29.)

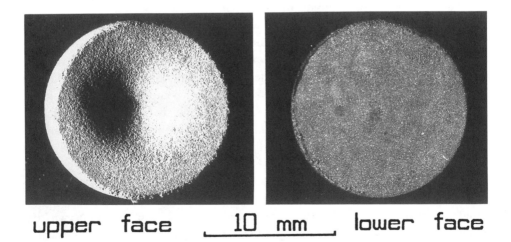

upper face   **10 mm**   **lower face**

Figure 6.28: *Gravity-induced slumping of the IC-218 compression-molded*
specimen of Figure 6.27 after debinding in an alumina wicking medium.
(Reprinted with permission from Ref. 29.)

There are numerous problems associated with the embedding of the complex shapes in a wicking medium and also the proper removal of the parts from the medium after debinding. Thus, the wicking process is currently not in commercial use.

C. Solvent Debinding

The slow debinding rates usually associated with pure thermal debinding have led to the concept of solvent extraction in multicomponent binder systems. In the process of solvent extraction, the primary component of the multicomponent binder is removed by preferentially taking it into solution in some chemical solvent. This form of solvent extraction is widely used in binder systems that use oil as a major portion of the binder. The binders are usually made up of oil and one or two other polymers that are not taken into solution by the chemical solvent. Thus, the fundamental principle of solvent debinding is again to remove a large volume fraction of the binder, which will result in a large amount of open pore structure, while retaining some polymers within the part for shape rigidity. The retained polymers can then be rapidly debound by thermal debinding processes. Usually the oil and polymer should be insoluble in one another. Hydrogenated vegetable oils, fish oils, palm oil, coconut oil, etc., have been commonly used in binder formulations that are suitable for the process of solvent extraction. Examples of such binders include 70% peanut oil and 30% polypropylene; 50% vegetable oil, 45% polystyrene, and 5% polyethylene; 30% mineral oil, 15% vegetable oil, and 55% polystyrene.

In its general form, the first stage of debinding involves the immersion of the green part in the desired chemical solvent for sufficient time to dissolve the primary volume fraction of the binder. The common solvents used in the process of debinding include acetone, carbon tetrachloride, trichloroethylene, and ethanol. The duration of the solvent extraction step usually varies from a few hours to a few days depending on the sample thickness [52]. This process has been used in the InjectAMAX Process [53], where the solvent-extracted porous green compact is directly introduced into the high-temperature sintering furnace. The residual binder is rapidly removed at the presintering temperature by thermal pyrolysis. Several thousand parts of various shapes and sizes have been successfully processed by this technique. Smaller-sized parts like 25-mm small-caliber fins, around 35.6 mm (1.4 inch) long and weighing around 8.9 g, have been debound in 10 800 s (3 hours), and larger-sized fins approximately 101.6 mm (4 inches) long and weighing around 187 g have been fully debound in 18 000 s (5 hours). A similar process has also been used by Technology Associates Corporation to process tungsten heavy alloys. In their experiments, after solvent extraction by immersion of green samples for 43 200 to 64 800 s (12 to 18 hours), the samples were dried by hot air fans. The remaining thermoplastic binders were removed during the presintering stage. The tungsten heavy alloys produced by this process exhibited properties comparable to those of similar alloys processed by the conventional pressing and sintering route [54].

There are other systems in which naphthenic or paraffin-type waxes serve as the primary binder, which is removed by solvent extraction. In this case, the binder to be solvent-extracted is in the solid state. The solvents required to extract these higher-molecular-weight waxes are usually stronger than those required to remove the oil-based binders.

The solvents often disrupt the chains of the waxes and dissolve them in solution. Higher the molecular weight of the waxes, the higher should be the molecular weight of the solvents. The dissolution rate or the reaction rate is increased by use of hot solvents instead of solvents at room temperature. A 10 K rise in the solvent temperature increases the reaction rate tenfold [55]. Apart from the fast debinding times, the solvent extraction technique also prevents the sample oxidation usually associated with thermal debinding, which is normally carried out in an atmosphere that has a partial pressure of oxygen.

The solvent becomes saturated with the dissolving organic material after sufficient loads have been solvent-extracted and it is necessary to replenish the solvent tank with fresh solvent after a period of use. The use will depend on the volume of binder being removed during solvent extraction versus the volume of the solvent present in the tank. As a rough rule, the solvent should be replenished after 12 to 13 days of use [55]. The contaminated solvent can be recycled by distillation.

One of the problems associated with the solvent extraction technique is the fact that some polymeric materials can swell when in contact with a solvent. This could lead to compact cracking. Another problem is the associated health hazard presented by a number of solvents presently in use. Thus, careful management of the solvent vapors should be implemented in the work environment.

A variant of the solvent extraction technique is the use of hot solvent vapors. In this form of debinding, the green sample, instead of being immersed in a tank filled with the solvent, is held in an atmosphere of heated vapors of the solvent that come into contact with the green part and extract the binder. In another form of this hot solvent vapor extraction technique, the heated solvent vapor that has dissolved the binder is condensed at the top, and then recirculated back into the pool of solvent below, thus creating a continuous supply of the solvent in the system. This process has been described by Lin and German [56]. The setup for solvent debinding by vapor condensation is shown in Figure 6.29. In this setup a sample chamber in which the solvent is contained is partially immersed in a water bath. The sample sits on a porous substrate

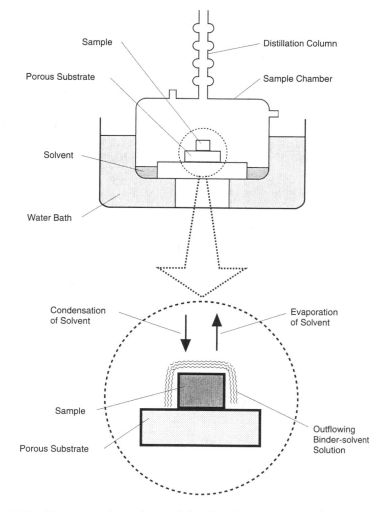

Figure 6.29: The setup for solvent debinding by vapor condensation.

that is placed inside the solvent tank at a level higher than the solvent level. The sample tank is attached to a distillation column as shown. When using a binder with the composition of 20 wt.% polypropylene, 10 wt.% carnauba wax, 69 wt.% paraffin wax, and 1 wt.% stearic acid, the vapor-phase solvent extraction can be carried out using heptane vapor operating at 358 K (85°C) [56]. A small amount of the thermoplastic material is retained, and has to be removed later by thermal debinding. In another variation of the solvent extraction process, the green parts are first subjected to a vapor-phase solvent extraction followed by their immersion in the liquid solvent. This patented solvent extraction process is being used commercially by Parmatech Corporation, California.

A form of solvent and thermal debinding technique has been used in a new binder development study by BASF [17]. In these experiments the green part is placed in an oven that is first purged with nitrogen and then back-filled with approximately 0.5 to 2.0 vol.% of gaseous acid. Under the influence of the gaseous acid, the polyacetal type of binder depolymerizes from the outside to the core. The shrinkage of the binder phase is homogeneous and proceeds at a rate of approximately 3×10^{-4} to 6×10^{-4} mm/s (1 to 2 mm per hour) in case of fine ceramics. After this form of acid vapor-induced depolymerization, some of the residual binder is rapidly pyrolyzed.

Another form of solvent extraction is known as supercritical binder extraction. In this process the green compact to be debound is subjected to simultaneous application of both temperature and pressure in a solvent atmosphere. In this process, the temperature and pressure are adjusted to a point where the vapor and the liquid become indistinguishable and have equal densities. During this form of extraction there is usually no compact cracking as the part is under positive pressures that can typically be as high as 20 MPa. The common solvents used in supercritical extraction are propane and carbon dioxide [6].

D. Debinding Equipment

Intelligent design of the debinding sequence for a particular feedstock is usually dependent on a number of factors. Typically the weight loss data with temperature, as measured by TGA, provides an insight into the critical temperature range where one or more binder component is pyrolyzed and removed from the green compact. It is important to design debinding cycles that will provide long hold times at those critical temperatures in order to allow the particular species being pyrolyzed to be completely removed from the part. The continuous measurement of the carbon potential of the atmosphere is a difficult monitoring task. This can be accomplished in furnaces specially fitted with Fourier transform infrared (FTIR) and oxygen probes. Analysis of the process of optimization has been described in several papers [57,58] and it is conceivable that intelligent debinding furnaces may become a reality in the future. However, the economics versus the advantages gained by the use of intelligent debinding furnaces will be the ultimate deciding factor.

The reactor vessels used for the debinding process can take a variety of shapes and sizes, usually depending on number of factors such as the volume of debinding that will be carried out in a single batch, the chemistry of the binders, the debinding process selected (thermal versus solvent extraction), whether the processing is batch or continuous type, and so on. Usually debinding is a batch type of process as the process control is usually much better. Thus, the reactors discussed in this section will be based on the batch-type debinding process.

For thermal debinding in a batch type of reactor, the green samples are often taken in a tray and loaded into the reactor chamber, which is usually a solid steel retort. The samples in the trays can be loaded into the steel retort either from the top or from the front; the larger batch-type retorts will rely on front loading. The retort can be sealed off from the outside atmosphere by shutting the doors or by closing a lid from the top. Gas lines usually lead into the reactor chamber and the atmosphere within the reactor can be changed. A simple top-loading type of reactor will normally have a gas inlet and a gas outlet on the top lid. This, however, is not the ideal gas flow condition. It is preferable to have an internal fan to agitate and homogenize the atmosphere within the reactor chamber. The compact support trays may have built-in heaters for uniform temperature control, or the retort may have heaters built in at the side. The problem with the gas inlet and gas outlet being in the top lid portion of the furnace is that the heavier molecules tend to settle or sink below, resulting in uneven furnace atmosphere control. Also, it has been found that unless the top exit tube is kept heated, the evaporated waxes and polymers often tend to condense and clog the gas exit pipe, which essentially removes the degraded products. Thus, the exit for the gas should preferably be at the bottom and should have a cold trap where the degraded products can collect, leaving the passage clear for the flue gases to exit. It is best if a sweeping gas flow is maintained to carry the plastic degradation products to a cold trap. It is good practice to allow the exit gases to be burnt off at the mouth of the exit tube as some of the exiting products are hydrocarbons like methane. A constant pilot flame to facilitate the burnout can be incorporated to achieve this end. Sophisticated reaction chambers can have sensors to monitor the exiting gases and adjust the debinding conditions within the reactor chamber accordingly.

Solvent extraction is usually carried out in a large tank; trays containing the green parts are immersed in the tank filled with the chemical solvent. Thousands of green parts can be solvent extracted together in one batch type of immersion operation. The solvent temperature should be carefully controlled in order to achieve proper solvent extraction. A distillation chamber may be used to separate out the binder from the solvent and recycle the solvent. The solvent tank should be built in such a way that it can be easily drained and refilled with fresh solvent. It should be remembered that the solvent extraction results in some residual plastic binder being retained after the solvent leaching process. This residual plastic binder can be removed at moderate temperatures in normal thermal debinding reactors or directly in a sintering furnace where the resulting pyrolyzed products can be safely removed without contaminating the furnace.

From the prior discussion on the thermal debinding process, it is clear that the removal of the binder can be achieved using a vacuum atmosphere with a partial pressure of reactive gases to cleave the binders. A special reactor that uses vacuum atmosphere for the binder removal and the collection of the degraded products has been described by Finn [59].

Direct heating of a reactive chamber to the debinding temperature usually results in the deposition of evaporated waxes and polymers on the cold furnace walls. Thus, from the earlier experience, a sweeping gas type of vacuum system was developed and modified to suit the debinding process. In this system, a graphite box with coated graphite shelves is used to carry the parts into the reactor chamber. The arrangement of the shelves is such that a sweeper gas introduced from one end of the shelves will flow through all the shelves and eventually exit from the other end. The sweeper gas essentially takes a tortuous serpentine path and carries all the evaporated binder products along with it directly into the pump, which is usually situated at the bottom of the retort. A view of the special retort, known as Injectavac, with the sweeper gas flow path is shown in Figure 6.30. A continuous pressure differential is maintained between the sweeper gas entry point and the vacuum pump exit port. The rate of binder removal has been found to be dependent on pressure and the component shape factor [60]. The evaporated wax can be collected by a modified "wax pot" introduced in the pump port side before it reaches the mechanical pumping system. Some of the wax reaching the mechanical pumping system resulted in breakdown of the pump; also, when the degradation product of the plastics reaches the pumping system, it also soon results in pump breakdown. One solution was to build a series of condensers to condense out the gaseous product produced by the polymer degradation. However, this would result in a drastic drop in the pumping efficiency. A "once through oiling" (OTO) type of pump, which is commercially available, can be used for the vacuum pumping system in a debinding retort. The use of this special pump has eliminated the need for on-line condensers, and has thereby allowed increased pumping capacity. It is claimed that tests have proved that this form of pumping system can handle nearly all the binders that are currently in use in the PIM industry. A new system, known as Vacuum Industries Binder Removal System, provides a single unit that can carry out the functions of debinding, sintering, and heat treatment [59]. A reactive sweeper gas that can help in breaking up the binders can be used to great advantage in this debinding system.

As more and more controls are added to the debinding reactor it becomes more and more expensive. Philosophically, it would be best to have closed-

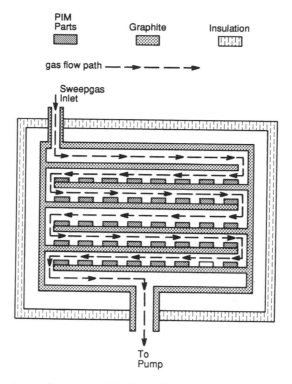

Figure 6.30: A view of the special debinding retort (Injectavac) showing the sweeper gas flow path. (After Ref. 59.)

loop feedback control to monitor and change the dew point, carbon potential, and some of the other important debinding parameters. However, the economics of obtaining such sophistication in the debinding reactor system have to be carefully balanced with the process economics. For example, systems that use the freeze drying technique and water as the primary binder can use a far less sophisticated vacuum reactor system for binder removal. In such a case, the special sensors in the reactor may represent serious "overkill."

E. Defects Generated during Debinding

Defects introduced during debinding should not be confused with the defects that were introduced during the mixing or injection molding stage. Usually the defects introduced during the earlier stages of PIM processing are not healed during the debinding stage but may be magnified. For example, ejection stress cracks that may be extremely fine in the green part can become fully developed cracks after debinding. Similarly, foldlines formed due to jetting may become very acute after debinding. However, a large number of defects are introduced into the injection molded part solely due to the debinding step. This part of the chapter will discuss only those defects generated due to the debinding step.

Some of the defects formed during the debinding step involve slumping and compact shape loss, surface blow holes, surface blisters, oxidation and discoloration of the parts, cracks, and carbon contamination. The importance of carefully designed multicomponent binders has been emphasized earlier. It is extremely difficult to produce a good part when a single-component binder is used, because of a combination of number of factors such as the formation of all the liquid at once as the binder usually has a narrow temperature range over which it can be melted or pyrolyzed. The compact essentially loses its rigidity and, depending on the powder characteristics and the volume of binder present, the part may either slump or lose its total shape. Also, during the pyrolysis of the binder the vapor phase formed inside the part could have very few avenues of exiting the part. The entrapment of these gases could cause sample swelling or even vapor-induced blowout of the part. Sometimes the buildup of a vapor pocket near the sample surface may cause the formation of pinholes if the vapor pressure built up in the pore near the surface is large enough to force open the pore to the sample surface, thereby venting the built-up pressure. In a slightly different situation, where the pressure generated within the pore is not high enough to blow open the pore and vent the built-up vapor, or the vapor pocket is formed a little distance away from the surface, there will be general blistering on the surface of the debound part instead of a vapor-induced blowhole.

It should be remembered that the thermal expansion characteristics of the binder and the particulate material are entirely different. Thus, during the heating of the part, stresses may develop due to the expansion mismatch. In case of a single binder, the binder softens at the elevated temperatures and has little yield strength, thus causing the part to deform. Gravity-induced viscous flow type of deformation also plays a major role during the debinding stage as the binder has very low yield strength during the debinding process. The problem of gravity-induced deformation becomes very acute in case of powders that have very low angle of repose, such as coarse spherical powders. This problem has been discussed in detail in the earlier sections.

Most of the problems discussed can be substantially reduced or almost eliminated by the selection of a well-designed multicomponent binder. The secondary binders that are retained after the primary binder has been effectively removed can hold the particles together while the primary component is being removed. Once open pore channels are introduced into the part, the problems associated with debinding are substantially reduced as the degraded binders can easily exit the part through the interconnected pores. The removal of the primary binder during the first stage of debinding in parts molded with multicomponent binders, can be achieved by various means. This step can be accomplished by heating the part very slowly and providing long holds during the critical temperature regime where the degradation of the primary binder has effectively started. An alternate route is to use the wicking approach to remove the primary binder (which usually has the lowest melting point) and open up the pore channels for the rapid removal of the remaining binder. The part can be removed from the wicking material after the primary binder has been removed as the remaining binders will provide sample rigidity. A third approach might be removal of the primary binder by a solvent extraction technique. However, one should remember that some of the polymers can swell in the presence of a solvent.

Cracks are a common defect related to incorrect debinding processes. Very fast heating rates during debinding will usually result in cracking of the part. Also parts that have too much binder (i.e., solid loading much lower than the critical loading) will allow rearrangement and generate areas of different packing densities that can lead to cracking [61]. Cracks can also occur in areas where a thick section meets a thin section.

The support of the parts during the process of debinding is an extremely critical step that can make the difference between a good distortion-free part and a badly warped part. Unless proper fixturing is provided during debinding, the parts can warp as the polymeric material becomes softer with temperature.

To facilitate the cleavage of the binders, oxidizing atmospheres are generally used during debinding. However, this can lead to an oxide layer building up on the particle surfaces, which can hinder the sintering process. A great deal of discoloration of the specimen surface could also occur. It may be extremely difficult to remove the discoloration during the subsequent sintering stage.

Too high debinding temperatures can cause the "cracking" of the polymeric binders into carbon, instead of the gradual gaseous pyrolysis products such as hydrocarbons, carbon dioxide, and carbon monoxide. This carbon retention in the part can create problems in materials for which carbon contamination is a problem. Such systems include invar, tungsten heavy alloys, and magnetic materials. However, in materials like carbides, this problem becomes insignificant. In the case of ferrous materials, higher carbon content could mean better strength of the material, but it is extremely difficult to control carbon additions via the debinding process.

4. Sintering

When a mass of powder in some form is heated to elevated temperatures, the powder particles tend to bond together by a process known as sintering. It is often assumed that a powder compact of some shape, when subjected to elevated temperature, will shrink and densify to result in a consolidated body that will be close to full density. Though this is often the case, it should be remembered that sintering can result in no shrinkage of the part or even in some cases in swelling of the part. Similarly, the green part, which before being sintered is porous in nature, may also undergo changes by which the actual porosity may decrease, remain nearly the same, or in some cases even increase with sintering. However, almost invariably, the process of sintering leads to the formation and growth of necks or weld joints at the particle–particle contacts. Sintering is often carried out at a temperature below 75% of the melting temperature of the major phase. Usually sintering does not involve the melting of the particulates, though in some special cases a small fraction of the particulate material may be in the liquid phase during the sintering process. The lowered processing temperature is one of the major advantages of particulate materials, so that even very high-temperature materials such as refractory metals, ceramics, etc., can be consolidated without the attainment of the extremely high melting temperatures of these materials.

The process of sintering forms the basis of powder metallurgy, and it is perhaps the most studied phenomenon in the area of particulate materials. Thus, it is not surprising that there is a tremendous wealth of literature available on sintering. It is not the purpose of this chapter to give extensive discussion on the extremely well-covered topic of sintering. This part will briefly touch on some of the fundamentals of sintering as related to particulate injection molding and point out some of the new developments.

A. Solid-State Sintering
During elevated-temperature exposure of a powder compact, the compact essentially tries to reduce the surface energy by transporting material from different areas by various mechanisms to eliminate the pores. There are a number of mechanisms by which mass transportation from the bulk of the sample to the pores can occur. They can be broadly divided into two categories, namely, surface and bulk transportation. The surface transportation mechanism can be further subdivided into mechanisms such as evaporation–condensation, diffusion–adhesion, and surface diffusion; while the bulk transportation mechanism includes processes such as plastic flow, grain boundary diffusion, and lattice diffusion. Figure 6.31 shows a schematic view of the transportation mechanisms occurring in a powder mass that has already formed the particle–particle necks. The detailed description of these individual mechanisms has been extensively covered in the powder metallurgy literature.

A brief description of the sintering stages should be given at this point, keeping in mind that for the majority of PIM parts the main aim is the attainment of high densities. The process of sintering can be roughly

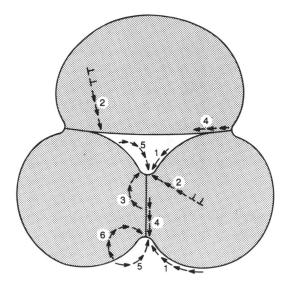

Figure 6.31: Schematic view of the transportation mechanisms in a powder mass that has already formed the particle–particle necks: 1, surface diffusion; 2, plastic flow; 3, volume diffusion; 4, grain boundary diffusion; 5, evaporation and condensation; 6, diffusion–adhesion.

categorized into three stages. The first stage is the point where rapid neck formation and neck growth occurs in the powder compact. At this initial stage, the pores are totally interconnected in nature and the pore shape is extremely jagged. The curvature gradients near the necks are responsible for the mass flow during this initial stage of sintering. The gradients associated with sintering are derived from Kelvin's equation, which gives the stress associated with a curved surface as follows:

$$\sigma = \left(\frac{1}{R_1} + \frac{1}{R_2} \right) \gamma \qquad (14)$$

where

σ is the stress associated with any curved surface (Pa);
γ is the surface tension (N/m);
R_1, R_2 are the principal radii of curvature (m).

The smaller the neck size, the higher is the curvature gradient, and therefore the faster is the sintering. As the neck grows, the curvature gradient is decreased and the sintering rate is decreased.

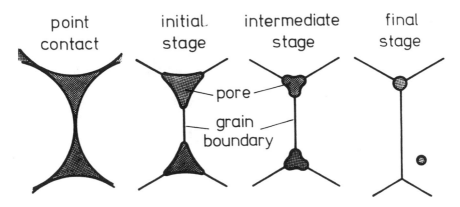

Figure 6.32: *Schematic view of the changes in pore morphology as sintering gradually progresses through the different stages. (Reprinted with permission from Ref. 62.)*

In the intermediate stage of sintering, when sufficient neck growth has occurred, the pore channels become more cylindrical in nature. The interfacial energy is the driving force during the intermediate stage of sintering. The pore channels are still interconnected at this stage. With continued sintering, these cylindrical pore channels become unstable and are gradually pinched off, resulting in isolated pore structures. This occurs at a pore concentration of approximately 8%, and represents the onset of the final stage of sintering.

The pinched-off, isolated pores are gradually smoothed out with continued sintering. The remaining 8% porosity that is retained in the microstructure has to be eliminated during this last stage of sintering, which is a very slow process. At this point, the isolated pores lose their jagged shapes and become more spherical in nature. Figure 6.32 [62] shows a schematic view of the changes in the pore morphology as the sintering gradually progresses through the different stages. The cylindrical pores, which were very effective in pinning the grain boundaries, gradually lose their pinning efficiency as the pores tend to isolate themselves.

It is important to realize that the gas that is entrapped in the isolated pores will tend to hinder the process of densification. This is usually the gas used in the sintering atmosphere. Generally, these gases have some solubility and diffusivity through the bulk material, and can be gradually removed. However, if the gas entrapped in the isolated pores has no solubility or diffusivity through the bulk material, it will remain in the pore, and densification will finally stop. This is treated in more detail in the discussion on liquid-phase sintering.

Apart from the gases entrapped in the closed pores, the pore–grain boundary interaction is another aspect that plays a vital role during the sintering of PIM parts. During sintering, mainly through bulk transportation, the neck sizes, density, and strength of the compact will increase. The rate of densification, which initially can be extremely fast, gradually slows down. The grain boundaries, which provide one of the fastest feeder routes for mass transfer to the pores, are generally attached to the pores during the initial and most of the intermediate stage of sintering. It is of vital importance to have the pores connected to the grain boundaries for as long as possible, so that continued mass transport to the pores can occur.

The pore–grain boundary interaction can vary due to different sintering conditions. In one scenario, the pores attached to the grain boundaries provide a pinning action to the grain boundary motion. As grain growth occurs, the pores can be dragged along with the grains. With increasing sintering temperature, the grain boundary motion is greatly increased, while the pore motion is comparatively sluggish. Thus, a situation is often encountered where the grain boundaries break away from the pores. This leaves the pores stranded within the grain boundaries, and the main arteries feeding the mass to the pores, namely the grain boundaries, have been separated. This results in a very sharp drop in the densification of the material and excessive grain growth. The two possible pore–grain boundary configurations are shown in Figure 6.33. When the pores are connected with the grain boundaries, the total energy of the system is lower compared to when the same pore is isolated within a bulk grain due to the fact that the pore actually reduces the total grain boundary area. It is therefore important that for PIM parts the sintering process is controlled so that the breakaway of the grain boundaries can be delayed as long as possible. It can be conceived that the incorporation within the microstructure of fine dispersoids that can pin the grain boundaries will result in faster densification. Nonisothermal sintering schedules can also be used to enhance densification during the sintering process.

B. Activated Sintering

All the discussion so far has been related to conventional solid-state pressure-less sintering of a monolithic material. However, there are other forms of sintering that can result in improved densification kinetics. The various modifications in the sintering process attempt to accomplish almost the same goal, which is rapid densification at lowered temperatures. However, the means by which it is achieved vary. In all cases, some form of activated sintering is used that actually lowers the activation energy for sintering. Some of the means of inducing rapid sintering are to blend a second additive material that can form a liquid at the sintering temperatures, to use chemical methods of coating a few monolayers of the additive on the primary powder, to use

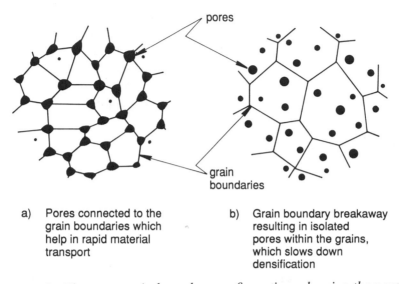

a) Pores connected to the grain boundaries which help in rapid material transport

b) Grain boundary breakaway resulting in isolated pores within the grains, which slows down densification

Figure 6.33: The pore–grain boundary configurations showing the pores located on the grain boundaries and pores located inside the grains. (After Ref. 62.)

external electric fields, to apply pressure with temperature, or to use additives that preferentially stabilize a phase in which rapid bulk diffusion is promoted.

One form of activated sintering relies on the presence of a minor additive through which the primary material can diffuse very rapidly. The segregation of the minor additive material to the interparticle boundary creates a short-circuit diffusion path through which the primary material rapidly diffuses, and densification is much faster than when the primary material is being sintered without the activating additive.

To promote this form of activated sintering, the additive should fulfill certain criteria as outlined by Zovas et al. [63]. The additive should form a low-melting phase during sintering and remain segregated at the interparticle interfaces throughout the sintering process. If the additive is highly soluble in the base material, its function as an activated sintering aid will be lost as it will be taken into solution in the base material and will not remain segregated at the interparticle boundaries. However, the additive should have a very high solubility for the base material as this will promote rapid diffusion of the base material through the additive phase. An example of the enhanced sintering capability of minute additives has been very well documented in the case of a 0.6-μm tungsten powder coated with a few monolayers of various additives. It has been observed that sintering of tungsten powders coated with nickel, platinum, or palladium results in much faster shrinkage compared to pure tungsten without the additive coating [62].

During activated sintering, the minor phase additive can be present as a liquid phase. This will be considered in a separate part of this chapter.

C. Pressure-assisted Sintering

In another form of rapid sintering, simultaneous application of pressure is used along with elevated temperatures. The external stress, which can be applied by various means such as hot pressing, hot isostatic pressing, hot extrusion, etc., promotes particle sliding that generates excess vacancies to aid rapid densification. The externally applied pressure also assists in the pore collapse and elimination of the closed pores formed during the final stages of sintering. In recent years, particulate injection molding has gained tremendously from the techniques in which simultaneous application of pressure and temperature has resulted in net-shaped parts with superior densities and final properties. Sintering furnaces are now available in which the material can be sintered and then subjected to moderately high pressure holds at the elevated temperatures.

This form of pressure-assisted consolidation, known as sinter/HIP or overpressure sintering, should not be confused with the case where an as-sintered (pressureless) material is subjected to conventional hot isostatic pressing. The sinter/HIP process combines sintering and postsintering densification with low gas pressure in a single cycle [64]. In the sinter/HIP process, the sintering of the material, which can be carried out under vacuum, is switched to a pressurized gas system in the final stage of sintering. This process of overpressure sintering has been used extensively by the hardmetal industry, which found it to be extremely beneficial. The process has also been applied to ceramics such as silicon nitride, where the use of high gas pressures during sintering prevents decomposition and the material can thus be sintered without any sintering aid [65]. Vacuum/overpressure sintering furnaces that combine the steps of hydrogen dewaxing, vacuum sintering, and relatively low-gas-pressure sintering (around 6 MPa) have been a boon for the hardmetal and cermet industries [66]. This form of consolidation could also benefit the PIM industry where application of pressure during sintering may cause rapid densification of the part and greatly increase its performance. The best advantage of pressure-assisted sintering in case of PIM parts can be achieved if the material is sintered to a stage where all the pores are closed off and pressure is then applied in the same furnace without having to cool the material. This has been the basis of sinter/HIP or vacuum/overpressure sintering developments.

Containerless, hot isostatic pressing (HIP) has also been used extensively by the PIM industry. In this process, the shaped part is first sintered to a density level where the remaining pores are isolated in nature. These parts are then subjected to simultaneous application of pressure and temperature, which

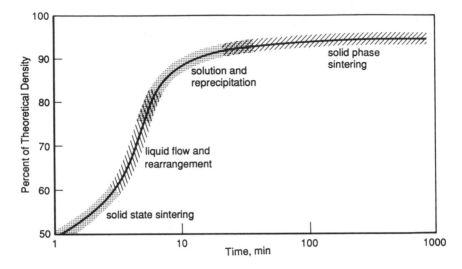

Figure 6.34: Schematic of the key stages of persistent liquid-phase sintering.

results in the elimination of the remaining porosity by pore collapse. However, this process will be expensive and not very energy-efficient, as the parts are heated to high temperatures twice.

D. Liquid-phase Sintering

Perhaps one of the most commonly used means of attaining rapid solidification is liquid-phase sintering. In this form of sintering, a portion of the mixed powder forms a liquid when heated to the sintering temperature. It must be realized that in a PIM part, or any powder-metallurgically produced part, there is a limit to the volume fraction of liquid that can be formed. Excessive liquid formation will result in the total slumping and distortion of the material. Thus, the total volume of liquid is usually not more than 20% to 25%.

In this case, the rapid densification that occurs is due to the effect of the liquid. In most of the popular liquid-phase-sintered materials, the liquid formed usually has some solubility for the solid. Once the liquid-phase formation temperature has been crossed, part of the material melts, takes the solid into solution, wets the solid, and spreads rapidly around the solid particles. A schematic drawing showing the key steps in this type of liquid-phase sintering, is shown in Figure 6.34. Assuming that the material was around 50% dense to start with, the application of temperature initially results in some degree of solid-state densification prior to the liquid formation. This will be especially true in systems that have considerable solubility for each other, such as W–Ni–Fe, WC–Co, etc. This first solid-state sintering stage should not be confused with the last stage of solid-phase sintering that continues in the presence of the liquid phase. The first solid-state sintering occurs at a time when the material is being heated to the sintering temperature, and liquid formation has not occurred. Once the liquid forms, there is a rapid increase in the density due to rearrangement, which gradually starts to taper off as the solution and reprecipitation stage takes over. During the first stage after liquid has formed, very rapid rearrangement of the solid particles occurs due to the capillary action of the liquid that is formed along with liquid flow. The liquid continues to dissolve the solid until it is saturated. If the temperature is still held above the liquid formation temperature, a process of solution and reprecipitation sets in, where the finer solid particles are preferentially dissolved and reprecipitated on the larger grains. This solution and reprecipitation event is the next stage of liquid-phase sintering. Density increase still occurs during this stage. However, the liquid volume or the liquid composition remains almost unchanged. The final stage of liquid-phase sintering is known as "solid-phase sintering." During this stage, grain coarsening occurs with shape accommodation in the solid particles. The densification rate during the final stage of sintering is very low and the solid-phase sintering continues in the presence of the saturated liquid.

The form of liquid-phase sintering described above is the most common. This can be termed "persistent liquid phase sintering," where the liquid phase is present throughout the sintering cycle. Here also there can be two different variations: in one case the liquid has some solubility for the solid material (such as W–Ni–Fe, WC–Co), while in the second case the liquid has very little solubility for the solid (W–Cu, Mo–Cu, W–Ag). In the latter case, the densification rate will be slow as the solution and reprecipitation stage will not be active in this type of system. Densification will primarily occur via the rearrangement and liquid flow. Persistent liquid-phase sintering has been used extensively for particulate injection molding of parts from material systems such as W–Ni–Fe, W–Mo–Ni–Fe, WC–Co, W–Cu, and Fe–Cu.

Another form of liquid-phase sintering, known as transient liquid-phase sintering, is also common in the conventional powder metallurgy area. In this form of sintering, the liquid that is formed is temporary in nature. This often happens when the liquid formed during the course of sintering has a high solubility for the solid phase. The liquid continues to take the solid into solution until it reaches a point where the solidus boundary is crossed, and the liquid resolidifies. Thus, in transient liquid-phase sintering, the liquid appears and then disappears. An example of such a system is the commonly used P/M material 90Cu–10Sn bronze, which is extensively used as a self-lubricating bearing material. When a compacted elemental powder mix of 90Cu–10Sn is being heated to the sintering temperature, which is around 1053 K (780°C), the

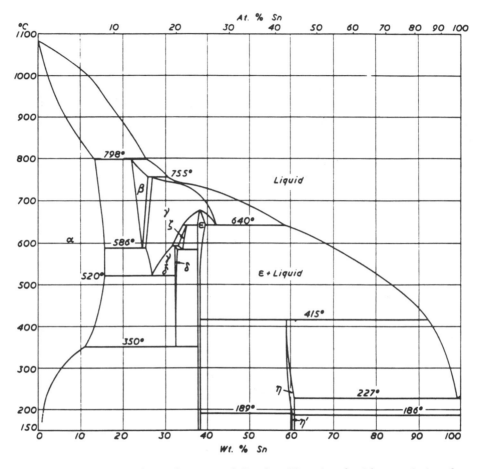

Figure 6.35: *Binary phase diagram of Cu–Sn. (Reprinted with permission from Ref. 67.)*

tin particles first melt, take copper into solution, and quickly resolidify. From the binary phase diagram shown in Figure 6.35 [67] it is clear that the 90Cu–10Sn material has a higher melting point than the sintering temperature. Thus, the liquid tin first forms, and disappears as the solidus line is again crossed.

E. Reactive Sintering
Another special form of rapid sintering technique, known as reactive sintering, has recently been used to process high-temperature materials. It should be pointed out that in a number of cases the reactive sintering is a form of very rapid transient liquid-phase sintering process. However, in other cases there is

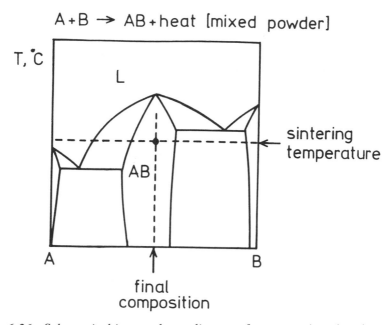

Figure 6.36: *Schematic binary phase diagram for a reactive sintering system, where a stoichiometric mixture of A and B powders is used to form an intermetallic compound product AB. (Reprinted with permission from JOM (formerly Journal of Metals), vol. 38, no. 8, p. 29 (1986), a publication of The Minerals, Metals, & Materials Society, Warrendale, PA 15086; Ref. 69.)*

no liquid formed during the process. Reactive sintering has been used to process various aluminide-type intermetallic compounds. Specifically, an interesting combination of particulate injection molding and reactive sintering has been used to process high-temperature composites of NiAl and Ni(Al,Si)-type intermetallic compound reinforced with chopped alumina whiskers. This will be discussed in more detail in a later section. It would, however, be appropriate to discuss the general process of reactive sintering when applied to systems where liquid formation can occur.

In general, reactive sintering involves a transient liquid phase [68]. The initial compact is composed of mixed powders that are heated to a temperature at which they react to form a compound product. Figure 6.36 shows a schematic binary phase diagram for a reactive sintering system, where a stoichiometric mixture of A and B powders is used to form an intermediate compound product AB [69]. The reaction occurs above the lowest eutectic temperature in the system, yet at a temperature where the compound is solid. Heat is liberated because of the thermodynamic stability of the high-melting-temperature compound. Consequently, reactive sintering is nearly spontaneous once the liquid forms. The liquid provides a capillary force on the structure, which leads to densification [68,70–72]. The liquid is transient as the process is conducted at a temperature below the melting temperature of the compound, typically near the eutectic. Reactive processing variants known as SHS, or gasless combustion synthesis, have been used by ceramists over a long period of time, and the Russian literature in this field is quite extensive. It is claimed that nearly all known ceramic compounds can be made by this reactive process [73]. Densification and compound formation are difficult to control, and are sensitive to processing parameters such as heating rates, green density, interfacial quality, and particle size.

As mentioned earlier, reactive processing with PIM has been used in Ni–Al systems. The system is characterized by five intermetallic compounds [74–76]. Reactive sintering of the Ni_3Al compound was first successfully carried out by sintering near 913 K (640°C), the lowest eutectic temperature [77]. Nickel and aluminum powders are randomly mixed in a stoichiometric ratio. This mixture is sintered under precise conditions of atmosphere, heating rate, time, and temperature. At the first eutectic temperature, liquid forms and rapidly spreads throughout the structure. The liquid consumes the elemental powders and forms a precipitated Ni_3Al solid behind the advancing liquid interface. Figure 6.37 shows a schematic of the reaction between the elemental nickel and aluminum to form the compound Ni_3Al [76]. Interdiffusion of nickel and aluminum is quite rapid in the liquid phase and the compound generates heat, which further accelerates the reaction [76]. If the reaction is controlled, the compound will be nearly fully dense and suitable for containerless hot isostatic compaction to full density.

The intermetallic compound NiAl has a much higher melting point than Ni_3Al. It has not been possible to ductilize NiAl, whereas the Ni_3Al compound can be extensively ductilized by the additions of small amounts of boron. However, the interest in processing the NiAl-based intermetallic is still great due to its very high melting temperature, low density, and oxidation resistance. One of the materials that is being considered involves reinforcement of NiAl with fibers or whiskers to improve the fracture toughness of the material, very much as in the ceramic matrix composites. Reactive processing of NiAl, however, produces excessive heat during the reaction, causing the material to melt. One way to overcome this problem is to dilute the reaction by the addition of prealloyed NiAl or some form of reinforcing material to form composites. Particulate injection molding has been used to form aligned fiber-reinforced composites of reactively processed NiAl compounds [78] and Ni(Al,Si) compounds where the silicon replaces some of the aluminum in the NiAl system [79]. The Ni(Al,Si)-based composite exhibits high compressive strength at elevated temperature.

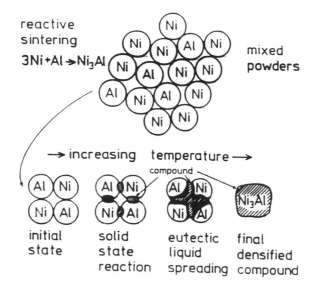

Figure 6.37: *Schematic diagram for the reactive sintering of Ni₃Al. (From Ref. 76.)*

F. Sintering Atmospheres

Since a major portion of particulate injection-molded composites often contain some metals, it is necessary to sinter the materials in the presence of a protective atmosphere. The sintering atmosphere can also serve the purpose of reducing the oxides that are often present in metallic powders. Thus, often the sintering atmosphere used contains a gas that is reducing in nature. The reaction products are usually water vapor in case of hydrogen or cracked ammonia-based reducing gases; or carbon dioxide when the reducing gas used is carbon monoxide. These are often removed as gaseous species.

Some of the commonly used sintering atmospheres are endogas (approximately 40% nitrogen, 40% hydrogen, 20% carbon monoxide), exogas (70% to 98% nitrogen, 2% to 20% hydrogen, and 1% to 6% carbon monoxide), dissociated ammonia (25% nitrogen and 75% hydrogen), nitrogen–hydrogen mixtures with 75% nitrogen and 25% hydrogen, and nearly pure hydrogen. Vacuum sintering atmospheres are also extensively used in the P/M industry. Vacuum sintering alone, however, cannot reduce most of the metallic oxides: a combination of presintering in a reducing atmosphere followed by a switch to a vacuum atmosphere can be the basis of an effective sintering cycle. Similarly, a hydrogen-based sintering cycle cannot be applied for materials that suffer from severe hydrogen embrittlement problems.

The development of a sintering schedule with the appropriate gas atmos-

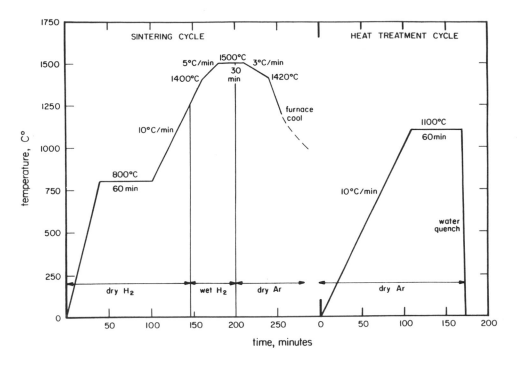

Figure 6.38: *The sintering cycle developed for liquid-phase sintering of tungsten heavy alloys.*

pheres is an extremely complex problem. The sintering schedule developed for liquid-phase sintering of tungsten heavy alloys illustrates the complexity of the design of a proper sintering schedule (which will vary for different materials). This example can also be used to demonstrate some of the fundamental principles of sintering.

A three-stage atmosphere cycle developed for sintering tungsten heavy alloys is shown in Figure 6.38 [80]. The sintering cycle in essence consists of a pre-reduction hold at 1073 K (800°C) in dry hydrogen followed by the heat-up to the sintering temperature, which is usually around 1773 K (1500°C). The pre-reduction hold step is provided to reduce as much of the oxides as possible before the liquid forms (around 1708 K; 1435°C). While heating up to the sintering temperature from 1073 K (800°C), a shift is made from dry hydrogen to wet hydrogen around 1523 K (1250°C), to prevent swelling due to the in-situ water vapor formation after the liquid appears. This water vapor entrapment within the system becomes more acute in case of materials with lower tungsten contents [80]. After 1200 s (20 minutes) hold at the sintering temperature, the wet hydrogen is changed to dry argon to remove the hydrogen embrittlement effect by biasing the diffusive flow of hydrogen outward [81]. The total sintering hold time is around 1800 s (30 minutes). The sample is then cooled at a slow rate (around 0.05 K/s) until the material has solidified. This is done to prevent solidification shrinkage pores [81]. On completion of sintering, the sintered samples are heat treated by water quenching from 1373 K (1100°C) after 3600 s (1 hour) hold at 1373 K (1100°C) in dry argon. This is done to suppress impurity segregation. Trace elements such as phosphorus and sulfur tend to segregate to the tungsten–matrix interface and severely degrade the properties of the tungsten heavy alloys. During the heat treatment step, the trace elements are resolubilized, and quenching from high temperature ensures that the elements do not segregate to the tungsten–matrix interfaces. Presently, the heat treatment is carried out as a separate step, but it would be advantageous if the heat treatment step could be incorporated into the sintering step itself. Thus, this sintering cycle addresses the various problems associated with the sintering of this particular material system, and generates a schedule that is expected to overcome most of the anticipated problems. Sintering schedules should be developed for various systems with such an outlook, and should be responsive to the ultimate property requirements of the material.

Consider what occurs in the case of an improper sintering cycle. If we consider the same sintering cycle described above and use dry hydrogen as the sintering atmosphere even after liquid formation occurs, in-situ water vapor formation can result. At the same time, rapid densification occurs, and all the porosity becomes isolated from the surface. According to Markworth's [82] model of the final stage of sintering, the densification rate, $d\rho/dt$, depends on the inverse of the pore radius and any gas pressure as shown in equation (15):

$$\frac{d\rho}{dt} = \frac{12\Omega D}{kTG^2}\left[\frac{2\tau}{r} - P_g\right]$$

(15)

where

ρ is the fractional density;
t is the time (s);
Ω is the atomic volume (m^3);
D is the diffusivity (m^2/s);
k is Boltzmann's constant (J/K);
T is the absolute temperature (K);
G is the grain size (m);
τ is the surface energy (J/m^2);
P_g is the gas pressure in the pore (N/m^2);
r is the remaining spherical pore radius (m).

The onset of the final stage of sintering occurs when pores have become closed. This occurs when the final porosity in the sample is around 8%. Due to liquid flow into the pore, the pore size will shrink, thus increasing the gas pressure within the pore. When the gas pressure within the pore is increased to a point where it balances the surface energy effect, no further densification will occur. This happens only for gases that have no solubility or diffusivity in the alloy matrix. If the gas has some diffusivity through the matrix, the densification rate will be slowed but will still continue as the gas entrapped gradually diffuses out. Thus, material will flow in and fill the pore. In the case of tungsten heavy alloys, if the sintering atmosphere used is dry hydrogen when the liquid formation occurs, some water vapor as well as hydrogen (which is the sintering atmosphere) may be trapped within the pores. A basic precept is that the hydrogen gas, which has some diffusivity through the matrix, will still allow the pore to be filled in with the liquid matrix, while the water vapor, which has no solubility or diffusivity in the matrix, will remain as a pore as densification occurs. The pore shape will usually be spherical in nature. Figure 6.39 [80] shows a schematic representation of the pore filling process when the gas entrapped in the pore has high solubility and diffusivity in the matrix, and when it has practically no solubility and diffusivity in the matrix.

Figure 6.40 [80] shows the microstructure of a 88W heavy alloy sintered at 1773 K (1500°C) in dry hydrogen, showing one pore that has been completely filled with the matrix while another pore, which probably had entrapped water vapor, is still retained in the microstructure.

From Markworth's model it can be observed that the use of vacuum as the sintering atmosphere will result in the fastest densification rate as no gas will be entrapped in the pores. German and Churn [81] have connected the pore depressurization rate to the diffusivity of various gases. Based on that model, the final sintered density variation with sintering time for a 95W heavy alloy, when sintered under different atmospheres, is as shown in Figure 6.41 [80]. The fastest densification, as expected, is in vacuum, closely followed by hydrogen, which has high solubility and diffusivity through the matrix alloy. However, sintering in an inert gas, argon (which has practically no solubility or diffusivity), does not result in complete densification. In fact, with increasing sintering time, the material may show a decrease in density due to the pore coalescence and growth phenomenon.

From the above it is clear that the design of a proper sintering schedule is extremely important for the attainment of the desired properties. Each individual material will have its unique set of problems that, unless carefully attended to, could result in poorly sintered materials. It should also be remembered that for a single material it is possible to have more than one sintering cycle that could provide adequate sintered properties.

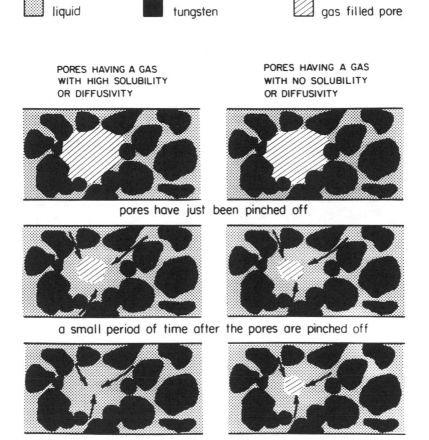

Figure 6.39: Schematic representation of the pore filling process when the gas entrapped in the pore has a high diffusivity in one case and practically no diffusivity in the other. (Reprinted with permission from Ref. 80.)

With this brief theoretical discussion on sintering, it would be best to provide readers with some interesting special cases where the PIM process in one form or another has been used to process the final product. The special cases of PIM that will be discussed in the following section will not only cover the PIM processing of specific materials (such as tungsten heavy alloys and titanium-based materials) but will also illustrate several unique processing variations of PIM. It may be argued that some of the processes (such as thixotropic alloys, aligned fiber composites, or freeze compression molding) discussed in the

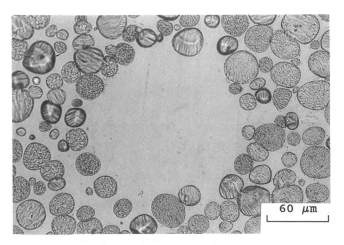

(a)

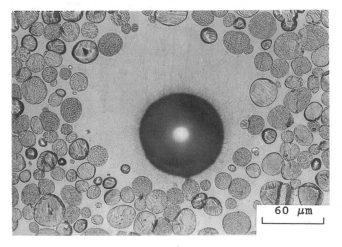

(b)

Figure 6.40: *Microstructure of a 88W heavy alloy sintered at 1773 K (1500°C) for 30 minutes in dry hydrogen, showing (a) one pore that has been completely filled with the matrix, and (b) another pore that probably had entrapped water vapor that is retained in the structure. (Reprinted with permission from Ref. 80.)*

following section, do not conform to the conventional PIM process. However, all the process variations discussed will at some point utilize one or more of the unique features of the PIM process. For example, the process of injection molding of thixotropic alloys starts with particulates (coarse chips) that are directly used in the injection molding machine where the particulate material is

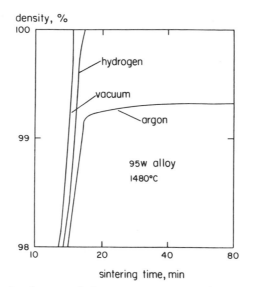

Figure 6.41: The final sintered density variation with sintering time for a 95W heavy alloy, sintered under different sintering atmospheres such as vacuum, hydrogen, or argon. (Reprinted with permission from Ref. 80.)

directly injection molded without any binder. Alternately, the processing of aligned fiber composites uses both particulates and binders, but the processing does not necessarily rely on the use of an injection molding machine. Similarly, the process of freeze compression molding, which uses both solid particulates and a binder, may or may not use an injection molding machine to process the materials. However, because these processes are so unique and interesting, and since they use some unique art of the PIM process in one form or the other, a brief discussion of these topics is considered almost mandatory.

Special Cases of Particulate Injection Molding

Tremendous developments have been occurring in the area of particulate injection molding in recent years. New and exciting alloys, novel techniques utilizing the PIM approach to produce aligned fiber-reinforced composites, new binderless injection molding techniques, injection molding of thixotropic alloys, and compression molding of water-based injection molding feedstock are presently being utilized. This part of the chapter discusses some of these exciting developments in the area of particulate injection molding. The common and popular PIM systems mainly based on iron, iron-nickel, various kinds of steels, and a number of common nonferrous alloys will not be discussed here.

Powder injection molding of tungsten heavy alloys will be one of the first topics discussed in this section. This alloy system will elucidate the role of physical metallurgy in providing materials that can have tremendous value-added impact when processed to near net-shape by particulate injection molding. However, before dealing with the heavy alloy modifications, it will be a good idea first to introduce this unique alloy system. A little of the alloy's sintering behavior has already been discussed in an earlier part of this chapter.

1. Tungsten Heavy Alloys

Tungsten heavy alloys are two-phase metal matrix composites having unique combinations of strength, ductility, and density. They are produced by persistent liquid-phase sintering of powder compacts comprised of mixed elemental powders generally consisting of 90 to 98 wt.% tungsten with nickel and other elements such as iron, cobalt, or copper. Usually a W–Ni–Fe alloy is the most popular alloy system. The final microstructure consists of a contiguous network of nearly pure tungsten grains embedded in a matrix of f.c.c. ductile tungsten–nickel–iron alloy. The typical mean tungsten grain size generally varies from 20 to 40 μm depending on volume fraction of tungsten, initial particle size, sintering temperature, and sintering time. Due to the unique property combination of the alloys, they are used extensively in kinetic energy penetrators, counterbalances, heavy-duty electrical contact materials, radiation shields, and numerous other applications in the defense industry [83–85]. The majority of the applications require a complex-shaped final form, which is generally obtained by machining. However, machining of these alloys results in the loss of expensive material. The use of particulate injection molding provides a tremendous economic advantage in the mass production of complex-shaped parts of tungsten heavy alloys. The success of powder injection molding of conventional tungsten alloys has generated a great deal of interest in producing net-shaped parts from this alloy system [54,86].

Powder injection molding of the conventional heavy alloy system will not be discussed at this point. Instead, injection molding of an alloyed tungsten heavy alloy system will be discussed, and the effect of and reasons for alloying will also be discussed.

Conventionally, higher strengths in tungsten heavy alloys are generated by postsintering mechanical deformation followed by aging. Usually, the tungsten heavy alloy rods are swaged and then aged to provide the increased strength and hardness that is a requirement in most of the defense-related applications. Figure 6.42 shows a schematic view of the conventional heavy alloy processing steps. It is not difficult to appreciate that the process of postsintered mechanical deformation does not lend itself to near net-shape forming

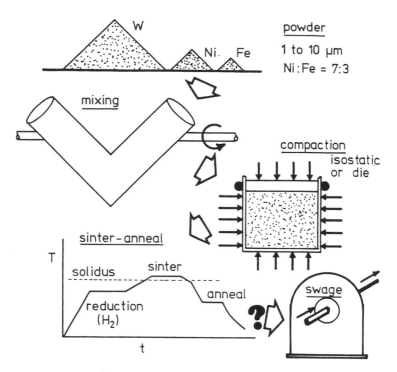

Figure 6.42: *Schematic view of the conventional heavy alloy processing steps.*

techniques such as particulate injection molding. The impetus for particulate injection molding of tungsten heavy alloys was generated by the success in improving the strength and hardness of the tungsten heavy alloys by alloying with other refractory metals [87–89]. The higher strengths of the new alloys are due to the combined effect of solid-solution strengthening and grain size refinement. Thus, the modified heavy alloys have the capability of utilizing the net-shape manufacturing potential of PIM, while exhibiting the increased strength and hardness obtainable by moderate amount of deformation of the classic tungsten heavy alloys.

Theory predicts, and it has also been verified experimentally, that molybdenum that goes into solution in tungsten, nickel, and iron will act as a good strengthening additive. Compared to the baseline properties with no molybdenum addition, the strength and hardness showed a continuous increase with molybdenum addition with a concomitant decrease in the sintered grain size and ductility [87].

The properties of tungsten heavy alloys are extremely sensitive to processing conditions, impurities, and postsintering treatments [80,81,90–94]. The metallurgy of this unique alloy system has also been studied extensively. In earlier

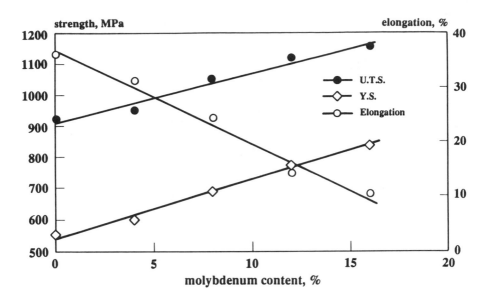

Figure 6.43: *Variation in the elongation, yield strength, and ultimate tensile strength with increasing molybdenum addition (samples pressed and sintered).*

experiments with pressed- and-sintered heavy alloys with molybdenum additions, it was found that with an increase in molybdenum content by the replacement of tungsten the theoretical density of the composite decreases. It was also observed that with increasing molybdenum additions the elongation decreases and the strengths (both yield and ultimate) increase monotonically (at least for the range of compositions used). Figure 6.43 shows the variation in the elongation, yield strength, and ultimate tensile strength with increasing molybdenum addition (samples pressed and sintered). The hardness also shows a similar monotonic increase with increasing molybdenum content. The increase in strength and hardness is due to the combined effect of solution hardening and grain refinement brought about by molybdenum additions [87].

The effect of grain refinement with increasing molybdenum addition is clearly demonstrated by the series of microstructures shown in Figure 6.44 [87]. Toward the high molybdenum contents the grains become slightly jagged, which suggests that a modification of the solution–reprecipitation step during sintering occurs in the presence of molybdenum. It is evident from the binary phase diagrams of Mo–Ni and Mo–Fe that molybdenum has a fair amount of solubility at 1573 K (1300°C) in both nickel ($\approx$38 wt.%) and iron ($\approx$24 wt.%). Tungsten and molybdenum exhibit complete solid solubility in each other over the whole range of compositions. Thus, addition of molybdenum to the classic tungsten heavy alloy system results in a situation where molybdenum is shared

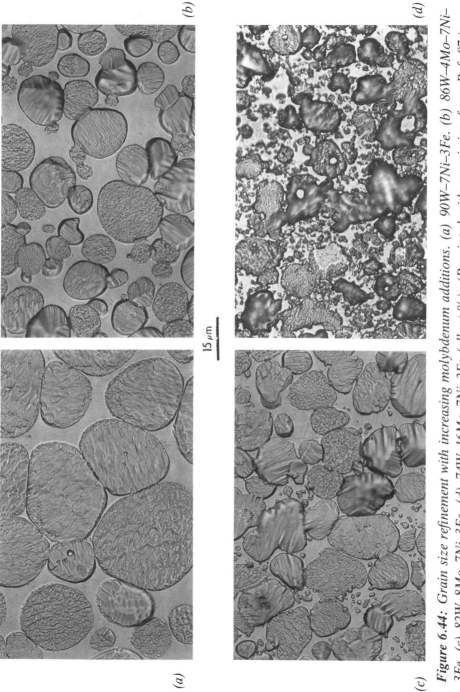

Figure 6.44: *Grain size refinement with increasing molybdenum additions. (a) 90W–7Ni–3Fe. (b) 86W–4Mo–7Ni–3Fe. (c) 82W–8Mo–7Ni–3Fe. (d) 74W–16Mo–7Ni–3Fe (all wt.%). (Reprinted with permission from Ref. 87.)*

between the tungsten grains as well as the matrix, and it can thereby strengthen both by solid solution. Due to the dissolution of molybdenum into the matrix alloy, the matrix can no longer dissolve the usual amount of tungsten. This in turn causes a modification in the solution–reprecipitation step that is responsible for grain growth in tungsten heavy alloys. During this stage, the smaller tungsten grains are preferentially dissolved and reprecipitated on the larger ones. With the molybdenum addition to the system, dissolution of tungsten is hindered, resulting in grain size refinement.

Molybdenum was therefore chosen as the alloying addition to the classic tungsten heavy alloy system. The alloy chosen for injection molding had a composition (wt.%) of 82W–8Mo–8Ni–2Fe, as preliminary investigations have shown that this alloy has an excellent combination of strength and ductility [87]. This composition also exhibited property combinations comparable to a moderately swaged classic tungsten heavy alloy (around 8% to 10% reduction in area). In the experiment on injection molding of molybdenum-doped tungsten heavy alloy, elemental tungsten, molybdenum, nickel, and iron powders were mixed in the desired ratio, which corresponded to a composition of 82W, 8Mo, 8Ni and 2Fe by weight [95]. The mixing was carried out in a turbula mixer for 3600 s (1 hour). Table 6.5 outlines the powder characteristics. Polyethylene and stearic acid were mixed in the ratio 90 : 10 by volume and a measured amount of this mixture was poured into the basket of a double planetary mixer. The temperature of the basket was maintained at 398 K (125°C) by hot oil circulation. The binder is liquefied and to it the desired amount of the premixed elemental powders is gradually added. The starting solid volume fraction loading was 0.4, which was gradually increased by adding more solid to the mix. The mixing was carried out in vacuum.

The loading curve was constructed using the experimental process outlined earlier. At 0.54 volume fraction of solid loading, the mixture became too viscous, and further solid loading was discontinued. Uniform mixing plays a key role in the proper realization of the optimum injection-molded properties. It was observed that mixing for 900 s (15 minutes) in the double planetary vacuum mixer resulted in mixture densities substantially less than the calculated values. When mixing under vacuum was carried out for 3600 s (1 hour), the calculated and the theoretical densities were almost identical. The loading curve for the modified heavy alloy composition with the binder is shown in Figure 6.45 [95]. High volume fraction of solid loading (0.6 or more) would not be possible due to the surface characteristics of the powders. The final volume fraction of solid loading was adjusted to 0.50.

Granulated feedstock of this material was used for all subsequent experiments. A small amount of the granulated feedstock was used in a capillary rheometer and the viscosity was measured as a function of strain rate at a

Table 6.5. Characteristics of various powders used in particulate injection molding of Mo-doped tungsten heavy alloys

Property	W	Ni	Fe	Mo
Vendor	GTE	INCO	GAF	GTE
Designation	M35	123	HP	Mo-638
Purity (%)	99.98	99.992	99.55	99.96
Fisher subsieve size (μm)	2.5	2.8	3.0	5.2
Mean size (μm)[a]	2.6	3.3	10.8	6.1
BET specific surface area (m²/g)	0.23	2.19	0.88	0.64
Apparent density (g/cm³)	2.57	2.15	2.20	2.03
Major impurities (ppm)	K(11)	Ca(10)	Ca(600)	Fe(28)
	Na(15)	Fe(30)	Al(600)	K(13)
	C(19)	Si(40)	Si(600)	C(10)
	O(770)		O(300)	W(120)
			Mn(2000)	Ni(7)

[a]Forward laser light scattering.

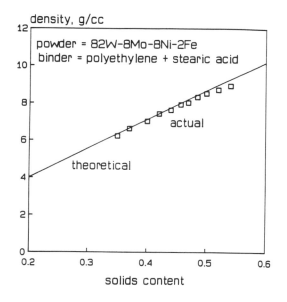

Figure 6.45: The loading curve for the Mo-doped heavy alloy. (Reprinted with permission from Ref. 95.)

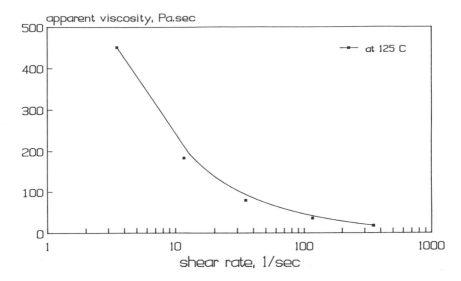

Figure 6.46: *The variation of apparent viscosity with shear rate of the 50 vol.% solid feedstock of Mo-doped heavy alloy.*

temperature of 398 K (125°C). The variation of the apparent viscosity with shear rate of the 50 volume fraction solid feedstock is shown in Figure 6.46. The apparent viscosity is seen to decrease sharply with high shear rates.

The granulated feedstock was then fed into the injection molding machine and small flat tensile bars and transverse rupture test bars were injection molded. The injection molding conditions selected are outliiined below:

Die temperature	300 K
Average melt temperature	372 K
Injection molding pressure	3.1 MPa

The injection-molded test bars were embedded in fine alumina (average particle size 1 μm) and thermally debound. The debinding of the fine powders proved to be quite difficult as the samples tended to crack and form pores under various debinding conditions. The samples were always embedded in alumina to allow faster debinding. The first case, where the samples were heated at 0.17 K/s (10 K/min) to 723 K (450°C) and held there for 3600 s (1 hour) before cooling, resulted in very fragile and porous material. The debinding in this case was obviously too rapid, which resulted in large pressure buildup in the sample, causing pores on the debound pieces. Through a few other debinding experiments, it was determined that a hold at around

573 K (300°C) was crucial for proper debinding of the specimens. Thus, the successful debinding cycle consisted of the following sequence:

0.17 K/s to 523 K, hold for 1800 s (30 minutes);
0.17 K/s to 573 K, hold for 1800 s (30 minutes);
0.17 K/s to 598 K, hold for 1800 s (30 minutes);
0.17 K/s to 623 K, hold for 1800 s (30 minutes);
0.17 K/s to 648 K, hold for 1800 s (30 minutes);
Furnace cools to room temperature.

This resulted in debound samples with no apparent flaws such as surface blisters, cracks, or pores. The hold time between 573 and 623 K (300 and 350°C) was critical for obtaining good quality debound samples. These debound samples were then presintered at 1173 K (900°C) for 3600 s (1 hour) in dry hydrogen atmosphere. The debound and presintered samples were removed from the embedding material and the surface alumina wicking material was carefully brushed off. The presintering step had provided the samples with good handling strength.

These samples were subjected to sintering that was carried out in a horizontal tube furnace programmed to control the heating and cooling rates as well as the hold temperatures. Sintering used the optimized three-step gas atmosphere cycle discussed earlier in this chapter and in more detail in a previous paper [80]. The sintered samples exhibited densities greater than 99.5% of theoretical, thus proving the success of the sintering schedule. The as-sintered microstructure of an injection-molded alloy of 82W–8Mo–8Ni–2Fe (wt.%) is similar to the pressed-and-sintered material of the same composition. The mechanical properties of the injection-molded samples are shown in Table 6.6.

As expected, the sintered strength increased and the elongation decreased with lower sintering temperatures. Thus, modification in the properties of the alloys is also possible by variation in the sintering conditions. For actual comparison, the properties of a 82W–8Mo–8Ni–2Fe (wt.%) alloy and a classic 90W–8Ni–2Fe (wt.%) alloy processed by the press-and-sinter method are also included in Table 6.6. It can be seen that the injection-molded properties compare very well with those of the pressed and-sintered alloys. Thus, it is possible to produce net-shape parts of high-strength heavy alloys by powder injection molding. With suitable alloying and adjustment of sintering conditions, it is possible to produce PIM heavy alloys that will have strength and hardness comparable to a moderately swaged and aged conventional tungsten heavy alloy. Since the alloyed material does not need any postsintering deformation, it is extremely suitable for producing near net-shape components.

Table 6.6. Mechanical properties of conventional and molybdenum-doped tungsten heavy processed by conventional pressing and sintering and the PIM route

Alloy and processing	Yield strength (MPa)	Ultimate tensile strength (MPa)	Hardness (HRA)	Elongation (%)
90W–8Ni–2Fe Press and sinter, 1500°C	551 ± 15	918 ± 6	63.5 ± 0.4	36 ± 2
82W–8Mo–8Ni–2Fe Press and sinter, 1500°C	688 ± 4	1048 ± 8	65.9 ± 0.4	24 ± 1
82W–8Mo–8Ni–2Fe Injection molded, 1475°C	775 ± 17	1144 ± 57	67.6 ± 0.1	8 ± 3
82W–8Mo–8Ni–2Fe Injection molded, 1500°C	700 ± 3	1115 ± 19	64.4 ± 1.4	20 ± 3
82W–8Mo–8Ni–2Fe Injection molded, 1530°C	688	1067 ± 6	64.4 ± 0.1	27 ± 1

The grain refinement of heavy alloys will also occur with the addition of other b.c.c. refractory metals that are soluble in both tungsten and matrix phases. Their additions can also provide solid-solution strengthening. The grain growth law for liquid-phase sintered heavy alloys is [81]

$$G^{2.8} = G_0^{2.8} + (\tfrac{4}{9}) K t \tag{16}$$

where G is the grain size (μm) after an isothermal hold for time t(s), with the initial grain size at $t = 0$ given by G_0. The parameter K (μm^3/s) depends on the material constants and can be expressed by [96]:

$$K = \frac{gD\Omega CS}{kT(1 - f^{1/3})} \tag{17}$$

where g is a numerical constant, D is the diffusivity of the solute in the liquid, Ω is the atomic volume, C is the solid solubility in the liquid, S is the interfacial energy, k is Boltzmann's constant, T is the temperature, and f is the volume fraction of solid.

The concept is that a change in the solubility of tungsten in the liquid by alloying would result in a smaller sintered grain size. Accordingly, the sintered strength would be improved for most b.c.c. refractory metal additions. It should be pointed out that high density is one of the desirable characteristics of

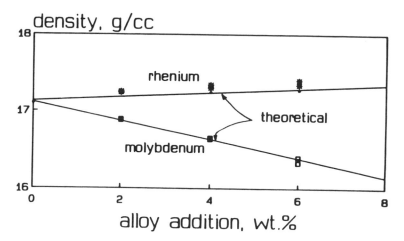

Figure 6.47: Variation of theoretical density of heavy alloys with molybdenum and rhenium additions. (Reprinted with permission from Ref. 89.)

tungsten heavy alloys. However, most of the common refractory b.c.c. elements, such as molybdenum, vanadium, tantalum, etc., have densities that are substantially lower than that of tungsten. In contrast, rhenium, which has a density greater than that of tungsten with a limited solubility in tungsten and a large solubility in the matrix, can serve the same purpose as molybdenum. The advantage of rhenium is the higher density and the greater solid-solution strengthening potential of rhenium over molybdenum, which is possibly due to atomic size difference. It can be seen from Figure 6.47 that as the molybdenum content is increased, the theoretical density of the composite decreases; while the theoretical density of the composite increases with the rhenium content [89] These opposite trends are due to the difference in the densities between molybdenum and rhenium. Figure 6.47 also shows the variation between the calculated and experimental densities with both rhenium and molybdenum additions. The sintered density values were in good agreement with the calculated densities. The effects of molybdenum, tantalum, and rhenium additions on the mechanical properties of tungsten heavy alloys are shown in Table 6.7. The increase in strength due to the rhenium addition is much larger than for the same weight fraction or atom fraction of molybdenum addition. The far superior strengthening potential of rhenium leads to the exciting possibility of processing rhenium-modified tungsten heavy alloys by powder injection molding. Another exciting possibility is that of obtaining coated powders, where the tungsten has been coated with small amounts of rhenium or molybdenum. In this case, substantial modification of the solution–reprecipitation step will occur compared to the case where elemental additions have been made. Injection molding of these powders could provide heavy alloys with novel property combinations with net-shape manufacturing capability.

2. Particulate Injection Molding of Thixotropic Alloys

Another extremely interesting development in the area of particulate injection molding is the Dow Chemical process for injection molding of thixotropic alloys. This relatively new technology has been demonstrated in the case of magnesium alloys.

A thixotropic material is a form of semisolid mass that becomes more fluid as it is subjected to high shearing action. An alloy, when it is being cooled through a zone between the solidus and the liquidus temperature, generally tends to forms dendrites. Subjecting this material in the mushy stage to high shearing often results in the formation of nearly spheroidal degenerate dendritic particles that are still dispersed in the liquid. Researchers at Massachusetts Institute of Technology found that the mushy semisolid thixotropic material could be molded or die cast. They also determined that the solidified alloys of the degenerate dendritic structure when reheated to the mushy state regained their thixotropic properties.

The processing of these materials in the semisolid state provides some advantages over the melt-and-cast method. As the temperature of the material is lower than in the melt and cast process (where sufficient superheat is required for proper flow), there is significant increase in the die life coupled with shorter cycle times. Since less than half the volume of the material is in the solid state, its solidification shrinkage is much lower than the shrinkage of the same material solidifying from a fully liquid state. Thus, the porosity in such a processing scheme is lower than in its cast counterpart.

This new process of injection molding of thixotropic alloys has the benefits of injection molding without its problems associated with the binders. The process can essentially transform granules of conventionally solidified magnesium alloys into as-cast metal parts without actually reaching the melting point of the alloy. The thixotropic nature of the material provides the flow characteristics and the aid of external binders is not required. This results in parts that are produced in just one shot, without going through the usual PIM processing steps of mixing, injection molding, debinding, and sintering. Also, the particulate size of the granules is not as critical as in conventional particulate injection molding.

Dow Chemical has developed a process for particulate (chips or very large particles) injection molding of magnesium alloys based on the thixotropic nature of this alloy system [97]. Chips of conventional alloy AZ91D (a popular

Table 6.7. Mechanical properties of tungsten heavy alloys with molybdenum, tantalum, and rhenium additions

Alloy composition (wt.%)	Sintering time (min)	Sintered density (g/cm³)	Percent theoretical density	Yield strength (MPa)	Ultimate strength (MPa)	Tensile elongation (%)	Hardness (HRA)
90W–7Ni–3Fe	30	17.09	99.6	535	925	31	62.8
88W–2Mo–7Ni–3Fe	30	16.89	99.7	570	945	28	63.2
86W–4Mo–7Ni–3Fe	30	16.64	100.1	625	980	24	63.9
84W–6Mo–7Ni–3Fe	30	16.37	99.9	650	1005	24	65.6
82W–8Mo–7Ni–3Fe	30	16.15	100.1	715	1030	20	67.8
78W–12Mo–7Ni–3Fe	30	15.69	100.1	835	1100	10	68.1
74W–16Mo–7Ni–3Fe	30	15.26	100.3	890	1145	7	68.8
86W–4Mo–7Ni–3Fe	120	16.60	99.9	570	920	29	63.8
78W–12Mo–7Ni–3Fe	120	15.68	100.0	670	1025	19	65.7
74W–16Mo–7Ni–3Fe	120	15.23	100.1	740	1065	14	66.8
85W–5Ta–7Ni–3Fe	30	16.20	95.0	740	1025	3	69.0
86W–4Re–7Ni–3Fe	30	17.18	99.9	732	1050	7	69.1
88W–2Re–8Ni–2Fe	60	17.24	100.1	703	1036	17	65.7
86W–4Re–8Ni–2Fe	60	17.32	100.5	766	1118	14	67.4
84W–6Re–8Ni–2Fe	60	17.37	100.6	815	1183	13	69.0
82W–8Mo–8Ni–2Fe[a]	30	16.07	99.4	700	1115	20	64.4

[a]Injection-molded, debound, and sintered at 1773 K (1500°C) for 30 minutes.

alloy of magnesium) are introduced through a hopper into the barrel of a specially designed injection molding machine. The particulates can be large as their size is only limited by the requirement that they fit into the flights of the screw. The granules are heated in the barrel of the injection molding machine. The barrel is heated by a combination of resistance and induction heating and the barrel is maintained under argon in order to prevent oxidation.

Inside the barrel, the magnesium alloy particles are heated to a desired temperature while being sheared and advanced to the front of the barrel. The combination of elevated temperature and high shear converts the material to a thixotropic state. The content of solid, which is composed of the spheroidal degenerate dendrites, is usually as high as 60 vol.% , with the rest being liquid. The thixotropic material transported to the front of the screw is ready to be injected. The temperature at the exit end is carefully controlled at 791 K (518°C) with an allowable deviation of 2 K. The next phase is similar to the normal particulate injection molding step where the material is shot into the die at high speeds. The velocity during this injecting stage can be as high as 3.8 m/s. The die cavity is filled within a matter of a second and almost pore-free parts with good dimensional stability are produced.

Two different profiles have been developed for producing the thixotropic solid from the AZ91D magnesium alloy particulates. In one case, known as the "hot" profile, the granules are taken to a temperature above the liquidus temperature of the alloy, 869 K (596°C). As the melt slowly cools into the mushy zone, the high shear action generated by the screw produces the desired thixotropic material. The second profile, known as the "cold" profile, only heats the particulates to a temperature that is below the liquidus but above the solidus, which is at 741 K (468°C). The shearing action of the screw again results in the formation of the desired thixotropic material. Interestingly, the properties of the semisolid material formed by either profile are almost identical. The advantage of the cold profile is that the complete material is never melted.

The microstructure of the material resembles that of a two-phase composite in which nearly spheroidal particles of degenerate dendritic material are dispersed in a matrix that was liquid when the material was in the thixotropic state.

The specially designed injection molding machine posed tremendous technical challenges. The temperature control at the exit end had to be within 2 K. The screw material had to have sufficient mechanical properties at the high operating temperatures such that the tolerance could be maintained. High expansion could potentially result in the seizing of the screw within the barrel. The hydraulic system and the thrust bearing had to be such that the high-speed injection molding could be easily accomplished. Coupled with the increased die

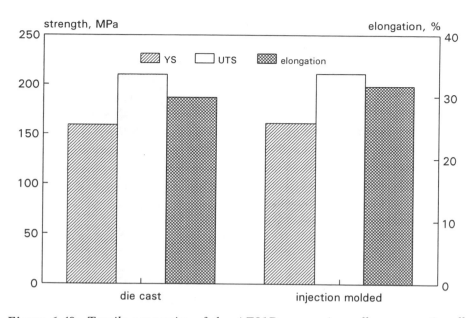

Figure 6.48: *Tensile properties of the AZ91D magnesium alloy conventionally die cast and injection molded by the process developed by Dow Chemical.*

life due to reduced thermal fatigue, the possibility of obtaining material with lower porosity would make this novel particulate injection molding process very attractive for producing high-quality magnesium alloy parts directly from particulate materials. The process is expected to be cost-effective in spite of the high initial investment required for the injection molding machine. Figure 6.48 compares some of the tensile properties of the AZ91D magnesium alloy that has been conventionally die cast and that has been injection molded by the process developed by Dow Chemical. It can be observed that the properties are similar for both the processes. The typical porosity level obtained in the injection-molded material is around 1.7 vol.% compared to 3.2 vol.% in the die-cast materials. The injection-molded material also has excellent corrosion resistance.

3. Aligned Fiber Composites

An interesting use of particulate injection molding is its possible application in processing of aligned fiber-reinforced composites. Composites consist of at least two distinct phases that may be totally different in their physical and mechanical properties. High-temperature composites may often contain a high-temperature metal, ceramic, or intermetallic compound with the reinforcing phase being a nonmetallic fiber or whisker.

A common reinforcing phase is a fiber, which may be continuous or discontinuous. In this part of the chapter we will deal only with discontinuous-fiber reinforcements. One of the major problems associated with fiber-reinforced composites is the random distribution of the fibers. Fibers that are perpendicular to the load-bearing axis are usually detrimental since they serve as flaws and decrease the strength of the material. However, if the fibers can be aligned in the direction of the loading axis (usually within 22°), the strength of the material can be improved. The fabrication of composites with aligned fibers is a real engineering challenge. A modified form of particulate injection molding has been used to process such fiber-reinforced composites.

Advanced nickel aluminide was among the first materials used to produce such a composite, with chopped alumina fiber being used as the reinforcing phase. The advanced nickel aluminide has great potential as a high-temperature material. The process of binder-assisted flow used to obtain the fiber alignment is quite generic and can be applied to produce a variety of composites including metal matrix or ceramic matrix composites with the reinforcing phase being either chopped short fibers or whiskers.

Binder-assisted extrusion of a 5 vol.% alumina fiber-reinforced advanced nickel aluminide-based (IC-218 alloy developed by Oak Ridge) composite was carried out to achieve fiber alignment within the matrix. Polyethylene wax acts as a binder for the phases and aids in attaining alignment during the forming process. The mixing of the fibers with the powder and polyethylene, and the loading curve characteristics of the composite, have been discussed earlier in this chapter. Mixing is important for attaining the desired fiber dispersion within the matrix and for the eventual orientation while forming the shape. A solid volume fraction of 0.66 was chosen for this study.

To demonstrate the generic concept of binder-assisted flow, the mix of polyethylene, IC-218 powder, and chopped short alumina fiber was granulated into small pieces and then placed inside a die with an opening of 17.6 mm. A specially designed nozzle that results in a gradual reduction in area was used as the lower punch. The special nozzle insert was constructed in such a way that the outer diameter of the insert just fitted into the die opening of 17.6 mm. After the composite feedstock was introduced, the top punch was inserted. The whole die assembly with the special insert, the composite feedstock, and the top punch was heated to 383 K (110°C), at which temperature the feedstock is a viscous fluid. Once the temperature was attained, the die assembly was quickly transferred to a hydraulic press and pressure was applied slowly to the top punch. The pressure on the top punch is transmitted to the viscous feedstock of the composite material, which is then slowly squeezed out through

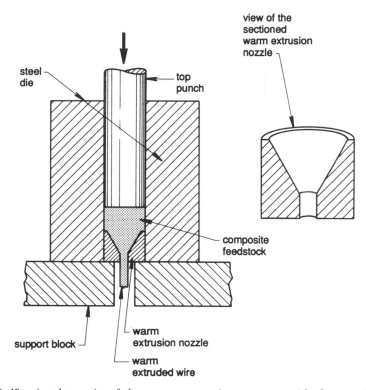

Figure 6.49: *A schematic of the warm extrusion process with the special nozzle insert. (Reprinted with permission from Ref. 79.)*

the special nozzle insert. The construction of the nozzle is such that the composite feedstock undergoes a gradual reduction in area, until it is finally squeezed out through the front orifice of the nozzle in the form of a thin rod. As the thin rod is squeezed out, the outside temperature cools it sufficiently that the feedstock solidifies, resulting in a smooth rod wire of the composite feedstock. A schematic of the warm extrusion process with the special nozzle is shown in Figure 6.49.

The early experiments used a nozzle that gave a wire diameter of 1.5 mm. It was expected that the fibers would be aligned parallel to the direction of extrusion. The hypothesis is that the viscous flow coupled with the gradual reduction in area, will align the fibers along the flow direction.

Wires made from the IC-218-based composite feedstock was debound and presintered. The fracture surface of a warm extruded IC-218 + 5 vol.% alumina fiber after debinding and presintering is shown in Figure 6.50 [98]. It can be observed that although there is some degree of alignment, the fibers are still quite randomly oriented (not within 22°). This is due to the powder

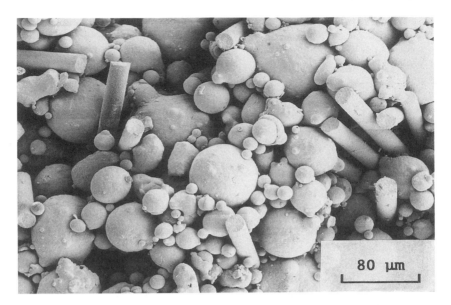

Figure 6.50: *Fracture surface of a warm extruded IC-218+5 vol.% alumina fiber after debinding and presintering. (Reprinted with permission from Ref. 98.)*

particle size of IC-218, which is quite large (70 μm) compared to the diameter of the fiber (round 20 μm). It was theoretically determined that the use of a finer powder would help in aligning the fibers.

To demonstrate this concept, fine BASF iron powder (particle size 5 μm) was mixed with 5 vol.% of the short alumina fiber and polyethylene wax. The binder-assisted flow experiments were conducted in the same manner as described in the earlier case of the IC-218-based composite. In this case, two different nozzles of different exit diameters were used. One resulted in wires of 1.5 mm in diameter and the other in wires 4 mm in diameter. The samples were debound by wicking with fine alumina and presintered using the following schedule:

To 433 K at 0.066 K/s, hold for 7200 s (2 hours).

To 723 K at 0.066 K/s, hold for 7200 s (2 hours).

To 1373 K at 0.17 K/s, hold for 3600 s (1 hour).

Then furnace cools.

After the warm extrusion, the specially designed nozzle with the material inside was allowed to cool to room temperature. The remaining material solidified inside the die and a replica of the inside of the special nozzle was

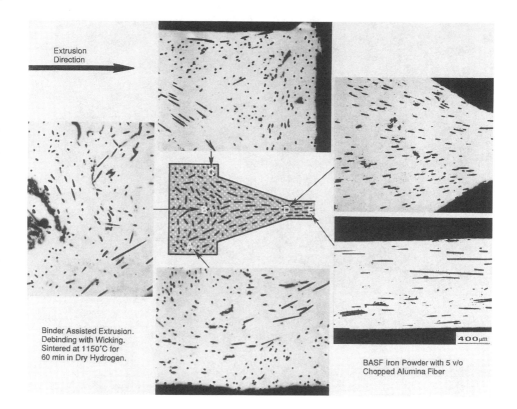

Figure 6.51: *The cross-section of the stub solidified inside the nozzle (like the nozzle shown in Figure 6.49) and microstructures taken from various regions of the material solidified within the nozzle.*

formed with the feedstock. The stub formed from the nozzle was also debound and sintered under the same conditions described earlier. This stub proved helpful in visually examining how the fibers were gradually orienting themselves when the feedstock was being squeezed out through the nozzle. The cross-section of the stub solidified inside the nozzle (like the nozzle shown in Figure 6.49) and the microstructures taken from various regions of the cross-sectioned stub are shown in Figure 6.51. This figure clearly shows the ability of the binder-assisted extrusion to align the fibers within the matrix material. An SEM photomicrograph of the fracture surface shown in Figure 6.52 [98] also shows the excellent fiber alignment that is attained with this fine powder.

It can be envisioned that the nozzle size and shape can vary, but the underlying principle of gradual reduction in area in binder-assisted flow is of vital importance in obtaining the desired fiber alignment within a matrix. The

Figure 6.52: The excellent alignment of the alumina fibers in the iron matrix, as seen on the fracture surface. (Reprinted with permission from Ref. 98.)

mode of application of pressure may vary, and the fluidized mix may be extruded by various methods such as slow-speed injection molding or hydraulic pressure-assisted or even air-pressurized extrusion. It is also envisioned that in the case of an expanding flow the fibers may be aligned perpendicular to the direction of flow, while in the case of a contracting flow the fibers will be aligned parallel to the direction of flow. This concept is illustrated schematically in Figure 6.53 [98].

The crux of the process is to have a streamlined reduction in area through which the fluid consisting of the powders, fibers, and the fluid material (usually a wax) is ejected. On ejection, the material must quickly solidify. The binder material has to be removed (preferably at low temperatures) and the remaining mass of fiber-dispersed powder sintered to attain the desired density.

The process can be applied to any composites in which fiber alignment is desired to improve the mechanical properties. This process has been used to produce reactively sintered alumina fiber-reinforced NiAl-based and Ni(Al,Si)-based composites [78,79]. The intermetallic compound NiAl has a much higher melting point and lower density than the intermetallic compound Ni_3Al. However, the material has practically no fracture toughness. Thus, attempts are in progress to reinforce NiAl with fibrous materials to increase its fracture toughness.

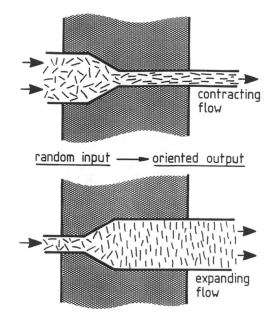

Figure 6.53: Schematic of fiber alignment in expanding or contracting flows. (Reprinted with permission from Ref. 98.)

For the developmental work based on warm extrusion and reactive processing of alumina-reinforced NiAl, elemental nickel and aluminum powders taken in the ratio of 51Ni : 49Al (atomic percent) were mixed in a turbula-type mixer for 3600 s (1 hour). To the elemental powder blend, 15 vol.% of alumina fibers was added and this mixture was used to form a PIM feedstock with 35 vol.% of binder. This feedstock served as the raw material for the binder-assisted extrusion for good fiber alignment. Since it had been shown that finer powders can provide better fiber alignment, the nickel powder used had a particle size less than 10 μm and the aluminum powder had an average particle size of 3 μm [98]. The warm extrusion was carried out in the apparatus described earlier. In this case the special extrusion die was gradually tapered from 12.7 mm diameter at the top to 1.5 mm at the exit end. The special extrusion die was inserted into another, larger, die and acted as the bottom punch. The composite feedstock was introduced into the larger die and the top punch was inserted. The die along with the punch and feedstock was heated to 363 K (90°C) and extruded by applying force to the top punch through a hand press. The resultant 1.5-mm-diameter wires were cut to approximately 30-mm-long pieces, carefully bunched together, and inserted into a soft polyurethane die and cold isostatically pressed at 208 MPa pressure. Cold isostatic pressing produced a specimen around 12.7 mm in diameter and 25.4 mm in length. The

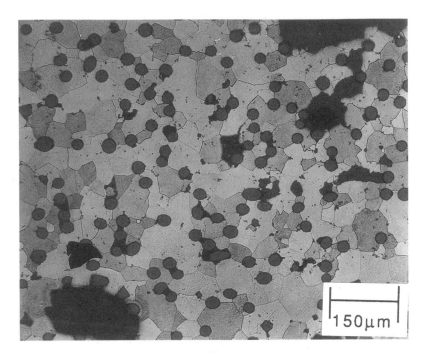

Figure 6.54: *Microstructure of a warm-extruded, cold isostatically pressed, and reactively sintered NiAl with chopped alumina fibers. Microstructure shows the large porosity and excellent fiber alignment. (Reprinted with permission from Ref. 78.)*

binder was removed by thermal debinding at 723 K (450°C) for 14400 s (4 hours).

For the consolidation process, a novel form of sintering termed reactive sintering was used [76,99]. The process of reactive sintering (described earlier in the Chapter) has been used to process high-density Ni_3Al-based intermetallic compounds. The process can also be used for NiAl-based intermetallic compounds, but the reaction in this case is much more vigorous and difficult to control. Thus, it is often necessary to dilute the exothermic reaction by using some volume fraction of prealloyed NiAl compound instead of using totally mixed elemental powders. In this case, however, the short alumina fibers provide the diluent for the reaction. The material was reactively sintered at 973 K (700°C) for 900 s (15 minutes) in a vacuum of 10^{-8} Pa. The material fabricated by this process was porous in nature, with the residual pores being around 150 to 200 μm. This can be seen in the microstructures shown in Figure 6.54 [78]. It can be observed from the figure that nearly perfect alignment of

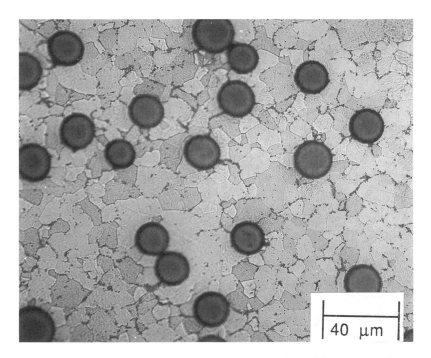

Figure 6.55: Microstructure of a warm-extruded, cold isostatically pressed, reactively sintered, and hot isostatically pressed NiAl with chopped alumina fibers. The microstructure is fully dense with good fiber alignment. (Reprinted with permission from Ref. 78.)

the fibers was achieved. The reactively sintered porous composite was hot isostatically pressed by first vacuum encapsulating the material in a stainless-steel can, followed by HIP at 1473 K (1200°C), for 3600 s (1 hour) using a pressure of 172 MPa. The samples obtained were fully dense, as seen in the photomicrographs in Figure 6.55 [78]. Initial testing on the composites by observing the interaction of the cracks with the alumina fibers suggests that the alumina fiber additions might improve the toughness of the NiAl compound. The addition of 15 vol.% of alumina fibers increases the Vickers hardness of the material to 350 compared to 277 VPN for the monolithic reactively sintered NiAl. Thus, this process could be a viable method for producing aligned reinforcements for toughening extremely brittle materials. A similar process has also been used to process a Ni(Al,Si)-based matrix material. This material has been found to exhibit very high compressive strength at an elevated temperature.

4. Freeze Compression Molding

Use of water as a medium for providing shape to fine ceramic powders is an important development in the area of particulate injection molding. Japan has been the forerunner in this area. In this section we discuss some of the advances in the area of shaping alumina with primarily water as the vehicle that provides the desirable moldability. It should be remembered that in injection molding of fine ceramic particulates the volume fraction of binder material is generally around half the total volume of feedstock. Thus, removal of all the binder through debinding, especially in thick sections, is an extremely slow and time-consuming process. The use of water as a replacement for the organic binders could lead to faster debinding times and lowered sample distortion or cracking during the debinding process.

In the process of freeze drying, a small amount of organic binder is usually added along with water to lend the compact some rigidity after the water has been removed. This technique relies on the use of a chilled mold to obtain a frozen green compact. The shape is usually imparted by compression molding the feedstock into the die cavity. The frozen green shape that is ejected from the die cavity is introduced directly into a vacuum furnace and then sintered. A major part of the binder is removed in the solid state to prevent crack formation during drying. This technique has been termed freeze compression molding [100].

The spiral flowability test was applied to determine the moldability of the alumina–water mixture. The alumina powder was submicrometer in size, with an average particle diameter of 0.2μm. The relationship between the water content and the spiral flow length is shown in Figure 6.56. The flowability is seen to increase rapidly when the water content is increased from 25 to 30 wt.%. The water–alumina mixture could be extruded continually through a narrow nozzle to produce very smooth wire with good shape retention, and the wire extrusion could be successfully carried out without the addition of any organic binder to the feedstock.

During molding, around 2 wt.% of PVA was added as the organic binder. The volume fraction of water used is around 45 to 50%. A wet powder billet is compression molded into shapes inside a die cavity that is cooled to a temperature between 258 and 253 K (-15 and $-20°$C). The shaped compact, after being held in the die for approximately 120 to 180 s (2 to 3 minutes), can be ejected by ejector pins without breakage. The frozen compact is introduced directly into a vacuum furnace and most of the debinding is achieved in the solid state. The small quantity of organic binder helps in retaining the compact shape after drying. The freeze drying is usually carried out at temperatures around 473 K (200°C) in vacuum. The drying time, which is usually between

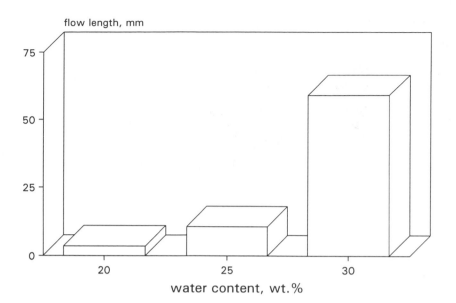

Figure 6.56: Relationship between the water content and the spiral flow length. (Data from Ref. 26.)

3600 and 14 400 s (1 and 4 hours) for a specimen 30 mm × 30 mm × 20 mm, can be decreased by increasing the freeze drying temperature. The material is then sintered in the usual manner.

It was determined that by using powders with a wider size distribution the flowability and formability of the mixture could be increased to some extent. This means that mixes produced using smaller amounts of water can also be formed. A large number of simple cup shapes with a bottom were processed from various materials such as alumina, silicon carbide, and silicon nitride. Some of the shapes obtained from the various materials are shown in Figure 6.57. The tool used for producing the desired shape is similar to a cold forging tool and the compression molding sequence in the tool is shown in Figure 6.58. It was found that 30 to 40 vol.% of the water-based binder was sufficient and that with decreasing water content the compact rigidity was also increased; in some cases the compact could be ejected without the freezing procedure.

Interestingly, it was found that the application of vibration could significantly increase the flowability of the ceramic powders in a water binder [101]. With application of vibration of frequency 15 Hz and an amplitude 0.25 mm, the density of compacts with only 8 wt.% water was significantly increased after drying, compared to ones pressed without vibration. Figure 6.59 illustrates graphically the effect of vibration on the increase of density of alumina with 8 wt.% water (binder)-containing compact after drying. Vibratory compression

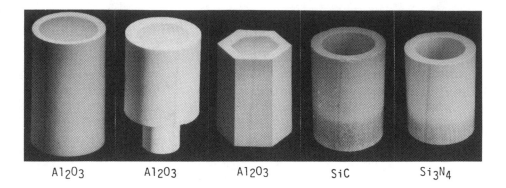

| Al$_2$O$_3$ | Al$_2$O$_3$ | Al$_2$O$_3$ | SiC | Si$_3$N$_4$ |

Figure 6.57: *Different ceramic shapes obtained by compression molding. (Reprinted with permission from Ref. 26.)*

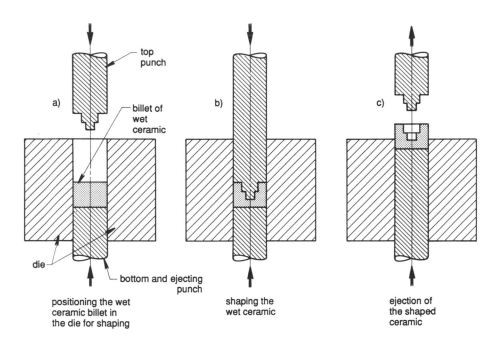

Figure 6.58: *The tool used for producing the desired shape by the compression molding and freeze drying technique. The tool is similar to a cold forging tool and the compression molding sequence is shown.*

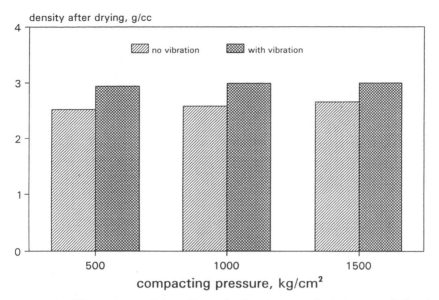

Figure 6.59: *Illustration of the effect of vibration on the increase of density of a compact of alumina with 8 wt.% water (binder) after the process of drying.*

molding can be used to form shapes that require smaller amounts of water as the binder, and it is also suitable for processing thinner wall sections.

By changing the setup shown for the compression molding tool and using a diamond nozzle with a small orifice, it is possible to extrude continuous ceramic wires. In this case, the volume fraction of water can be as high as 40 to 50% with 1 to 2 wt.% of PVA being used as the organic binder. Wires with diameters ranging from 0.1 mm to 0.5 mm have been successfully processed by this technique.

5. Particulate Injection Molding of Titanium

Particulate injection molding of aluminum, titanium, and other reactive metals poses a great technological challenge due to the inherent tendency of the materials to react. Thus, the choice of binder, which is a source of contaminant, is very difficult for these materials. Among the reactive materials, titanium is used extensively in the aerospace industry. Unfortunately, it is also a difficult material to machine. Thus, any development in the near net-shape processing of titanium or titanium alloys would be a big step in the processing of advanced materials. Recent Japanese investigations have shown that titanium can be injection-molded and sintered to densities as high as 95 to 97% [102]. The development used a novel binder that is essentially a copolymer of poly(butylmethacrylate) and ethylene–vinyl acetate, mixed with atactic poly-propylene, a plasticizer (dibutyl phthalate), and wax.

In this investigation [102], a commercial titanium powder with an oxygen content of 0.34 wt.% and a titanium hydride powder with 3.94 wt.% hydrogen were used as the particulate materials. Both the powders had an average particle size around 30 μm, which is coarser than typical injection molding grade powders. The powders were separately mixed with the above binder at 413 K (140°C) for 1800 s (30 minutes) and then injection molded into a cylindrical die with 10 mm diameter and 100 mm length. For both the mixes around 85 wt.% of the powder was used. The debinding step, which was carried out in air, is as follows:

From room temperature to 373 K at 0.045 K/s.

From 373 K to 593 K at 0.1 K/s.

Hold at 593 K for 7200 s (2 hours).

The debound material was sintered in vacuum at 1573 K (1300°C) for 3600 s (1 hour). The material obtained from titanium powder was around 95% dense, while the sample produced from the titanium hydride was around 97% dense. The microstructure represented a two-phase material in which the second phase that was dispersed in the matrix was determined to be TiC. This was confirmed by X-ray diffraction, electron spectroscopy for chemical analysis (ESCA), and Auger electron spectroscopy (AES). The volume fraction of TiC dispersion was greater in the sample prepared from commercially pure titanium powder.

The injection-molded materials and an as-cast bulk material that was also heat treated in vacuum at 1573 K (1300°C) for 3600 s (1 hour), were tested in compression. The injection-molded material processed from the titanium hydride powder had a compressive strength of 1800 MPa, while the sample processed from injection-molded titanium powder had a strength of 1300 MPa. The as-cast and heat-treated material had the lowest strength but the highest ductility of 35%. The low ductility of 4% exhibited by the sample made from the titanium hydride powder was probably due to hydrogen embrittlement. The sample processed from titanium powder exhibited a ductility of 16%.

Thus, it is possible to produce near net shapes from titanium by the process of particulate injection molding. However, TiC particles are found to be dispersed within the matrix. It is not known at this time whether the volume fraction of TiC particles can be controlled and also whether they can be consistently reproduced. Whether particulate injection molding of titanium will be a commercial success will depend to a great extent on the reproducibility of the microstructure and properties of the parts produced by PIM.

6. Particulate Injection Molding plus Hot Isostatic Pressing

The combination of injection molding with hot isostatic pressing is an extremely potent technique in the area of materials processing. This development could ultimately lead to an economic process for production of alloy steel parts by hot isostatic pressing. The processing concept is extremely interesting and should not be confused with the postsintering HIP treatment that is sometimes carried out to remove the remaining pores or heal defects.

According to the authors, if the sintered green body produced by injection molding could be processed by HIP to attain properties comparable to forged machine parts, the advantages would be tremendous. However, conventional HIP technique would not work as the process of encapsulating the material in a can is mainly suitable for simple shapes such as rods or bars. Containerless HIP would not work either as the HIP medium, which is a gas, would enter the pores and prevent densification. Clearly, a medium is necessary that will not penetrate the pores of the material. ASEA has developed a process for producing high-performance silicon nitride parts such as gas turbine wheels by the injection molding/HIP process.

The process consists of mixing the silicon nitride powder with a plasticizer to form the desired injection molding feedstock. The feedstock is then injection molded to produce a green shape such as a turbine impeller. The green part is debound to remove the plastic constituent, resulting in a porous shaped part of silicon nitride. The shaped part is then enveloped in a special glass powder and hot isostatically pressed to full density. The glass probably melts and forms a viscous fluid that conforms to the exact shape of the material undergoing HIP. The pressure is then transmitted through the fluidized glass to the porous green part. Lower temperatures are required to consolidate silicon nitride to full density. Thus, excessive grain growth is not a problem in this process and the silicon nitride has extraordinarily high strength, such as a bend strength of 700 MPa at room temperature. Also, the material properties are found to be isotropic. This process provides a method of combining the best attributes of the two processes of particulate injection molding and hot isostatic pressing.

7. Binderless Pressureless Injection Molding

In another development, Gorham Advanced Materials Institute has described a binderless pressureless injection molding of P/M parts. The process, with the trade name BPIM, claims that the powder without the aid of a binder can be injected into a mold cavity and then preformed by an electromechanical process. The triaxial thermomechanical energy that is applied results in the development of metallurgical bonds for powder particles that are in contact. The preform is then removed from the mold and consolidated to full density via any number of processes such as sintering, pressure-assisted sintering, sinter/HIP, or sinter+isostatic forging. The advantage of the BPIM process is that it can use a wide range of engineering materials. It is claimed that it can use the inexpensive coarse powders, and the need for fine powders is still a handicap for the PIM industry. The major advantage, however, is that the process requires no binders, and thereby does not require the slow step of debinding. Also, larger parts are possible as the binder-removal problems associated with thicker parts are no longer a stumbling-block (since there is no binder present). The markets and applications for BPIM include high-speed steel cutting tools, automotive components, gun components, valve components, replacement of machined investment castings, and turbine engine components.

The Future

The future of particulate injection molding is indeed very bright. From all the predictions that are available in the market, a modest growth rate of 20% to 50% can be estimated for this technology.

Some of the problems that are hindering the growth of this process have been discussed. However, these problems are gradually being reduced or eliminated. The availability of trained personnel is gradually increasing thanks mainly to the efforts of various universities and institutes. There is also a trend of consolidation among the PIM manufacturers, with the small and undercapitalized companies either being absorbed by larger companies or going out of business. The companies that have the technology and the capital are expected to grow rapidly during the 1990s. This consolidation of the PIM industry is a healthy trend, and is expected to be beneficial for the overall industry.

A curious phenomenon, of companies having no idea of what applications they wish to pursue trying to acquire the PIM technology, has apparently subsided to some extent. According to Verduzco [14], it is important not only for an injection molding company to have the proper technology but also for it to decide what specific field of application it wishes to pursue. As a rough estimate, to produce a couple of million dollars' worth of injection-molded parts per year, the initial capital investment in machinery is around 1 million dollars. It is also considered that to take the operation to a breakeven point the working capital required may be around another million dollars. To run a successful particulate injection molding operation it is not only important to know the technical characteristics of the parts to be produced such as size, wall

(a)

(b)

(c)

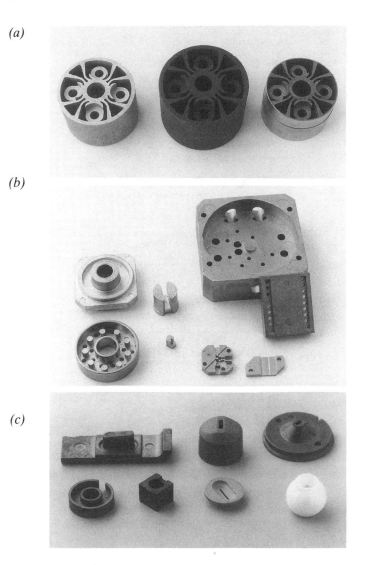

Figure 6.60: *Pictures of various injection-molded components. (Courtesy of Parmatech Corporation.)*

thickness, tolerances, mechanical property requirements, etc., but also extremely important to know what the competitive processes can do. The main competition for PIM comes from traditional processes such as investment casting and machining. Very rarely will a PIM part compete successfully with the highly automated metalworking processes such as forging, stamping, die casting, or even conventional press-and-sinter powder metallurgy processing.

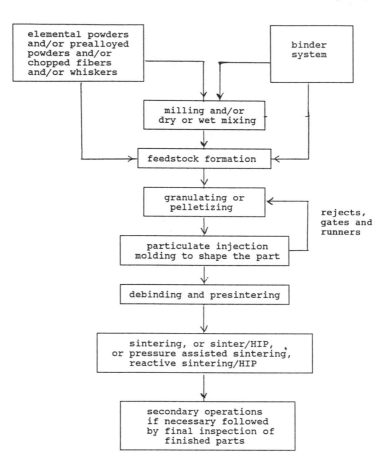

Figure 6.61: *Flow diagram of the PIM process providing a summary of the key steps of this shaping technique.*

Thus, the parts chosen for PIM should have a fair degree of shape complexity, and the material chosen should be difficult to machine and also difficult to melt and cast. The future expansion of the PIM market will rely more on value-added technology than the processing of large volumes of reasonably complex parts from materials that are easy to cast or machine.

To provide readers with a glimpse of some of the complex shapes and the variety of materials that can be used in the PIM process, photographs of some actual samples that have been produced by PIM are shown in Figures 6.60a through 6.60c. Figure 6.60a shows a complex-shaped controlled-expansion component produced from a 36Ni–Fe (wt.%) alloy. The central piece that is darkish in color is the green injection-molded part. The one on the left is the part obtained after sintering, while the piece on the right is after final machining. The complexity of the part and the nature of the material make this part almost ideal for PIM. Figure 6.60b represents some metal components made from 2Ni–Fe (wt.%), 50Ni–Fe (wt.%), 50Ni–Cu (wt.%), F-15 alloy, and Figure 6.60c represents some PIM ceramic components made from Mn–Zn ferrite, Ni–Zn ferrite, Ca-titanate, alumina, and Al_2O_3–Cr_2O_3.

It would be best to end this chapter at this point and allow readers to appreciate the shape complexity that can be achieved in the PIM process. The technological advancement that is achieved by being able to produce exotic and extremely difficult to machine materials into the complex shapes shown in the pictures will aid the materials community to a great extent. Figure 6.61, which shows a flow diagram of the PIM process, gives a Summary of the key steps of this unique process that is pushing the envelope of advanced materials-shaping technology further and further. By the turn of the century, it is expected that PIM will be the leading near net-shape processing technique for producing small and complex parts of advanced materials that are difficult to machine.

References

1. K. Schwartzwalder, *Ceramic Bulletin*, vol. 22, p.459, 1949.
2. R.P. Kuzy, *Metal-Filled Polymers*, ed. S.K. Bhattacharya, Marcel Dekker, New York, p.1, 1986.
3. L.F. Pease III, *International Journal of Powder Metallurgy*, vol. 24, p.123, 1988.
4. J.R. Merhar, *International Journal of Powder Metallurgy*, vol. 27, p.106, 1991.
5. P.K. Johnson, *International Journal of Powder Metallurgy*, vol. 27, p.163, 1991.
6. R.M. German, *Powder Injection Molding*, Metal Powder Industries Federation, Princeton, NJ, 1990.
7. R.D. Rivers, U.S. Patent 4,113,480, September 12, 1976.
8. R.E. Wiech, Jr., U.S. Patent 4,197,118, April 8, 1980.
9. R.E. Wiech, Jr., U.S. Patent 4,404,166, September 13, 1983.
10. R.E. Wiech, Jr., U.S. Patent 4,415,528, November 15, 1983.
11. R.E. Wiech, Jr., U.S. Patent 4,602,953, July 29, 1986.
12. B.F. Rosof, *Journal of Metals*, vol. 41, p.13, 1989.
13. D.G. White, *Advances in Powder Metallurgy*, compiled by E.R. Andreotti and P.J. McGeehan, Metal Powder Industries Federation, Princeton, NJ, vol. 1, p.1, 1990.
14. B. Verduzco, *Advances in Powder Metallurgy*, compiled by T.G. Gasbarre and W.F. Jandeska, Metal Powder Industries Federation, Princeton, NJ, vol. 3, p.127, 1989.
15. W.R. Mossner, paper presented at the Powder Injection Molding International Symposium, Albany, NY, July 1991.
16. R.J. Merhar, paper presented at the Powder Injection Molding International Symposium, Albany, NY, July 1991.
17. M. Bloemacher, J. Ebenhoech, J.H. ter Maat, and H.J. Sterzel, paper Presented

at the Powder Injection Molding International Symposium, Albany, NY, July, 1991.

18. D. Lee, paper Presented at the Powder Injection Molding International Symposium, Albany, NY, July, 1991.

19. J.R. Gasperevich, *International Journal of Powder Metallurgy*, vol. 27, p.133, 1991.

20. C.I. Chung, B.O. Rhee, M.Y. Cao, and C.K. Liu, *Advances in Powder Metallurgy*, compiled by T.G. Gasbarre and W.F. Jandeska, Jr., Metal Powder Industries Federation, Princeton, NJ, vol. 3, p.67, 1989.

21. C.C. Chou and M. Senna, *Ceramic Bulletin*, vol. 66, p.1129, 1987.

22. S. Maron and P. Pierce, *Journal of Colloid Science*, vol. 11, p.80, 1956.

23. B. Rhee, Rensselaer Polytechnic Institute, private communication.

24. C.I. Chung, paper Presented at the Powder Injection Molding International Symposium, Albany, NY, July, 1991.

25. N. Takahashi, U.S. Patent 4,740,352, April 26, 1988.

26. T. Nakagawa, L. Zhang, H. Noguchi, N. Takahashi, and K. Suzuki, *Modern Developments in Powder Metallurgy*, ed. P.U. Gummeson and D.A. Gustafson, Metal Powder Industries Federation, Princeton, NJ, vol. 20, p.763, 1988.

27. M.J. Edirisinghe and J.R.G. Evans, *British Ceramic Society Transactions and Journal*, vol. 86, p.18, 1987.

28. J. Shah and R.E. Nunn, *Powder Metallurgy International*, vol. 19, p.38, 1987.

29. A. Bose and R.M. German, *Modern Developments in Powder Metallurgy*, compiled by P.U. Gummeson and D.A. Gustafson, Metal Powder Industries Federation, Princeton, NJ, vol. 18, p.299, 1988.

30. C. Khipput and R.M. German, *International Journal of Powder Metallurgy*, vol. 27, p.117, 1991.

31. A. Bose and R.M. German, *Advanced Materials and Manufacturing Processes*, vol. 3, p.37, 1988.

32. I.F. Snider, Jr., paper presented at the Powder Injection Molding International Symposium, July, Albany, NY, 1991.

33. R.M. German, *Particle Packing Characteristics*, Metal Powder Industries Federation, Princeton, NJ, 1989.

34. A. Bose, paper presented at the Powder Injection Molding International Symposium, July, Albany, NY, 1991.

35. D.I. Lee, *Journal of Paint Technology*, vol. 42, p.579, 1970.

36. A.R. Dexter and D.W. Tanner, *Nature Physical Science*, vol. 230, p.177, 1971.

37. *Powder Metallurgy Standards*, Metal Powder Industries Federation, Princeton, NJ.

38. J. Warren and R.M. German, *Modern Developments in Powder Metallurgy*, ed. P.U. Gummeson and D.A. Gustafson, Metal Powder Industries Federation, Princeton, NJ, vol. 18, p.391, 1988.

39. C.J. Markhoff, B.C. Mutsuddy, and J.W. Lennon, *Forming of Ceramics*, ed. J.A. Mangels and G.L. Messing, American Ceramic Society, Columbus, OH, p.246, 1986.

40. R.T. Fox and D. Lee, *International Journal of Powder Metallurgy*, vol. 26, p.233, 1990.

41. M.L. Thorpe, *Progress Report — Sprayed Metal Faced Plastic Tooling*, Technical Bulletin 2.5.1, TAFA Inc., Concord, NH.

42. W.A. Zanchuk and L.J. Grant, *Advances in Powder Metallurgy*, compiled by T.G.

Gasbarre and W.F. Jandeska, Metal Powder Industries Federation, Princeton, NJ, vol. 3, p.25, 1989.

43. K.F. Hens, D. Lee, and R.M. German, *International Journal of Powder Metallurgy*, vol. 27, p.141, 1991.

44. J. Kubat and A. Szalanczi, *Polymer Engineering and Science*, vol. 14, p.873, 1974.

45. B.C. Mutsuddy and L.R. Kahn, *Forming of Ceramics*, ed. J.A. Mangels and G.L. Messing, ACS, Columbus, OH, p.251, 1984.

46. P.A. Willeermet, R.A. Pett, and T.J. Whalen, *Ceramic Bulletin*, vol. 57, p.744, 1978.

47. C.I. Chung, M.Y. Cao, G.B. Kupperblat, and B.O. Rhee, *Advances in Powder Metallurgy*, compiled by E.R. Andreotti and P.J. McGeehan, Metal Powder Industries Federation, Princeton, NJ, vol. 3, p.193, 1990.

48. B. Haworth and P.J. James, *Metal Powder Report*, vol. 41, p.146, 1986.

49. G.R. Youngquist, *Flow Through Porous Media*, American Chemical Society Reprint Publication, Washington, DC, p.58, 1970.

50. R.M. German, *International Journal of Powder Metallurgy*, vol. 23, p. 237, 1987.

51. C.Y. Liu, K. Murakami, and T. Okamoto, *Acta Metallurgia*, vol. 34, p.159, 1986.

52. K. P. Johnson, *Near Net Shape Manufacturing*, ed. P.W. Lee and B.L. Ferguson, ASM International, Metals Park, OH, p.201, 1988.

53. K.P. Johnson, *Advances in Powder Metallurgy*, compiled by T.G. Gasbarre and W.F. Jandeska, Metal Powder Industries Federation, Princeton, NJ, vol. 3, p.17, 1989.

54. A. Bose, R.J. Dowding, and G. Allen, *Powder Injection Molding Symposium*, ed. P.H. Booker, J. Gaspervich, R.M. German, Metal Powder Industries Federation, Princeton, NJ, p.261, 1992.

55. H.E. Amaye, *Advances in Powder Metallurgy*, compiled by E.R. Andreotti and P.J. McGeehan, Metal Powder Industries Federation, Princeton, NJ, vol. 3, p.233, 1990.

56. S.T. Lin and R.M. German, *Powder Metallurgy International*, vol. 21, p.19, 1989.

57. D.R. Bankovic and R.M. German, *Advances in Powder Metallurgy*, compiled by E.R. Andreotti and P.J. McGeehan, Metal Powder Industries Federation, Princeton, NJ, vol. 3, p. 223, 1990.

58. H. Zhang, R.M. German, and A. Bose, *International Journal of Powder Metallurgy*, vol. 26, p.217, 1990.

59. C.W. Finn, *International Journal of Powder Metallurgy*, vol. 27, p.127, 1991.

60. S.W. Kennedy and C.W.P. Finn, *Powder Metallurgy Science and Technology*, vol. 1, p.37, 1990.

61. A.R. Cooper, *Ceramic Processing Before Firing*, ed. G.Y. Onoda and L.L. Hench, Wiley, New York, p.261, 1978.

62. R.M. German, *Powder Metallurgy Science*, Metal Powder Industries Federation, Princeton, NJ, p.150, 1984.

63. P.E. Zovas, R.M. German, K.S. Hwang, and C.J. Li, *Journal of Metals*, vol. 35, p.28, 1983.

64. A. Frisch, W.A. Kaysser, and G. Petzow, *World Conference on Powder Metallurgy*, Institute of Metals, London, vol. 1, p.237, 1990.

65. C. Greskovick, *Journal of the American Ceramic Society*, vol. 64, p.725, 1981.

66. R.E. Bauer and M. Poniatowski, *World Conference on Powder Metallurgy*, The Institute of Metals, London, vol. 3, p.112, 1990.

67. Brandes, E.A., ed., Cu-Sn Phase Diagram, *Smithells Reference Book*, 6th ed., Butterworths, London, pp.11–231, 1983.
68. R.M. German, *Liquid Phase Sintering*, Plenum Press, New York, Chapters 7 and 8, 1985.
69. R.M. German, *Journal of Metals*, vol.38, p.26, 1986.
70. W.H. Baek and R.M. German, *International Journal of Powder Metallurgy*, vol. 22, p.235, 1986.
71. W.H. Baek and R.M. German, *Powder Metallurgy International*, vol. 17, p.273, 1985.
72. R.M. German and J.W. Dunlap, *Metallurgical Transactions*, vol. 17A, p.205, 1986.
73. L.M. Sheppard, *Advanced Materials and Processes*, vol. 2, p.25, 1986.
74. M. Hansen and K. Anderko, *Constitution of Binary Alloys*, 2nd ed., McGraw-Hill, New York, 1958.
75. I.M. Robertson and C.M. Wayman, *Metallographia*, vol. 17, p.43, 1984.
76. A. Bose, B.H. Rabin, and R.M. German, *Powder Metallurgy International*, vol. 20, p.25, 1988.
77. R.M. German, A. Bose, and D.M. Sims, U.S. Patent 4,762,558, August 9, 1988.
78. D. Alman and N.S. Stoloff, *International Journal of Powder Metallurgy*, vol. 27, p.29, 1991.
79. A. Bose, D.E. Alman, and N.S. Stoloff, *Advances in Powder Metallurgy and Particulate Materials*, compiled by J.M. Capus and R.M. German, Metal Powder Industries Federation, Princeton, NJ, vol. 9, p.209, 1992.
80. A. Bose and R.M. German, *Metallurgical Transactions*, vol. 19A, p.2467, 1988.
81. R.M. German and K.S. Churn, *Metallurgical Transactions*, vol. 15A, p.747, 1984.
82. A.J. Markworth, *Scripta Metallurgia*, vol. 6, p.957, 1972.
83. E.I. Larsen and P.C. Murphy, *Canadian Mining Metallurgy Bulletin*, vol. 58, p.413, 1965.
84. J.F. Kuzmick, *Modern Developments in Powder Metallurgy* vol. 3, ed. H.H. Hausner, Plenum Press, New York, p.166, 1966.
85. F.V. Lenel, *Powder Metallurgy Principles and Applications*, Metal Powder Industries Federation, Princeton, NJ, p.284, 1980.
86. T.S. Wei and R.M. German, *International Journal of Powder Metallurgy*, vol. 24, p.327, 1988.
87. A. Bose and R.M. German, *Modern Developments in Powder Metallurgy*, ed. P.U. Gummerson and D.A. Gustafson, Metal Powder Industries Federation, Princeton, NJ, vol. 19, p.139, 1988.
88. A. Bose and R.M. German, *Metallurgical Transactions*, vol. 19A, p.3000, 1988.
89. A. Bose, G. Jerman, and R.M. German, *Powder Metallurgy International*, vol. 21, p.9, 1989.
90. H.K. Yoon, S.H. Lee, S.J.L. Kang, and D.N. Yoon, *Journal of Material Science*, vol. 18, p. 1374, 1983.
91. M. Yodoagawa, *Sintering—Theory and Practice*, ed. D. Kolar, S. Pejovnik and M.M. Ristic, Elsevier Scientific, Amsterdam, The Netherlands, p.519, 1982.
92. B.C. Muddle, *Metallurgical Transactions*, vol. 15A, p.1089, 1984.
93. D.J. Jones and P. Munnery, *Powder Metallurgy*, vol. 10, p.156, 1967.
94. D.V. Edmonds and P.N. Jones, *Metallurgical Transactions*, vol. 10A, p. 289, 1979.
95. A. Bose, H. Zhang, P. Kemp, and R.M. German, *Advances in Powder*

Metallurgy, compiled by E.R. Andreotti and P.J. McGeehan, Metal Powder Industries Federation, Princeton, NJ, vol. 3, p.401, 1990.

96. P.W. Voorhees and M.E. Glicksman, *Metallurgical Transactions*, vol. 15A, p.1081, 1984.

97. P.S. Fredrick, N.L. Bradley, and S.C. Erickson, *Advanced Materials and Processes*, vol. 10, p.53, 1988.

98. R.M. German and A. Bose, *Materials Science and Engineering*, vol. A107, p.107, 1989.

99. D.M. Sims, A. Bose, and R.M. German, *Progress in Powder Metallurgy*, compiled by C.L. Freeby and H. Hjort, Metal Powder Industries Federation, Princeton, NJ, vol. 43, p.575, 1987.

100. T. Nakagawa, N. Takahashi, H. Noguchi, and L. Zhang, *Journal of the Japan Society for Powders and Powder Metallurgy*, vol. 34, p.383, 1987.

101. T. Nakagawa, L. Zhang, H. Noguchi, and K. Suzuki, *Proceedings Spring Conference of Japanese Society of Powders and Powder Metallurgy*, 1988.

102. K. Ameyama, Y. Kaneko, H. Iwasaki, and M. Tokizane, *Advances in Powder Metallurgy*, compiled by T.G. Gasbarre and W.F. Jandeska, Metal Powder Industries Federation, Princeton, NJ, vol. 3, p.121, 1989.

Index